printreading
for welders

Thomas E. Proctor
Jonathan F. Gosse

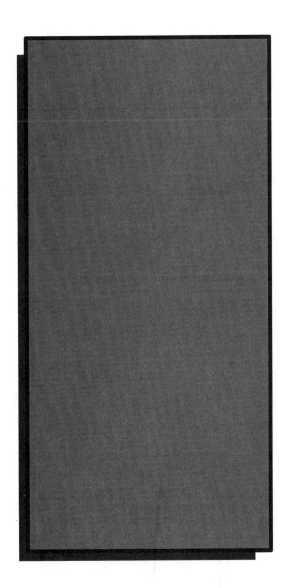

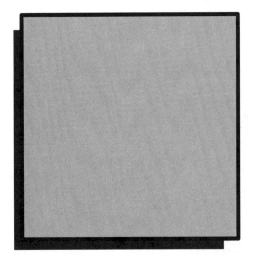

AMERICAN TECHNICAL PUBLISHERS, INC.
HOMEWOOD, ILLINOIS 60430

Acknowledgments

The authors and publisher are grateful to the following companies and organizations for providing technical information and assistance. Selected charts and prints have been modified for educational purposes.

Allied/Gary Safe & Vault Co., Inc.
American Institute of Steel Construction, Inc.
The American Society of Mechanical Engineers
American Welding Society
AMETEK, Inc., Houston Instrument Division
The Bodine Corp.
Calkins Manufacturing Company
CVI Incorporated
EZ Loader Boat Trailers, Inc.
The Flxible Corporation

Gasdorf Tool & Mach. Co.
General Electric Co.
Hobart Brothers Company
Ivey's Mechanical Inc.
Koh-I-Noor Radiograph, Inc.
Komar Industries, Inc.
Toledo Scale Corporation
VME AMERICAS INC.
Worthington Industries

1 2 3 4 5 6 7 8 9 - 92 - 9 8 7 6 5 4 3 2 1

Printed in the United States of America

ISBN 0-8269-3025-5

CONTENTS

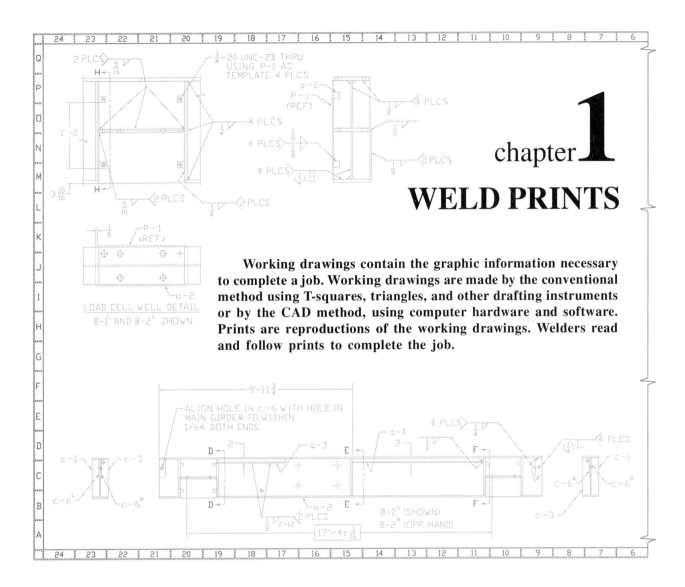

chapter 1

WELD PRINTS

Working drawings contain the graphic information necessary to complete a job. Working drawings are made by the conventional method using T-squares, triangles, and other drafting instruments or by the CAD method, using computer hardware and software. Prints are reproductions of the working drawings. Welders read and follow prints to complete the job.

PRINTS

Prints are reproductions of working drawings. A *working drawing* is a set of plans that contains the information necessary to complete a job. *Specifications* are documents that supplement working drawings with written instructions giving additional information. Originally, prints were referred to as blueprints because the process used to make them produced a white line on a blue background. Any number of copies or blueprints could be made from working drawings by using a process similar to the process used for making prints of photographs.

Today, diazo prints are generally preferred over blueprints because of their white background and dark lines. Electrostatic prints are becoming increasingly popular due to their advantage of easy enlargement or reduction. See Figure 1-1.

Blueprints

The use of blueprints began in 1840 when a method was discovered to produce paper sensitized with iron salts that would undergo a chemical change when exposed to light. *Translucent paper* is paper that allows light to pass through. Drawings made on translucent paper were placed over the sensitized paper in a glass frame used to hold the paper firmly. The frame was then exposed to sunlight. A chemical action occurred wherever the light was permitted to strike the sensitized paper. When the blueprint paper was washed in water, the part protected by the pencil or ink lines on the tracing would show as white lines on a blue background. A fixing bath of potassium dichromate, a second rinse with water, and print drying completed the process. Blueprints are not common today, although they are used in

1

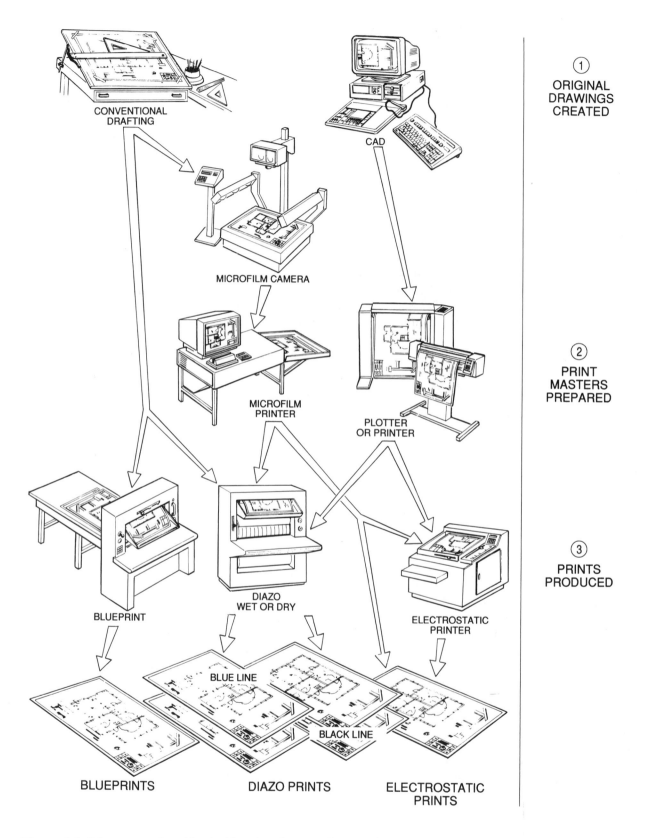

CONVENTIONAL DRAFTING

CAD

MICROFILM CAMERA

MICROFILM PRINTER

PLOTTER OR PRINTER

BLUEPRINT

DIAZO WET OR DRY

ELECTROSTATIC PRINTER

BLUE LINE

BLACK LINE

BLUEPRINTS

DIAZO PRINTS

ELECTROSTATIC PRINTS

① **ORIGINAL DRAWINGS CREATED**

② **PRINT MASTERS PREPARED**

③ **PRINTS PRODUCED**

Figure 1-1. Prints are produced by the blueprint, diazo, or electrostatic process.

some industries such as oil production because they fade less rapidly in sunlight than prints produced by the diazo process.

Diazo Prints

The majority of prints used today have blue or black lines on a white background. Blue line prints are generally preferred by engineers while black line prints are preferred by architects. These prints are made by the diazo process. This process provides excellent reproductions with very good accuracy because the paper has not been soaked with water and then dried. Diazo prints, with their white background, are easier to read than blueprints. Additionally, the white background provides a convenient area for writing field notes or making emergency changes.

Two types of sensitized paper are used in the diazo process, one for each development method. These papers are coated with a chemical that, when exposed to ultraviolet light, becomes a part of a dye complex. The original drawing, or a copy on translucent material, is placed over a sheet of the sensitized paper (yellow side) and is fed by a belt conveyor into the print machine. The two sheets revolve around a glass cylinder containing an ultraviolet lamp and are exposed to the light. The sensitized paper is exposed through the translucent original in the clear areas but not where lines or images block the light. The sheets are separated, the original is returned to the operator, and the sensitized paper is transported through the developing area. It is then developed by either a wet diazo or dry diazo process.

Wet Development Method. In the wet development method, the sensitized paper passes under a roller which moistens the exposed top surface completing the chemical reaction to bring out the image. Prints made by this method have black or blue lines on a white background.

Dry Development Method. In the dry development method, the sensitized paper is passed through a heated chamber in which its surface is exposed to ammonia vapor. The ammonia vapor precipitates the dye to bring out the image. Prints made by this method have black or blue lines on a white background. This method is most commonly used today since high-quality reproduction can be achieved on mylar (plastic film) or sepia (brown line) copies.

Electrostatic Prints

Electrostatic prints are produced by the same process used by office copiers. Full-size working drawings are exposed to light and are projected through a lens onto a negatively charged drum. They may be photographed with a camera and reduced to a 35 mm ($1\frac{3}{8}''$) frame.

The film, after processing, is inserted into an aperture card which can be keypunched for computer sorting, filing, and retrieval. The microfilm in the aperture card is then exposed to light and projected through the lens onto the negatively charged drum. The drum is discharged by the projected light from the nonimage areas but retains the negative charge in the unexposed areas. The drum then turns past a roller where black toner particles are attracted to the negatively charged image areas on the drum surface. As the drum continues to turn in synchronization with the positively charged copy paper, toner particles are attracted to the paper and fused to it by heat and pressure. See Figure 1-2.

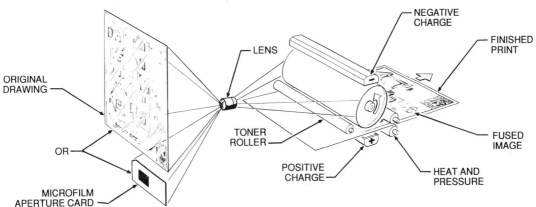

Figure 1-2. Electrostatic prints are produced as light is projected through a lens onto a negatively charged drum which offsets the image to positively charged paper.

Prints made by this method have black lines on a white background. The advantages of the electrostatic process include easy enlargement and reduction of drawings, small storage size, quick retrieval and duplication, and reduced shipping costs. The major disadvantage is the potential for distortion by projection through a lens.

CONVENTIONAL DRAFTING

Working drawings for prints may be made using conventional drafting practices or computer-aided design. *Conventional drafting practices* are a language of standard lines, symbols, and abbreviations used in conjunction with drafting principles so that drawings are consistent and easy to read. See ANSI Y series for additional information.

Basic tools are used to produce working drawings by the conventional method. These tools include T-squares, triangles, drafting instruments, scales, and pencils. See Figure 1-3.

Drafting machines (combination T-square, scale, and triangles) and parallel straightedges (combination drafting board and modified T-square) are commonly used in production situations. Drawings are begun after taping the drafting paper to the drafting board. All line work is done with construction lines that are darkened to produce the final drawing.

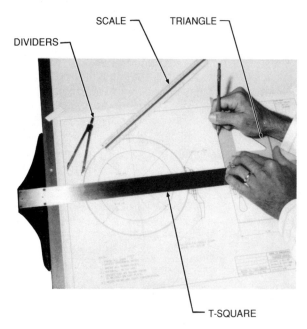

Figure 1-3. Conventional drafting tools include T-squares, triangles, instruments, scales, and pencils.

T-square

The T-square is used to draw horizontal lines and as a reference base for positioning triangles. The head of the T-square is held firmly against one edge of the board to ensure accuracy. T-squares are made of wood, plastic, or aluminum and are available in various lengths. The most popular T-squares are 24″ to 36″ in length.

Triangles

Triangles are used to draw vertical and inclined lines. The base of the triangle is held firmly against the blade of the T-square to ensure accuracy. Two standard triangles, 30°-60° and 45°, are commercially available in a variety of sizes. Triangles are commonly made of clear plastic. The 30°-60° triangle is used to produce vertical lines and inclined lines of 30° to 60° sloping to the left or right. The 45° triangle is used to produce vertical lines and inclined lines of 45° sloping to the left or right. The triangles may be used together to produce inclined lines every 15°. See Figure 1-4.

Drafting Instruments

Although a wide variety of precision drafting instruments is available, the compass and dividers are the most commonly used. Each of these is available in a variety of sizes. See Figure 1-5.

The compass is used to draw arcs and circles. One leg of the compass contains a needlepoint that is positioned on the centerpoint of the arc or circle to be drawn. The other leg contains the pencil lead used to draw the line. Two types of compasses are center-wheel and friction. The radius of the arc on the center-wheel compass is changed by adjusting the center wheel. Arcs of various radii are obtained on the friction compass by opening or closing the legs. Center-wheel compasses are the most popular and most accurate.

Dividers are used to transfer dimensions. Each leg contains a needlepoint to assure accuracy. Two types of dividers are center-wheel and friction. Friction dividers are the more useful for general work.

Other drafting instruments include irregular (french) curves and templates. Irregular curves are used to draw curves that do not have consistent radii. Templates are used to save time when drawing standard items such as fasteners, springs, etc.

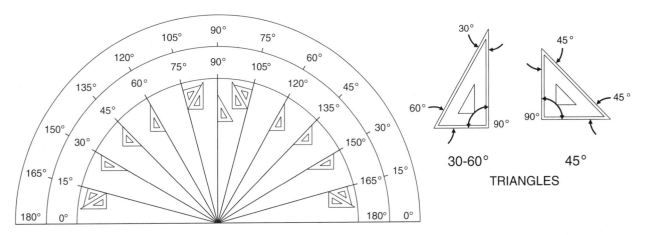

Figure 1-4. The 30°-60° and 45° triangles are used to draw lines 15° apart.

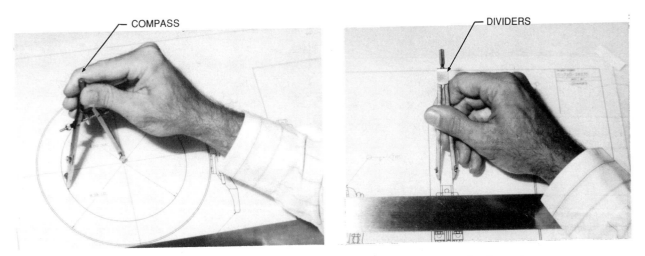

Figure 1-5. The compass is used to draw arcs and circles. The dividers is used to transfer dimensions.

Scales

Scales are used to measure lines and reduce or enlarge them proportionally. The three types of scales are the architect's scale, civil engineer's scale, and mechanical engineer's scale. A variety of sizes is commercially available.

An architect's scale is used when making drawings of buildings and other structural parts. A common type of scale is triangular in shape. One edge of the scale is a standard ruler divided into inches and sixteenths of an inch. The other edges contain 10 scales that are labeled 3, 1½, 1, ¾, ½, ⅜, ¼, 3/16, ⅛, and 3/32. The ¼ scale means that ¼″ = 1′-0″, and so forth. For larger scale drawings, the 1½″ = 1′-0″, or 3″ = 1′-0″ scales are used.

The civil engineer's scale is used when making maps and survey drawings. Plot plans also may be drawn using this scale. The civil engineer's scale is graduated in decimal units. One inch units on the scale are divided into 10, 20, 30, 40, 50, or 60 parts. These units are used to represent the desired measuring unit such as inches, feet, or miles. For example, a building lot line that is 100′-0″ long drawn with the 20 scale (1″ = 20′-0″) measures 5″ on the drawing.

The mechanical engineer's scale is used when drawing machines and machine parts. This scale is similar to the architect's scale except that the edges are limited to fractional scales of ⅛, ¼, ½, and 1 related to inches. Decimal scales are also available. See Figure 1-6.

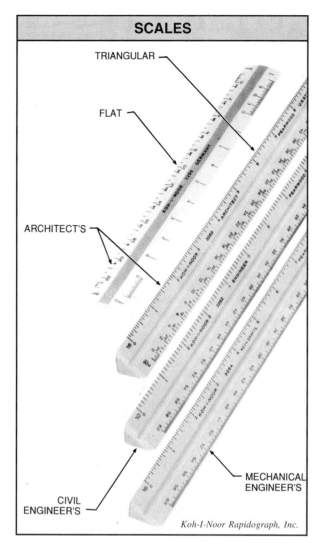

Figure 1-6. Three types of scales are used to produce scaled drawings.

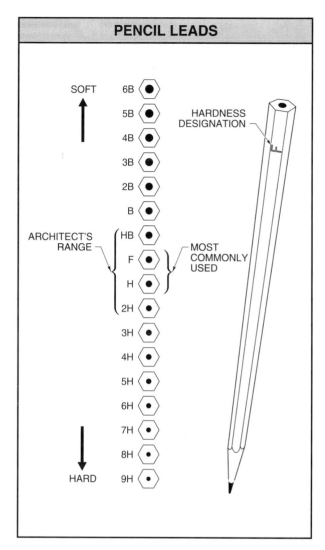

Figure 1-7. Pencil leads range from extremely soft to exceptionally hard.

Pencils

Wooden or mechanical pencils are used to draw lines. Wooden pencils contain a stamp near one end indicating the hardness or softness of the lead. The lead of desired hardness is inserted into mechanical pencils.

Hard leads are used to draw fine, precise lines. Medium leads are used to draw object lines. Soft leads are used primarily for sketching. Grades of lead range from 6B (extremely soft) to 9H (exceptionally hard). The engineer's range is HB, F, H, and 2H. F and H are most used for producing drawings. See Figure 1-7.

COMPUTER-AIDED DESIGN (CAD)

Computer-aided design (CAD) is the generation and reproduction of line drawings and prints with computers. It is also known as computer-aided drafting, or computer-aided drafting and design (CADD). This system is popular in architectural and engineering offices.

Tradesworkers reading CAD-generated plans benefit from the quality and consistency of line work, symbol representation, and lettering. Designers and engineers benefit from increased drafting productivity achieved in the planning, design, drafting, and reproduction of prints. See Figure 1-8.

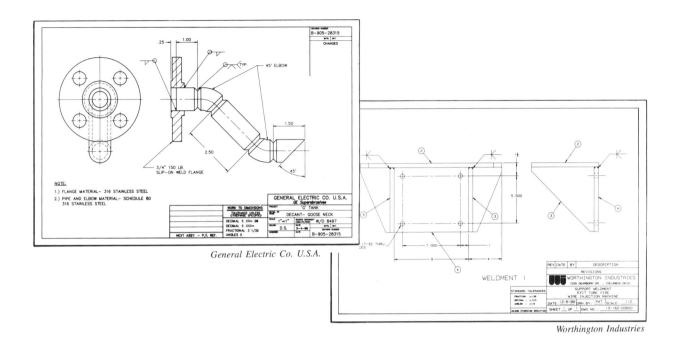

General Electric Co. U.S.A.

Worthington Industries

Figure 1-8. CAD systems produce excellent quality drawings consistently.

Six primary factors contributing to increased productivity are

Consistency - constant sameness in line width, symbol depiction, and representation of drawing components.

Changeability - revisions, additions, and deletions are easily made.

Layering - a method, similar to using overlays on conventionally generated drawings, in which base work is used to generate additional drawings.

Modeling - viewing of the complete part in pictorial form and subjection to stress tests.

Storage - drawings, stored on magnetic tapes or small magnetic disks, require minimal space.

Repeatability - an unlimited number of sheets may be reproduced, each with original quality.

CAD systems use hardware and software to generate drawings. *Hardware* is the physical components of a computer system, including the input devices, central processing unit (CPU), and output devices.

A large variety of commercially available components are used to control, manage, and process information in CAD systems. See Figure 1-9. *Software* is the operating system of a computer, on magnetic tape or disks, that provides operational instructions for capturing and formatting keystrokes and generating lines. Software allows system hardware components to interact in the production of drawings. See Figure 1-10.

Figure 1-9. Hardware components of a CAD system are the input devices, central processing unit, and output devices.

Input Devices

Input devices are hardware used to enter information into a computer system. The input devices are interfaced (connected) with the central processing unit, which controls the CAD system. Information may be input by the use of electronic or electromechanical devices. Electronic devices use electronic signals to relay information to the CPU.

Electromechanical devices use mechanical actuation to input electronic information. Common input devices include the graphics tablet, digitizing tablet, keyboard, and light pen. Tablet accessories include the stylus, mouse, joystick, thumbwheel, and trackball.

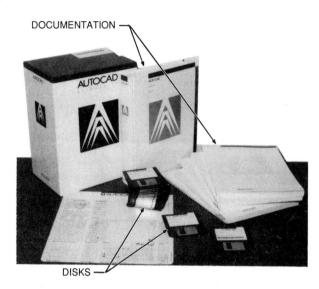

Figure 1-10. Software is the operating system that provides instructions to the hardware components.

Graphics Tablet. A graphics tablet is a common input device used with a CAD system. It consists of a drawing area and a menu. See Figure 1-11. The menu is a group of commands. Menu selections may consist of simple commands such as inserting circles and erasing lines, or more complex operations such as drawing tangent lines and dimensioning objects.

An electronic or electromechanical device is used with the graphics tablet to choose a function from the menu and "draw" the object in the space provided. For example, a CAD operator may choose "Seam Weld, Other Side" from the Compugraphx menu with a stylus (pen-like device). The stylus is placed in the drawing area and a cursor on the display screen shows its present location. A *cursor* is the solid or flashing pointer indicating position of work. As the stylus is moved to its desired location, the cursor moves accordingly. Once in position, the size of the circle may be input by using the keyboard or stylus.

In addition to standard commands, the graphics tablet may display a library of symbols. Common symbols include weld symbols, surface finish, geometric tolerance, ANSI dimension, fasteners, set screws, rivet pins, etc. A symbol that is fre-

quently used, but not part of the original symbols library, can be created and stored. When needed again, the symbol can be recalled from the library and inserted in its desired location. For example, a reoccurring examination symbol could be created, stored, recalled, and inserted as often as needed.

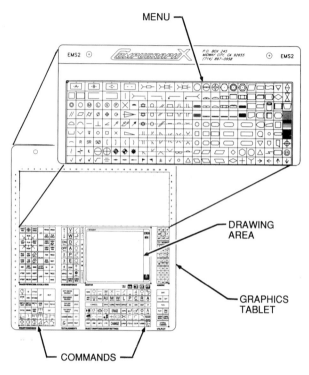

Figure 1-11. A graphics tablet consists of a group of commands and a drawing area.

Digitizing Tablet. A larger version of the graphics tablet, known as a digitizing tablet, can be used to convert existing drawings to CAD without reinputting all drawing components. An existing print is placed on the digitizing tablet and digitized (traced) with a stylus or mouse. The digitized drawing is then treated as other CAD drawings and may be revised to required specifications. See Figure 1-12.

Keyboard. A *keyboard* is an electronic device that sends signals to the CPU. The keyboard is the most common input device. The number and letter arrangement is similar to a typewriter. For example, the standard QWERTY letter arrangement is used in addition to function keys. A function key is a key that performs a particular function when depressed. For example, a function key labeled Delete will delete highlighted keystrokes.

The keyboard has the capability of inputting notes and dimensions as well as positioning the cursor on the display monitor. A keyboard may be the sole input device or it may be used in conjunction with other input devices.

Figure 1-12. A digitizing tablet is used to convert existing drawings to a CAD format.

Light Pen. A *light pen* is a photosensitive electronic device used to enter data into the computer. Data (information) is input when the tip of the light pen is pressed directly on the video display screen. The operator may chose various commands listed along the edge of the screen in the same manner as when using a graphics tablet. An advantage of using the light pen is that the operator is working directly on the drawing rather than indirectly, as with the graphics tablet.

Tablet Accessories. A stylus, mouse, joystick, thumbwheels, and trackballs are common accessories used in conjunction with tablets to input information. These accessories are either electronically or electromechanically operated.

A *stylus* is an electromechanical device used to input information into the computer. The stylus is used to select commands from the menu and position the cursor in the required area. When the desired command or position is located, the tip of the stylus is depressed against the surface of the tablet. As the tip is depressed, the stylus relays an electronic signal to the CPU.

A *mouse* is an electronic device used to input information into the computer. A mouse may be interfaced with a tablet or used separately. When used separately, the mouse controls the cursor on the display screen through movement on a hard surface. Menu selections are made by moving the cursor to the edge of the display screen and selecting the desired command.

A *joystick* is an electromechanical device used to control the cursor on the display screen and enter information into the computer. It may be used to choose a command from the display screen and locate its position on the drawing area.

A *thumbwheel* is an electromechanical device used to control the position of the cursor in vertical and horizontal planes. Two thumbwheels are required. One thumbwheel controls the vertical movement and the other controls the horizontal movement of the cursor.

A *trackball* is an electromechanical device similar to a thumbwheel in that it is only used to control cursor movement. A single trackball controls both the horizontal and vertical movement of the cursor and also has the capability of moving it diagonally. The cursor is moved by rolling the trackball in the desired direction.

Central Processing Unit

The *central processing unit (CPU)* is the control center of the computer. It receives information through the input devices and produces an output image. A CPU is classified by its memory capacity and the speed at which it carries out commands. A larger memory capacity generally has a greater capability of producing quality drawings. The central processing unit stores information through the use of magnetic tapes, disks, and internal memory. CPUs may be dedicated for CAD systems only or may also run other software, such as word or data processing programs.

Output Devices

Output devices are hardware that either display or generate drawings. The basic types of output devices are monitors (screens), printers, and plotters.

Monitor. A *monitor* is a video display terminal. A monitor is a necessity for all CAD stations. The monitor displays the drawing that the operator is developing. Monitors are available in many sizes and are chosen based upon the application. A large monitor may be required when using a light pen because the large working surface enables an operator to work more accurately with the drawing. On some types of CAD systems two monitors are used. One monitor displays the drawing and the other displays the menu.

Printer. Printers produce drawings on paper. They provide a fast and convenient method of checking the placement of drawing features. The drawing generated by a printer consists of small dots (dot matrix) or a laser-generated image. The copy produced by a printer is generally of less-than-desirable quality. Consequently, final drawings are seldom produced by a printer.

Plotter. A *plotter* is an output device that generates finished drawings with pens. Plotters are commercially available with single-color or multiple-color pens. See Figure 1-13.

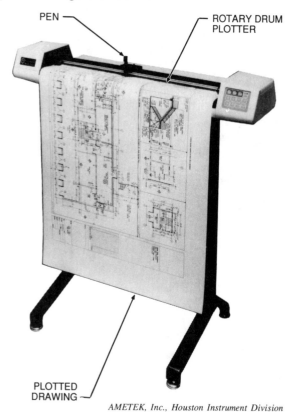

PEN — — ROTARY DRUM PLOTTER

PLOTTED DRAWING —

AMETEK, Inc., Houston Instrument Division

Figure 1-13. A plotter produces high-quality drawings.

The quality of a plotted drawing is much better than a printed drawing. Plotters vary in size and are chosen based upon the application. Larger plotters have the capability of plotting E-size sheets of paper or film. Plotters are available in two major styles: rotary-drum and flatbed plotters.

On a rotary-drum plotter, the paper is mounted on a drum and moves with the drum's rotation. The pen moves parallel with the length of the drum. Vertical lines are drawn with the drum remaining in a fixed position while the pen moves along the sheet of paper. Horizontal lines are drawn with the pen remaining stationary and the drum rotating. A flatbed plotter allows the piece of paper to lie flat on its bed. The pen moves along the width and height of the paper while the paper remains stationary.

WELD PRINTS

Working drawings are commonly produced on film or paper. The working drawings are copied to produce prints. The two basic materials used for working drawings are polyester and vellum. Prints are produced on various sizes of paper depending upon the scale used and the complexity of the part being drawn. By using standard size paper for drawings, basic formats, title blocks, revision blocks, parts list, supplementary blocks, and drawing numbers can be utilized per ANSI standards.

Polyester

Polyester film provides good surfaces for drafting requirements. Ink lines are smooth, black, and consistent. Pencil lines are uniform in density and continuity. Polyester film is available in sheets and rolls punched for specific plotters. Polyester film may have a matte finish on one or both sides. A *matte finish* is a dull finish that will accept and hold pencil and ink lines well. Polyester film may be purchased in 20 yd and 50 yd rolls and standard size flat sheets. The thickness varies from .002″ to .007″.

Vellum

Vellum is a translucent paper made from a rag base. A *translucent* paper allows light to pass through. Vellum does not yellow, become brittle, or deteriorate with age. It can be used for pencil or ink lines. Erasures are easily made and do not reproduce. Vellum is available in sheets, rolls, and rolls punched for specific plotters. It may be purchased in 10, 50, or 100 flat-sheet packages and in 20 yd and 50 yd rolls. The thickness varies from .0025″ to .0030″.

Paper Sizes

USA flat and roll sheets are standardized by ANSI Y14.1. Flat sheets are designated by the letters A through F. Rolls are designated by the letters G, H, J, and K. Margins are sufficiently large enough to allow reproductions of drawings made to these standard sizes and to international paper sizes. See Figure 1-14.

USA PAPER SIZES		
FLAT SHEETS		
SIZE	WIDTH	LENGTH
A	8.5	11
B	11	17
C	17	22
D	22	34
E	34	44
F	28	40
ROLLS		
SIZE	WIDTH	LENGTH MIN	LENGTH MAX
G	11	22.5	90
H	28	44	143
J	34	55	176
K	40	55	143

All measurements are in inches.

Figure 1-14. Standard USA paper sizes are designated by letters.

International paper sizes are designated by letter-number combinations. The length of an international sheet is found by multiplying the width by the $\sqrt{2}$. The $\sqrt{2}$ is 1.414 (1.414 × 1.414 = 1.999 = 2). For example, the length of an A3 size sheet is 420 mm (297 × 1.414 = 419.958 = 420 mm). See Figure 1-15.

Basic Formats

Drawing paper is oriented with its base horizontal along the long dimension with the exception of A

size vertical paper. The borderline is thick (approximately .030″). The vertical margins between the edges of the paper and the border line vary from .25″ to 1.00″ depending on the paper size. The horizontal margins vary from .38″ to 1.00″ depending on the paper size. See Figure 1-16.

INTERNATIONAL PAPER SIZES			
SIZE	WIDTH	LENGTH	NEAREST USA SIZE
A4	210	297	A
A3	297	420	B
A2	420	594	C
A1	594	841	D
A0	841	1189	E

FINDING LENGTH
$L = W \times \sqrt{2}$
where
L = length W = width $\sqrt{2}$ = 1.414
What is the length of an A4 sheet?
$L = W \times \sqrt{2}$
$L = 210 \times 1.414 = 296.94 = 297$
$L = \textbf{297 mm}$

All measurements are in millimeters.

Figure 1-15. Standard international paper sizes are designated by letter-number combinations.

Lettering. Lettering size and style is per ANSI Y14.2M. Uppercase, single-stroke gothic letters are used. They may be inclined or vertical. The preferred slope for inclined letters is 68° with the horizontal. The height of freehand lettering varies from .120″ to .290″ depending on the paper size.

Zoning. Drawing paper in flat sheets and rolls may include zones for reference purposes. Zones are referenced by alphabetical and numerical entries in margins. Commonly, the numerical entries are along the base of the paper and the alphabetical entries are along the right and left hand margins. The sizes of the individual zones vary based upon the paper size.

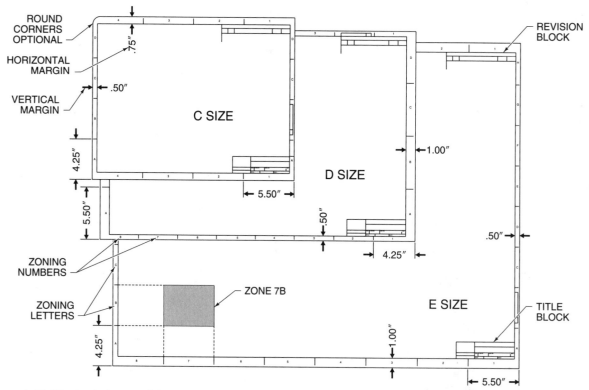

Figure 1-16. The basic format of drawing sheets includes the title block, margins, and zoning.

Title Blocks

Title blocks are located in the lower right-hand corner of the drawing sheet. Title blocks are often customized to the individual company. Information common to all drawings is located in the title block. Two basic title blocks are given; one for sheet sizes A, B, C, and G and the other for sheet sizes D, E, F, H, J, and K. The basic difference between the two sizes is the dimensions of the various blocks. See Figure 1-17.

Revision Block. Revision blocks are located in the upper right-hand corner of the sheet. They are extended downward as required. The revision symbol, description of the change authorization document, date, and approvals are included. A zone column is added if required.

Parts List. The parts list is also known as a bill of materials. It is located in the lower right-hand corner of the sheet above the title block. Supplementary parts lists may be located to the left of the original parts list. The parts list contains the name of the part, the identifying number if required, and how many are required.

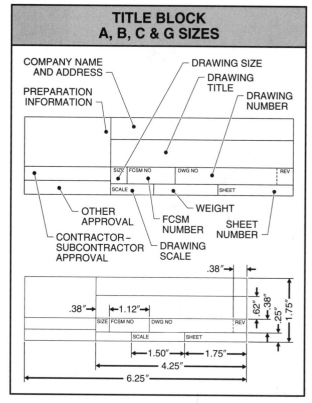

Figure 1-17. Title blocks are located in the lower right-hand corner of the sheet.

Review Questions

Name_____ Date _____

Matching — Angles

_____ 1. 135°

_____ 2. 120°

_____ 3. 165°

_____ 4. 75°

_____ 5. 30°

_____ 6. 45°

_____ 7. 90°

_____ 8. 150°

_____ 9. 60°

_____ 10. 15°

Completion

_____ 1. Electrostatic prints have _____ lines on a white background.

_____ 2. T-squares, triangles, drafting instruments, etc. are used to produce drawings by the _____ method.

_____ 3. Triangles may be used together to produce inclined lines every _____°.

_____ 4. The _____ is an instrument used to draw arcs and circles.

_____ 5. The _____ engineer's scale is used when drawing machines and machine parts.

_____ 6. _____ is the operating system of a computer.

_____ 7. The keyboard is an electronic device that sends signals to the _____ of a computer.

_____ 8. _____ drawings are commonly produced on film or paper.

_____ 9. A(n) _____ finish on paper is a dull finish.

_____ 10. The preferred slope for inclined letters is _____° with the horizontal.

_____ 11. _____ are used to draw vertical and inclined lines.

_____ 12. The _____ engineer's scale is used when making maps and survey drawings.

_____ 13. A thumbwheel controls the vertical and horizontal movement of the _____.

_____ 14. Vellum may be purchased in 20 yd or _____ yd rolls.

_____ 15. Lettering size and style on drawings is established per _____ Y14.2M.

Multiple Choice

_____ 1. _____ supplement working drawings with written information necessary to complete a job.
A. Revision blocks
B. Title blocks
C. Specifications
D. none of the above

_____ 2. Blueprints are used in some industries today because they _____ than other prints.
A. are less expensive
B. are easier to produce
C. fade less rapidly in sunlight
D. all of the above

_____ 3. The most popular T-squares are _____ in length.
A. 12″ to 24″
B. 18″ to 24″
C. 18″ to 30″
D. 24″ to 36″

_____ 4. The _____ scale contains a standard 12″ ruler and 10 scales.
A. architect's
B. civil engineer's
C. drafter's
D. assembler's

_____ 5. Vertical margins between the edges of drawing paper and the border line vary from _____ depending on the paper size.
A. .10″ to .20″
B. .15″ to .25″
C. .25″ to 1.00″
D. .25″ to 1.25″

_____ 6. The computer keyboard contains function keys and the standard _____ letter arrangement.
A. POINT
B. XYZ
C. ALPHA
D. QWERTY

7. USA flat and roll sheets of paper are designated for size by _____.

    A. letters
    B. numbers
    C. both A and B
    D. neither A nor B

8. International flat and roll sheets of paper are designated for size by _____.

    A. letters
    B. numbers
    C. both A and B
    D. neither A nor B

9. Zones on drawing paper are referenced by _____ entries in the margins.

    A. alphabetical
    B. numerical
    C. both A and B
    D. neither A nor B

10. The length of an international sheet of paper is found by _____.

    A. adding 250 mm to the width
    B. multiplying the width by 3
    C. multiplying the width by the $\sqrt{2}$
    D. none of the above

11. Blue line prints on a white background are made today by the _____ process.

    A. diazo
    B. electrostatic
    C. both A and B
    D. neither A nor B

12. Dividers are drafting instruments used to _____.

    A. draw arcs
    B. draw circles
    C. transfer dimensions
    D. transfer letters

13. The civil engineer's scale is graduated in _____.

    A. inches and sixteenths of an inch
    B. decimal units
    C. both A and B
    D. neither A nor B

14. _____ is the physical components of a computer system.

    A. Software
    B. Hardware
    C. CAD
    D. CADD

15. The _____ of a CAD system is a group of commands.

    A. output device
    B. digitizer
    C. cursor
    D. menu

True-False

T   F   **1.** Prints are originals of working drawings.

T   F   **2.** Diazo prints have dark lines on a white background.

T   F   **3.** The T-square is a drafting instrument used to draw horizontal lines.

T   F   **4.** A translucent sheet stops light from passing through.

T   F   **5.** Hard pencil leads are used to draw object lines.

T   F   **6.** Information may be input into computers by the use of electronic or electromechanical devices.

T   F   **7.** A monitor is an input device.

T   F   **8.** Vellum is made from a rag base.

T   F   **9.** The parts list on a print is located above the title block.

T   F   **10.** Plotters generate finished drawings by a photographic process.

T   F   **11.** The majority of prints used today have blue or black lines on a white background.

T   F   **12.** Friction compasses are more accurate than center-wheel compasses.

T   F   **13.** A pencil lead labeled 6B is exceptionally hard.

T   F   **14.** A digitizing tablet is an input device.

T   F   **15.** Revision blocks may be extended downwards if necessary.

Matching — Paper Sizes

_____   **1.** A sheet

_____   **2.** A3 sheet

_____   **3.** A4 sheet

_____   **4.** G roll

_____   **5.** H roll

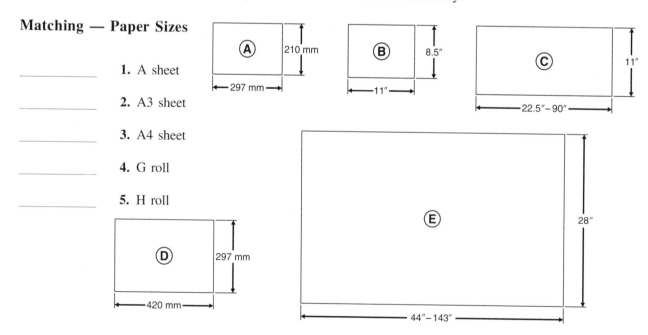

Trade Competency Test

Name_____ Date _____

DRAFTER RJH	DATE 3-23-92	PIERCE FIXTURE COMPANY CHICAGO, ILLINOIS			
CHECKER TAJ	DATE 3/25/92	CENTERING FIXTURE			
APPROVAL CT	DATE 3-26-92				
		SIZE C	FCSM NO NA	DWG NO 316915	REV
		SCALE ¼"=1'-0"		SHEET 1 OF 3	

TITLE BLOCK 1

Refer to Title Block 1 on page 17.

_____ 1. The scale of the drawing is _____" = 1'-0".

_____ 2. There are _____ sheets in this set of prints.

_____ 3. The drawing number is _____.

T   F   4. The drawing was drawn and checked by the same person.

_____ 5. Prints for this drawing are reproduced on size _____ sheets.

T   F   6. The FCSM number is required for this drawing.

T   F   7. Approval of the drawing was made by TAJ.

_____ 8. Approval of the drawing was completed on _____-92.

_____ 9. The title of the part to be fabricated is _____.

T   F   10. Pierce Fixture Co. is located in Detroit, Michigan.

_____ 11. This sheet is Sheet _____.

T   F   12. This particular drawing has been revised.

_____ 13. The drafter for the drawing was _____.

_____ 14. The drawing was checked on _____-92.

T   F   15. Checking and approval of the drawing were completed on consecutive days.

Refer to the Flange Assembly print on page 18.

_____ **1.** The drawing was drawn by _____.

_____ **2.** Drawing approval was completed on _____-89.

_____ **3.** The tolerance for decimal dimensions on the print is ±_____".

_____ **4.** Part 2 is a(n) _____.

T   F **5.** There is no scale specified on the print.

_____ **6.** One _____ and one flange are required.

T   F **7.** The drawing has been revised three times.

T   F **8.** All sharp edges are to be broken.

T   F **9.** The general note pertains to welding specifications.

_____ **10.** The next assembly is part number _____.

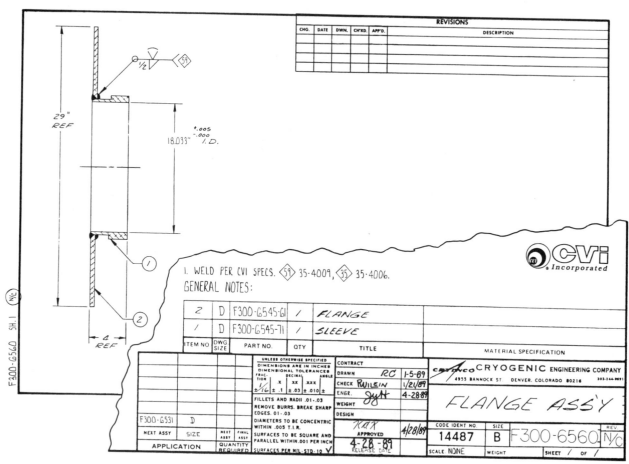

FLANGE ASSEMBLY

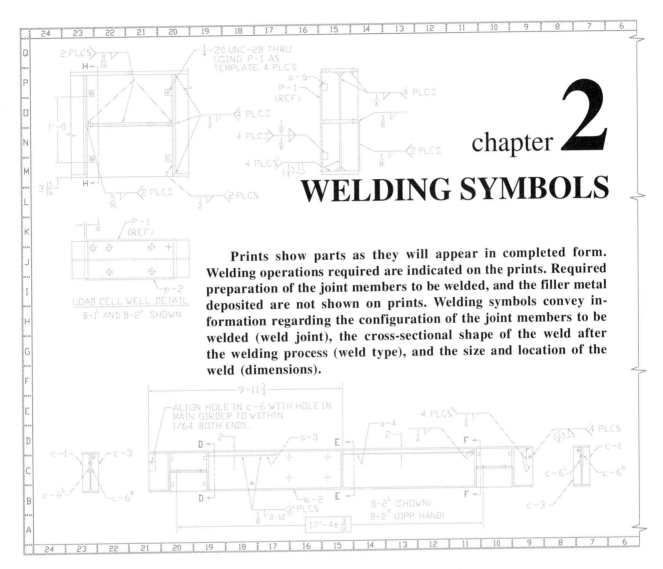

chapter 2
WELDING SYMBOLS

Prints show parts as they will appear in completed form. Welding operations required are indicated on the prints. Required preparation of the joint members to be welded, and the filler metal deposited are not shown on prints. Welding symbols convey information regarding the configuration of the joint members to be welded (weld joint), the cross-sectional shape of the weld after the welding process (weld type), and the size and location of the weld (dimensions).

WELDING SYMBOL USE

Welding symbols are symbols developed by the American Welding Society (AWS). Welding symbols are used to specify information about a weld on a print. The weld requirements are considered and specified by engineers on prints using welding symbols. Welders interpret these welding symbols and other weld joint information on the print to fabricate the part to required specifications. Standards pertaining to welding symbols are specified in the publication "Symbols for Welding, Brazing, and Nondestructive Examination" (ANSI/AWS A2.4) developed by the American Welding Society. This standard has been developed in accordance with the American National Standards Institute's rules for development and approval of American National Standards.

WELD JOINTS

A *weld joint* is the physical configuration of the joint members to be joined. The weld joint is the recipient of the filler metal deposited. Different weld joints are used on a fabrication to meet design requirements. The weld joint is selected by location, preparation required, welding equipment used, and the application of the weld joint. Joint penetration and strength of the filler metal determine the strength of the weld joint.

Filler metal is the metal deposited during the welding process. Filler metal in the weld joint is usually matched to equal the strength of the base metal. Weld joints are commonly specified by the engineer and are indicated on prints. Basic weld joints include butt, lap, T, edge, and corner joints. See Figure 2-1.

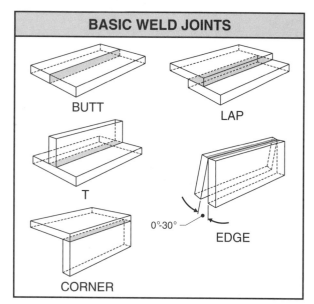

Figure 2-1. The five basic weld joints are the butt, lap, T, edge, and corner weld joints.

Butt Joint

A *butt joint* is a weld joint formed when two joint members, located approximately in the same plane, are positioned edge to edge. Butt joints can be used with or without preparation on joint members having the same or different thicknesses. Butt joints are commonly used in subassemblies and during fabrication and repair operations.

Lap Joint

A *lap joint* is a weld joint formed when two joint members are lapped over one another. A lap joint is stronger than a butt joint, but results in the thickness of one of the joint members added to the overall thickness. Lap joints are commonly welded on both sides. They are commonly used during repair operations and to extend standard material lengths to required specifications.

T-Joint

A *T-joint* is a weld joint formed when two joint members are positioned approximately 90° to one another in the form of a T. If possible, the T-joint is welded on both sides for maximum strength. T-joints are commonly used in fabricating support structures where the load is transferred to different planes at approximately 90°.

Edge Joint

An *edge joint* is a weld joint formed when the edges of two joint members are joined. The edges are parallel to one another. Edge joints are commonly used to join support structures and short lengths of structural steel. They are also used to combine the strength of two joint members with exposed edges.

Corner Joint

A *corner joint* is a weld joint formed when two joint members are positioned at an approximate 90° angle with the weld joint at the outside of the joint members. Corner joints are commonly used in tank and pressure vessel construction. Filler metal may or may not be required depending on the design and function of the joint.

WELD TYPES

A *weld type* is the cross-sectional shape of the filler metal after welding. Weld types are different from weld joints as weld joints are the configuration of the joint members. The configuration of joint members and the edge preparation have an effect on the weld type.

The weld joint is shown on prints by the description of the object. The shape of the joint member after welding is not shown on the prints. The weld type specified for the weld joint is indicated by welding symbols on the prints. Weld types include groove, fillet, plug, slot, stud, spot, projection, seam, back or backing, surfacing, and flange welds. See Figure 2-2. Different weld types can be used with different weld joints as necessary in the fabrication process.

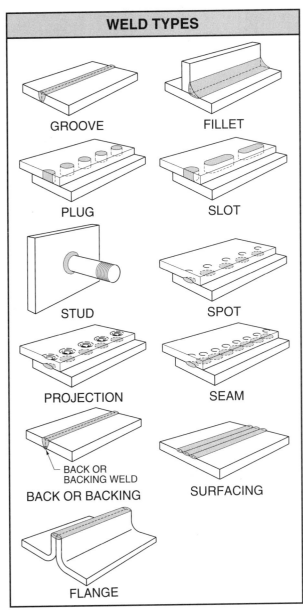

WELD TYPES

GROOVE   FILLET

PLUG   SLOT

STUD   SPOT

PROJECTION   SEAM

BACK OR
BACKING WELD

BACK OR BACKING   SURFACING

FLANGE

Figure 2-2. Weld types are classified by the cross-sectional shape of the weld.

Groove Weld

A *groove weld* is a weld type made in the groove of the welded pieces. Groove welds can be modified for specific weld joints. Parts of the groove weld include the base metal, filler metal, weld face, weld toes, weld root, face reinforcement, root reinforcement, weld interface, and root surface.

Base metal is the material to be welded. Filler metal is added to the weld joint during the welding process. The *weld face* is the exposed surface of

the weld, bounded by the weld toes of the side on which the welding was done. The *weld toe* is the intersection of the base metal and the weld face. The *weld root* is the area where the filler metal intersects the base metal opposite the weld face. *Face reinforcement* is filler metal which extends above the surface of the joint member on the side of the joint on which welding was done. *Root reinforcement* is filler metal which extends above the surface of the joint on the opposite side of the joint on which welding was done. *Weld interface* is the area where the filler metal and the base metal mix together (interface). The *root surface* is the surface of the weld on the opposite side of the joint on which welding was done.

The shape of a groove weld is determined by the preparation of the edge or edges and the orientation of the joint members. See Figure 2-3. Groove welds are classified as square-groove, V-groove, bevel-groove, U-groove, J-groove, flare-V-groove, and flare-bevel-groove depending on the edge preparation performed.

Joint penetration in a groove weld is determined by the distance between the joint members, the thickness of the joint member, and the edge preparation. Edge preparation permits a greater amount of joint penetration and fusion of filler and base metal. This results in greater strength of the joint when compared with the square-groove weld. *Fusion* is the melting together of filler metal and base metal.

The welding process used is also a factor on which groove weld type is best suited for the joint. For example, submerged arc welding (SAW) has greater joint penetrating characteristics than shielded metal arc welding (SMAW). Joints welded using submerged arc welding require less edge preparation than joints welded with shielded metal arc welding.

Fillet Weld

A *fillet weld* is a weld type in the cross-sectional shape of a triangle. Fillet welds can be used on joint members of right angle welds such as T-joints and lap joints. A fillet weld without additional information positions the joint members at 90°. Weld joints requiring more or less than 90° must be specified on the print.

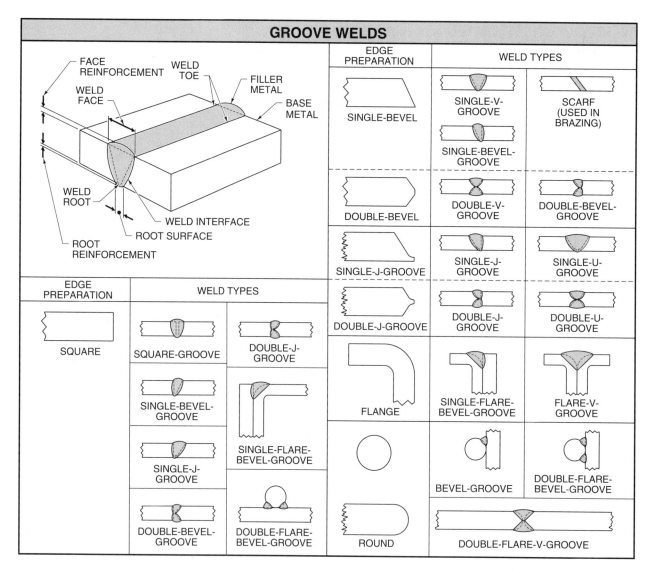

Figure 2-3. Groove welds are classified by the edge preparation of the joint members.

Fillet welds are the most popular type of welds. They require little or no edge preparation. *Single-fillet welds* are fillet welds that have filler metal deposited on one side. They are limited to smaller loads than double-fillet welds. *Double-fillet welds* are fillet welds that have filler metal deposited on both sides. The weld on both sides provides additional strength. See Figure 2-4.

Plug and Slot Welds

A *plug weld* is a weld type in the cross-sectional shape of a hole in one of the joint members. A *slot weld* is a weld type in the cross-sectional shape of a

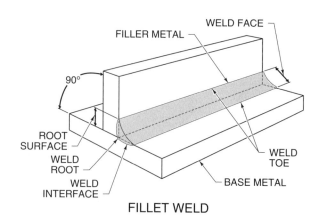

Figure 2-4. Fillet welds are triangular in cross-sectional shape. They transfer loads to joint members positioned at 90° to one another.

slot (elongated hole) in one of the joint members. These welds are made through the opening of the hole or slot to the other joint member. The round hole, which is filled or partially filled with filler metal during the welding operation, becomes the plug weld. The elongated hole, which is filled or partially filled with filler metal during the welding operation, becomes the slot weld. See Figure 2-5.

Plug and slot welds are used for similar applications as the lap joint, except when the edges of the joint members cannot be welded. Slot welds are stronger than plug welds because of the larger weld surface area. Plug and slot welds are not to be confused with the fillet weld as the base of the hole or slot is filled. A fillet weld would only deposit weld material in a triangular shape on the perimeter of the hole or slot.

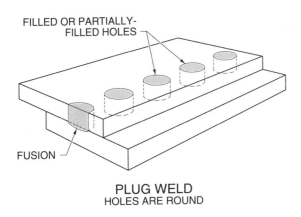

PLUG WELD
HOLES ARE ROUND

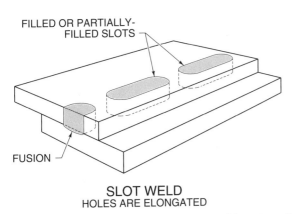

SLOT WELD
HOLES ARE ELONGATED

Figure 2-5. Plug and slot welds provide strength without affecting the edges of the joint members.

Stud Weld

A *stud weld* is a weld type produced by joining threaded studs with other parts using heat and pressure. During the welding process part of the stud is melted, providing weld reinforcement at the base of the stud. After welding, the stud is permanently joined to the part. See Figure 2-6.

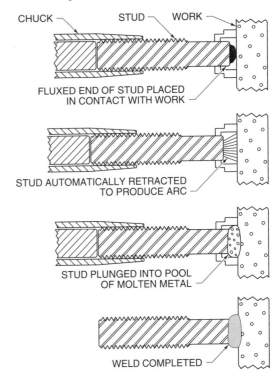

Figure 2-6. Stud welds permanently join threaded fasteners to parts using heat and pressure.

Spot, Projection, and Seam Welds

Spot, projection, and seam welds are welds produced by confining the fusion of molten base metal using heat and pressure. A weld nugget is formed where the fusion occurs. Electric current directed through the joint members is commonly used to produce the heat required. See Figure 2-7.

A *spot weld* is a weld type produced by confining the fusion of molten base metal using heat and pressure without preparation to the joint members. Joint members are in contact at the faying surface prior to welding. *Faying surface* is the part of the joint member which is in full contact prior to welding.

A *projection weld* is a weld type produced by confining the fusion of molten base metal using heat and pressure with a preformed dimple or projection in one of the joint members prior to welding. The joint members are in contact at the projection of the joint member prior to welding. The projection is melted during welding and becomes part of the weld nugget.

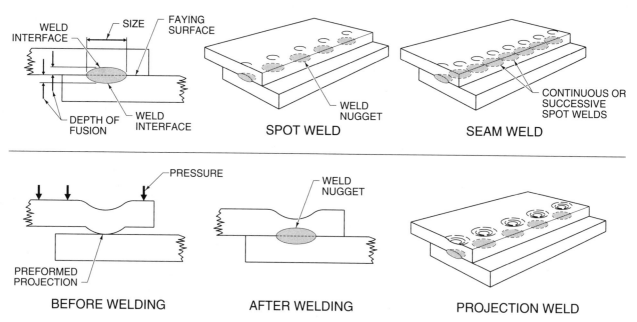

Figure 2-7. The size of spot and projection welds is the diameter of the weld nugget. The size of seam welds is the diameter of the weld nugget and length of the weld.

A *seam weld* is a weld type produced by confining the fusion of molten base metal using heat and pressure for a series of continuous or overlapping successive spot welds on the joint members. Continuous spot welds are made using rotary electrodes. Overlapping successive spot welds are made using conventional spot welding equipment.

these two welds is based on when the weld is deposited. Back welds are deposited after the weld on the opposite side of the part. Backing welds are deposited before the weld on the opposite side of the part. See Figure 2-8.

Surfacing Weld

A *surfacing weld* is a weld type in which weld beads are deposited on a surface to increase the dimensions of the part. Successive overlapping beads are deposited to form a layer of filler metal on the base metal.

Back or Backing Weld

A *back or backing weld* is a weld deposited in the weld root opposite the face of the weld on the other side of the joint member. The difference between

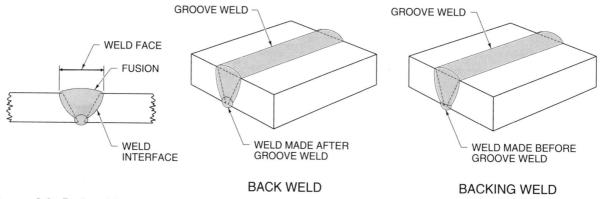

Figure 2-8. Back welds are deposited after the weld on the opposite side of the part. Backing welds are deposited before the weld on the opposite side of the part.

The filler metal can have the same or different characteristics as the base metal depending on the application. Surfacing welds are commonly used for building up worn parts or depositing abrasive-resistant metals. Layers of filler metal are deposited with additional layers deposited 90° to the prior layer. See Figure 2-9.

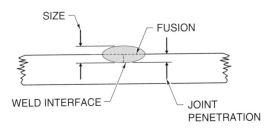

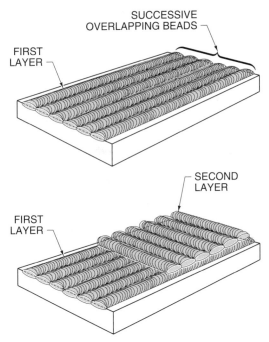

Figure 2-9. Surfacing welds are commonly used to build up worn parts.

Flange Weld

A *flange weld* is a weld type of light-gauge metal with one or both of the joint members bent at approximately 90°. An *edge-flange weld* is a flange weld with both joint members bent. A *corner-flange weld* is a flange weld with one joint member bent. In some instances, part of the joint members may be melted and become part of the filler metal. This allows joining of the two joint members without the addition of filler metal. See Figure 2-10.

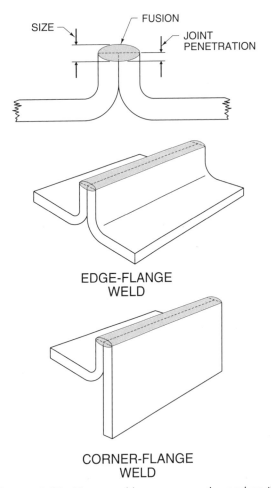

Figure 2-10. Flange welds are commonly used on thin materials where filler metal is not required.

WELDING SYMBOLS

A *welding symbol* is a graphic symbol that shows weld locations and specifications on prints. Object lines on the prints show the joint members before welding. The actual weld, however, is not commonly shown on conventionally drawn prints. With CAD drawings, more and more welds are being shown on the prints. The weld location and specifications, including type, size, process, and examination procedures, can be specified on the welding symbol.

Only information pertaining to the particular weld is included on the welding symbol. If welding requirements are the same on all welds on the print, a general note is included on the print. The parts of a welding symbol include the arrow, reference line, weld symbol, dimensions, and supplementary symbols. See Figure 2-11.

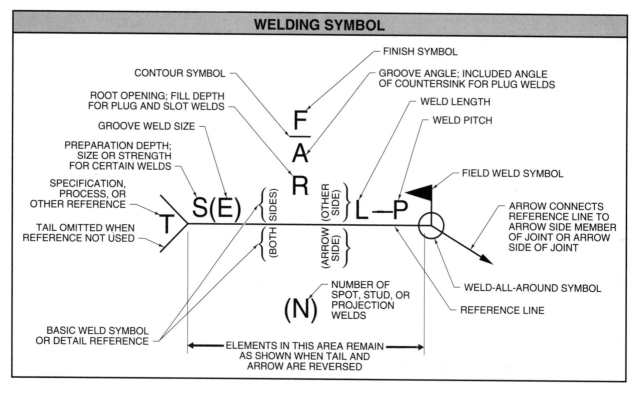

Figure 2-11. Welding symbols provide standardized information regarding welding and examination requirements.

Arrow

The *arrow* of the welding symbol identifies the location where the welding operation is to be performed. The arrow points to the joint and the tip of the arrowhead touches the object lines on the print. The arrow includes an arrowhead and a leader line which connects to the reference line of the welding symbol. More than one arrow can be used to specify the same weld required at different locations. See Figure 2-12.

Reference Line

The *reference line* of the welding symbol identifies the side of the joint to be welded. It is a horizontal line attached to the arrow. Information regarding weld type and specification is divided by the reference line into two parts, the arrow side and the other side. The side to which the arrow is attached determines arrow side and other side. Weld information located on the arrow side indicates the weld information pertaining to the view side of the weld joint nearest the welding symbol.

Weld information located on the other side of the arrow indicates the welding operation to be completed on the side of the weld joint furthest from the welding symbol. See Figure 2-13.

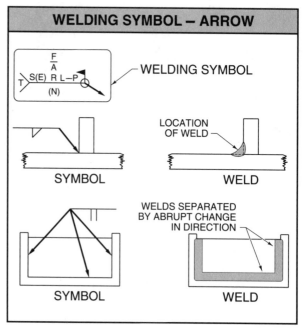

Figure 2-12. The arrow of the welding symbol points to the location of the weld specified.

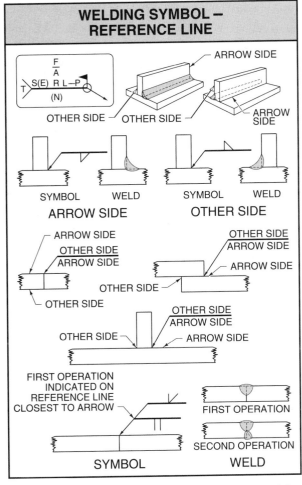

Figure 2-13. The reference line divides the welding symbol into the arrow side and the other side.

Multiple reference lines are used to indicate separate welding operations. The sequence of welding operations is determined by the proximity of the reference lines to the arrowhead. The operation indicated on the reference line closest to the arrowhead is completed first. Operations on other reference lines further from the arrowhead are then completed.

Weld Symbol

A *weld symbol* defines the cross-sectional shape of the weld. This is different from a welding symbol which contains all required information to complete the weld specified. Each weld type has a distinct weld symbol.

Weld symbols are located on the reference line of the welding symbol. If there is no significance on which side the weld is made, the weld symbol is centered in the reference line. Weld symbols with a vertical line always show the vertical line on the left side. More than one weld symbol may be required to specify a welding operation. See Figure 2-14.

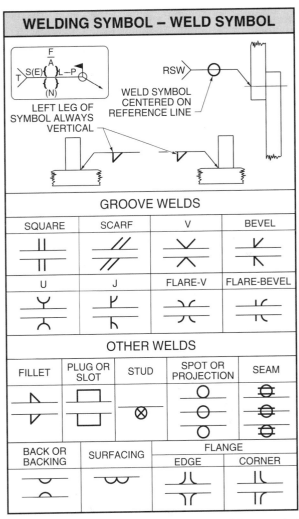

Figure 2-14. Weld symbols on the welding symbol indicate the type of weld required.

Dimensions

Dimensions are the part of a welding symbol that specify the weld size, number, and location. The location of dimensions for welds varies depending upon the weld type, joint, and application. See Figure 2-15. When no length dimension is specified on the welding symbol, the weld is continuous and extends the maximum length of the joint member.

Generally, inch marks and the millimeter designation (mm) are omitted on the welding symbol. A general note on the print indicates the unit of measurement. Degree symbols are also included with angular dimensions. Center-to-center dimensions are given for spacing of intermittent welds. The length of the weld increment should be increased to terminate a weld at the end of the joint.

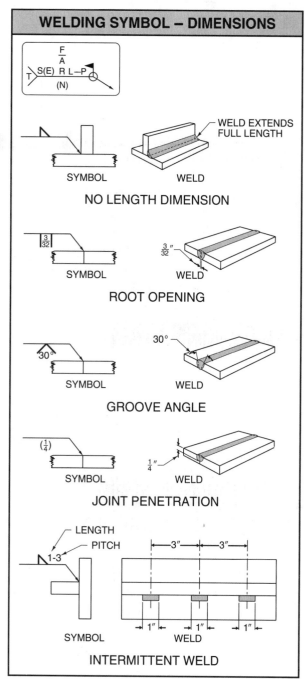

Figure 2-15. Dimensions included in the welding symbol specify weld size and location requirements.

Tail

The *tail* is the part of a welding symbol included when a specific welding process, specification, or procedure must be indicated. Special notations regarding detail and cross-sectional drawings are also included in the tail. For example, the letters GTAW in the tail specify the weld to be completed using the gas tungsten arc welding process.

Welding and allied processes are abbreviated with letter designations. The tail is omitted when no information is required in the tail of the welding symbol. See Figure 2-16.

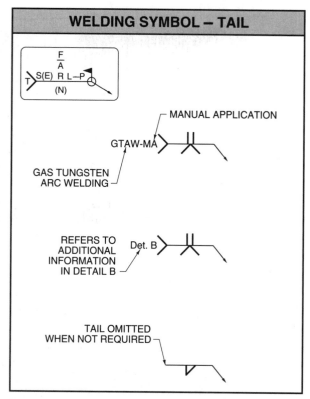

Figure 2-16. Information in the tail of the welding symbol pertains to weld specification, weld process, or other reference.

Supplementary Symbols

A *supplementary symbol* is a symbol that further defines the welding operation to be completed. Supplementary symbols used on welding symbols include weld-all-around, field weld, melt-through, consumable insert, backing, spacer, and contour. See Figure 2-17.

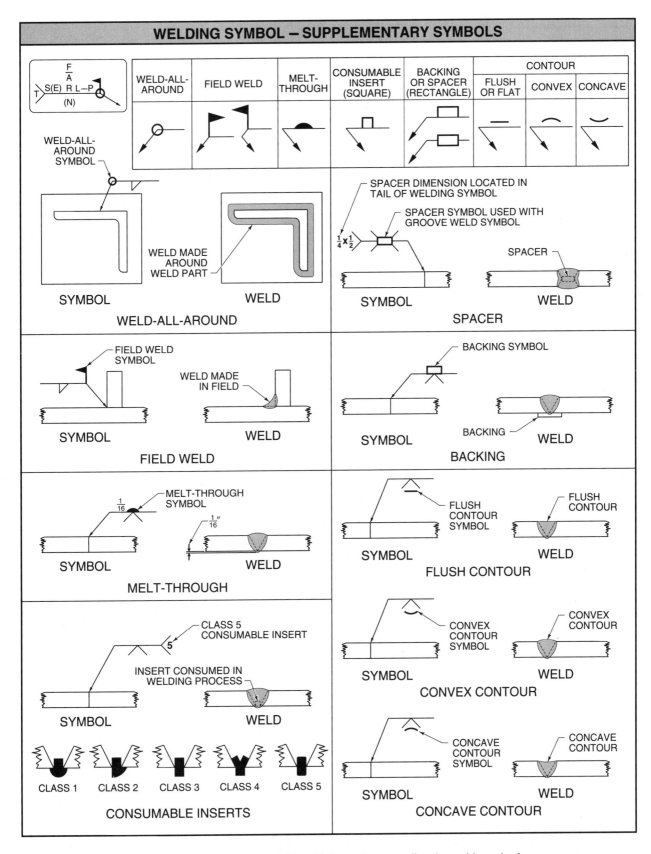

Figure 2-17. Supplementary symbols provide additional information regarding the weld required.

Weld-All-Around Symbol. The *weld-all-around symbol* is a supplementary symbol indicated by a circle at the intersection of the arrow and the reference line. The weld-all-around symbol specifies that the weld extends completely around the joint. The weld is continuous and does not have a break in joint penetration.

Field Weld Symbol. The *field weld symbol* is a supplementary symbol indicated by a triangular flag on a stem rising from the intersection of the arrow and reference line. The field weld symbol specifies that the welding operation is to be completed in the field at the location of final installation. The point of the flag always points toward the tail of the welding symbol.

Melt-Through Symbol. The *melt-through symbol* is a supplementary symbol indicated by a darkened radius on the reference line opposite the weld symbol specified. The melt-through symbol specifies that the filler metal deposited on one side must completely penetrate through to the other side of the weld.

Consumable Insert Symbol. The *consumable insert symbol* is a supplementary symbol indicated by a square on the opposite side of a groove weld on the reference line. A *consumable insert* is a spacer that provides the proper opening of the weld joint and becomes part of the filler metal during the welding process.

Information pertaining to the consumable insert is included in the tail of the welding symbol. The classification numbers 1, 2, 3, 4, and 5 refer to AWS classes of consumable inserts listed in the publication "Specifications for Consumable Inserts" (ANSI/AWS A5.30).

Backing Symbol. The *backing symbol* is a supplementary symbol indicated by a rectangle on the opposite side of the groove weld symbol on the reference line. Information pertaining to the composition and dimensions of the backing is included in the tail. The letter R is placed in the backing symbol if the backing is to be removed after welding.

Spacer Symbol. The *spacer symbol* is a supplementary symbol indicated by a rectangle centered on the reference line. Like the consumable insert, the spacer is melted and becomes part of the filler metal in the weld. Dimensions of the spacer are located in the tail of the welding symbol.

Contour Symbol. The *contour symbol* is a supplementary symbol indicated by a horizontal line or arc parallel to the weld symbol. The contour symbol specifies the shape of the weld after the weld is completed. A straight horizontal line specifies that the filler metal is to be flush with the joint member.

A radius with the open side down specifies that the weld is to have a convex contour. A radius with the open side up specifies that the weld is to have a concave contour. A letter above or below the contour symbol indicates the finishing operation for the weld. Letters used are C - chipping, H - hammering, G - grinding, M - machining, R - rolling, and U - unspecified.

EDGE PREPARATION

Edge preparation is indicated for weld joints which require weld reinforcement and/or access room to the weld area. The type of edge preparation required is determined by the engineer and is indicated by the welding symbol. The joint member requiring edge preparation is indicated on the welding symbol by a change in direction of the arrow. See Figure 2-18. The arrow always points to the joint member to be prepared. Edge preparation of the joint members is specified in the welding symbol but is not shown on the print.

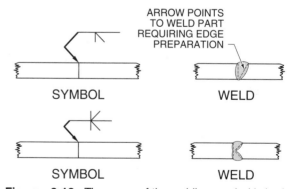

Figure 2-18. The arrow of the welding symbol is broken if a specific joint member requires edge preparation. If either joint member could be prepared, the arrow is not broken.

BRAZING SYMBOLS

A *brazing symbol* is a graphic symbol that shows braze locations and specifications on prints. Brazing symbols require no special symbols other than those used for specifying welds. However, the brazing process uses adhesion, instead of fusion, as in the welding process. *Adhesion* is the joining together of dissimilar metals by capillary action. See Figure 2-19.

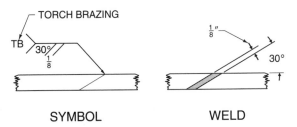

Figure 2-19. Brazing symbols include the brazing process designation in the tail.

For adequate strength, there must be sufficient surface area for the filler metal. For example, square-groove butt joints, or lap joints without the proper clearance, do not provide sufficient surface area for flow of filler metal. Edge preparation, joint member clearances, brazing process used, and required additional information are included in the brazing symbol.

NONDESTRUCTIVE EXAMINATION (NDE) SYMBOLS

Nondestructive examination (NDE) symbols are symbols that specify examination methods and requirements to verify weld quality. The method of examination required can be specified on a separate reference line of the welding symbol or as a separate NDE symbol.

When specified on a separate reference line, the order of operation is the same as for multiple welding operations. The reference line furthest from the arrowhead indicates the last operation to be performed. When used separately, NDE symbols include an arrow, reference line, examination letter designation, dimensions, areas, number of examinations, supplementary symbols, tail, and specifications and other references. See Figure 2-20.

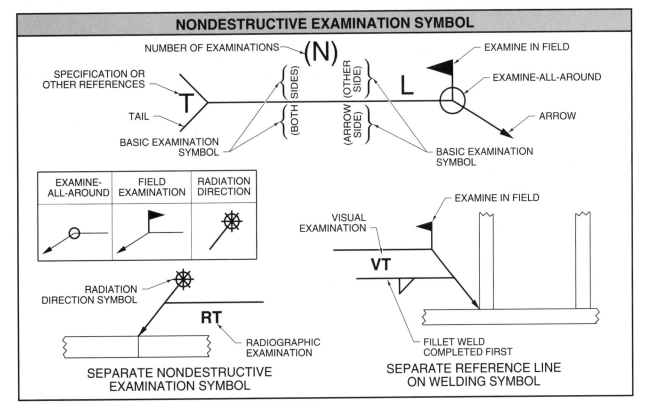

Figure 2-20. Nondestructive examination (NDE) symbols can be part of a welding symbol or a separate NDE symbol.

The location of the basic examination symbol is treated like a welding symbol. Examination required on the arrow side of the welded part is indicated by the basic examination symbol on the same side as the arrowhead. Examination required on the other side of the welded part is indicated by the basic examination symbol on the opposite side of the arrowhead. Examinations required on both sides are indicated by the basic examination symbol on both sides. When side has no significance, the basic examination symbol is located in the center of the reference line. Letter designations are used to indicate the method of examination. See Figure 2-21.

EXAMINATION METHODS	
METHOD	LETTER DESIGNATION
Acoustic emmision	AET
Electromagnetic	ET
Leak	LT
Magnetic particle	MT
Neutron radiographic	NRT
Penetrant	PT
Proof	PRT
Radiographic	RT
Ultrasonic	UT
Visual	VT

Figure 2-21. Letter designations indicate the method of examination.

The arrow of the NDE symbol can be used to indicate direction of radiographic examination. See Figure 2-22. *Radiographic examination* is the testing of welds for weld defects and strength requirements using X rays. The direction of radiographic examination can also be specified with a separate line in the required direction and by including the required angle on the print for clarity.

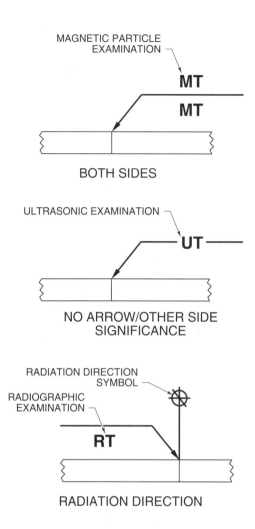

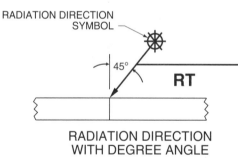

Figure 2-22. The location and direction of nondestructive examination are indicated on the NDE symbol.

Sketching

Name _____ Date _____

Sketching — Welding Symbols

Sketch the welding symbols. For example, weld-all-around.
All welds are arrow side unless otherwise indicated.

1. V-groove weld; flat contour	
2. V-groove weld; melt-through	
3. Fillet weld	
4. Fillet weld; other side	
5. Square-groove weld (first operation); back weld (second operation)	
6. Stud weld	
7. Scarf weld	
8. GTAW process	
9. Field weld	
10. J-groove weld; other side	

11. Class 5 consumable insert	
12. V-groove weld; flush contour	
13. J-groove weld; other side; concave contour	
14. Radiation direction	
15. Bevel-groove weld	
16. Bevel-groove weld; both sides	
17. Slot weld	
18. Spot weld; other side	
19. Seam weld	
20. Surfacing	

Sketching — Weld Symbols

Sketch the weld symbols. All welds are arrow side.

1. Square	
2. Scarf	
3. V	
4. Bevel	
5. Fillet	
6. Plug or slot	
7. Stud	
8. Spot	

9. U	
10. J	
11. Back or Backing	
12. Flare-V	
13. Flare-bevel	
14. Surfacing	
15. Corner-flange	

Sketching — Edge Preparation

Sketch the edges.

1. Square	
2. Double-bevel	
3. Single-J-groove	

4. Double-J-groove	
5. Round	

Review Questions

Name _____ Date _____

Matching — Weld Joints

_____ **1.** T-joint

_____ **2.** Edge joint

_____ **3.** Lap joint

_____ **4.** Corner joint

_____ **5.** Butt joint

True-False

T F **1.** Welding symbols are developed by the American Welding Society.

T F **2.** Joint penetrations and strength of the filler metal determine the strength of a weld joint.

T F **3.** A butt joint is generally stronger than a lap joint.

T F **4.** The weld type is shown on prints by the description of the object.

T F **5.** The weld toe is the intersection of the base metal and the weld face.

T F **6.** A fillet weld without additional information positions the joint members at 90°.

T F **7.** Slot welds are stronger than plug welds.

T F **8.** Back welds are deposited before the weld on the opposite side of the part.

T F **9.** The reference line of the welding symbol identifies the side of the joint to be welded.

T F **10.** Weld symbols with a vertical line always show the vertical line on the right side.

T F **11.** If no length dimension is specified on the welding symbol, the welder may make the weld any length desired.

T F **12.** Welding and allied processes are abbreviated in the tail with letter designations.

T   F   **13.** The melt-through symbol is a supplementary symbol.

T   F   **14.** If backing is to be removed after welding, the letters RAW are placed in the backing symbol.

T   F   **15.** The arrow of the welding symbol always points to the joint member to be prepared if edge preparation is indicated.

T   F   **16.** Brazing symbols are similar to welding symbols.

T   F   **17.** Edge joints are commonly used to join support structures and short lengths of structural steel.

T   F   **18.** The weld root is the exposed surface of the weld.

T   F   **19.** Fillet welds are the most popular type of weld.

T   F   **20.** Flange welds are commonly made on material over 6″ thick.

Matching — Weld Types

_____   **1.** Back

_____   **2.** Seam

_____   **3.** Fillet

_____   **4.** Slot

_____   **5.** Stud

_____   **6.** Plug

_____   **7.** Groove

_____   **8.** Projection

_____   **9.** Spot

_____   **10.** Surfacing

Completion

_____   **1.** A(n) _____ is the physical configuration of the joint members to be joined.

_____   **2.** A(n) _____ joint is a weld joint formed when two joint members, located approximately in the same plane, are positioned edge to edge.

_____   **3.** A weld _____ is the cross-sectional shape of the filler metal after welding.

4. The joint members of a spot weld are in contact at the _____ surface prior to welding.

5. A weld _____ defines the cross-sectional shape of the weld.

6. The _____ symbol specifies the shape of the weld after the weld is completed.

7. The _____ is the part of a welding symbol that is included when a specific welding process, specification, or procedure must be indicated.

8. _____ metal is the metal deposited during the welding process.

9. A(n) _____ weld is a weld type made in the groove of the welded pieces.

10. The _____ of the welding symbol identifies the location where the welding operation is to be performed.

Multiple Choice

1. _____ metal is the material to be welded.

    A. Filler
    B. Base
    C. both A and B
    D. neither A nor B

2. The field weld symbol is a supplementary symbol indicated by a _____.

    A. triangular flag
    B. rectangular flag
    C. circle
    D. half-circle

3. The arrow of the NDE symbol can be used to indicate the direction of _____.

    A. travel
    B. radiographic examination
    C. part orientation
    D. none of the above

4. A fillet weld is a weld type in the cross-sectional shape of a _____.

    A. square
    B. rectangle
    C. circle
    D. triangle

5. Plug and slot welds are used for similar applications as the _____ joint.

    A. butt
    B. lap
    C. edge
    D. corner

6. Surfacing welds are commonly used for _____.

    A. joining two edges together
    B. appearance over the weld
    C. increasing the thickness of a part
    D. none of the above

_____ **7.** _____ reinforcement is filler material which extends above the surface of the joint member on the side of the joint on which welding was done.
  A. Face
  B. Root
  C. Crown
  D. none of the above

_____ **8.** The consumable insert symbol is indicated on the reference line by a _____ on the opposite side of a groove weld.
  A. triangle
  B. circle
  C. square
  D. rectangle

_____ **9.** Joint penetration in a groove weld is determined by the distance between the joint members and the _____.
  A. thickness of the joint members
  B. edge preparation
  C. both A and B
  D. neither A nor B

_____ **10.** Lap joints are commonly welded on _____.

  A. one side
  B. both sides
  C. all surfaces
  D. none of the above

Matching — Welding Symbols

_____ **1.** Fillet weld; other side

_____ **2.** Bevel weld; arrow side

_____ **3.** Spot weld; other side

_____ **4.** Scarf weld; arrow side

_____ **5.** Fillet weld; both sides

_____ **6.** J-groove weld; other side

_____ **7.** Fillet weld; arrow side; Gas Tungsten Arc Welding

_____ **8.** Seam weld; other side

_____ **9.** Square-groove weld; both sides

_____ **10.** Surfacing weld; arrow side

Trade Competency Test

Name_____ Date _____

DRAWING NUMBER
B-905-28315

SHTS. | SHT.
CHANGES

.25 1.00

① ②

45° ELBOW

TYP.

1.50

2.50

45°

3/4" 150 LB.
SLIP-ON WELD FLANGE

NOTE:

1.) FLANGE MATERIAL— 316 STAINLESS STEEL

2.) PIPE AND ELBOW MATERIAL— SCHEDULE 80
316 STAINLESS STEEL

GENERAL ELECTRIC CO. U.S.A.
GE Superabrasives

	WORK TO DIMENSIONS	PROJECT	'G' TANK
	TOLERANCE UNLESS OTHERWISE SPECIFIED	DETAIL OR INST OF	DECANT— GOOSE NECK
	DECIMAL ± .XX= .06	SCALE 1"=1"	SOURCE DOCUMENT (CREATE/CHG) W/O 8497
	DECIMAL ± .XXX=	DRAWN D.S.	DATE 12-11-91	SHTS. SHT.
	FRACTIONAL ± 1/32	CHECKED	DATE	DRAWING NUMBER
NEXT ASSY. – P/L REF.	ANGLES ±		B-905-28315

'G' TANK

Refer to the 'G' Tank print on page 39.

_____ **1.** The tail designation on the welding symbol includes the letters _____.

T   F **2.** All welds on the print are specified for weld-all-around.

_____ **3.** Flange material to be used is _____.

_____ **4.** Weld 1 is a(n) _____ weld.

_____ **5.** Weld 2 is a(n) _____ weld.

_____ **6.** The manufacturer of the part is _____.

_____ 7. Tolerances specified for decimal dimensions are ±_____.

_____ 8. Tolerances specified for fractional dimensions are ±_____.

_____ 9. The scale for the drawing is 1″ = _____.

_____ 10. The drawings for the prints were drawn by _____.

_____ 11. Pipe and elbow material to be used is _____.

_____ 12. _____ 45° elbow(s) is (are) required to fabricate the part.
  A. One
  B. Two
  C. Three
  D. Four

_____ 13. _____ V-groove weld(s) is (are) specified.
  A. One
  B. Two
  C. Three
  D. Four

_____ 14. The drawing number of the part is _____.

_____ 15. The date the drawing was completed is _____.

Matching — Welding Symbols — 1

_____ 1. Seam weld

_____ 2. Flare-bevel weld

_____ 3. Stud weld

_____ 4. U-groove weld

_____ 5. Square-groove weld

_____ 6. V-groove weld

_____ 7. Weld-all-around symbol

_____ 8. Bevel-groove weld

_____ 9. Fillet weld

_____ 10. J-groove weld

ANCHOR BRACKET

CLIP ASSEMBLY

STUD MOUNT

Matching — Welding Symbols — 2

_____ 1. Back weld

_____ 2. Melt-through

_____ 3. Edge-flange weld

_____ 4. Field weld

_____ 5. Resistance welding

_____ 6. Corner-flange weld

_____ 7. Brazing symbol

_____ 8. Double-fillet weld

_____ 9. Spot weld

_____ 10. Flux cored arc welding

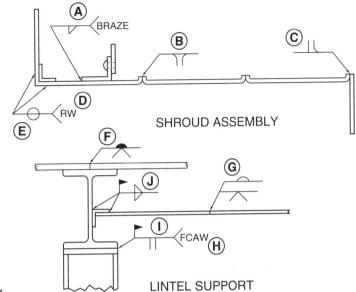

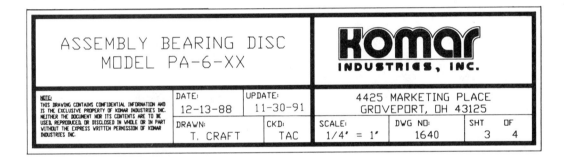

ASSEMBLY BEARING DISC MODEL PA-6-XX	**Komar** INDUSTRIES, INC.				
NOTE: THIS DRAWING CONTAINS CONFIDENTIAL INFORMATION AND IS THE EXCLUSIVE PROPERTY OF KOMAR INDUSTRIES INC. NEITHER THE DOCUMENT NOR ITS CONTENTS ARE TO BE USED, REPRODUCED, OR DISCLOSED IN WHOLE OR IN PART WITHOUT THE EXPRESS WRITTEN PERMISSION OF KOMAR INDUSTRIES INC.	DATE: 12-13-88	UPDATE: 11-30-91	4425 MARKETING PLACE GROVEPORT, OH 43125		
	DRAWN: T. CRAFT	CKD: TAC	SCALE: 1/4" = 1"	DWG NO: 1640	SHT 3 OF 4

TITLE BLOCK 1

Refer to Title Block 1 on page 41.

_____ 1. The drawing was updated on _____.

_____ 2. The total number of sheets in the drawing is _____.

    A. one
    B. two
    C. three
    D. four

_____ 3. The drawing was drawn by _____.

_____ 4. The scale for the drawing is _____ = 1".

_____ 5. The model number of the part is _____.

T   F   **6.** The drawing was checked by TAC.

T   F   **7.** The drawing number is 1640.

_____   **8.** The manufacturer of the part is _____.

_____   **9.** The manufacturer is located in _____.

T   F   **10.** The part is an Assembly Bearing Disc.

Matching — Welding Symbols — 3

_____   **1.** Length of weld

_____   **2.** Flare-bevel-groove weld

_____   **3.** Consumable insert

_____   **4.** Weld-all-around

_____   **5.** Flush contour

_____   **6.** Bevel-groove weld

_____   **7.** Backing

_____   **8.** Pitch of weld

_____   **9.** Plug weld

_____   **10.** Convex contour

VENT PIPE DETAIL

BRACE

FLOOR MOUNT

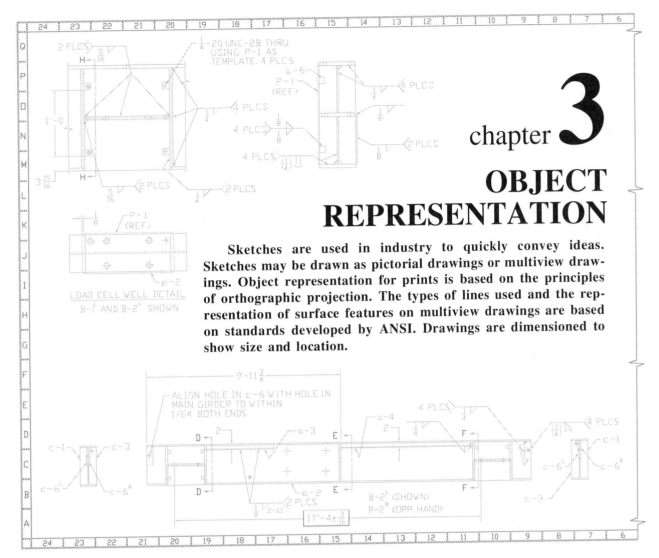

chapter 3

OBJECT REPRESENTATION

Sketches are used in industry to quickly convey ideas. Sketches may be drawn as pictorial drawings or multiview drawings. Object representation for prints is based on the principles of orthographic projection. The types of lines used and the representation of surface features on multiview drawings are based on standards developed by ANSI. Drawings are dimensioned to show size and location.

SKETCHING

Sketching is drawing without instruments. Sketches are made by the freehand method. The only tools required are a pencil, paper, and an eraser. See Figure 3-1.

Sketching pencils are either wooden or mechanical. Wooden pencils must be sharpened, and the lead must be pointed. Erasers are not commonly placed on sketching pencils.

One type of mechanical pencil contains a thick lead that is pointed with a file, sandpaper, or lead pointer. Another type contains a thin lead which does not require pointing. Softer leads, such as HB, F, and H, are commonly used for sketching.

The paper selected for sketching depends upon the end use of the sketch. Plain paper is commonly used. Tracing vellum is used if the sketch is to be duplicated on a diazo printer. Sketching papers and tracing vellums are available in pads or sheets in standard sizes designated A, B, or C. Size A is 8½″ × 11″. Size B is 11″ × 17″. Size C is 17″ × 22″. The paper is either plain or preprinted with grids to facilitate sketching. Preprinted paper is available in a variety of grid sizes for orthographic and pictorial sketches. A grid size of ¼″ is popular. Grids are commonly printed in light-blue, non-reproducing inks.

Erasers are designed for use with specific papers and leads. The eraser selected should be soft enough to remove pencil lines without smearing lines or damaging the paper. Pink pearl and white vinyl erasers are commonly used while sketching.

Electric erasers and battery-powered erasers are available. These light-weight erasers provide smooth easy operation with efficient cleanup.

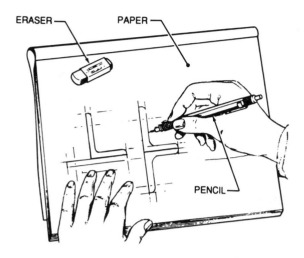

Figure 3-1. Sketching is drawing without instruments.

Sketching Techniques

The pencil point should be pulled across the paper while sketching. Pushing the pencil point can tear the paper. While pulling the pencil, slowly rotate it to produce lines of consistent width.

Shading techniques are not used with orthographic drawings. Pictorial drawings may be shaded. Horizontal, inclined, vertical, and curved lines are drawn to produce orthographic and pictorial drawings. See Figure 3-2.

Sketching Horizontal Lines. To sketch horizontal lines, locate the end points with dots to indicate the position and length of the line. For short lines, the end dots are connected with a smooth wrist movement from left to right (for a right-handed person). Long lines may require intermediate dots. If grid paper is used, intermediate dots are not required. For long lines, a full arm movement may be required to avoid making an arc.

The top or bottom edges of the paper or pad may be used as a guide while sketching horizontal lines. Light, trial lines are drawn first to establish the straightness of the line. The line is then darkened. With sketching experience, the trial lines may be omitted.

Sketching Vertical and Slanted Lines. To sketch vertical lines, locate the end dots and draw from the top to the bottom. The side edges of the paper

or pad may be used as a guide while sketching vertical lines.

Slanted lines are inclined lines. They are neither horizontal nor vertical. To draw slanted lines, locate end dots and draw from left to right (for a right-handed person). The paper may be rotated so that the slanted lines are in either a horizontal or vertical position to facilitate sketching.

SKETCHING LINES		
HORIZONTAL	INCLINED	VERTICAL

1. Locate end points.

2. Draw horizontal and inclined lines from left to right. Draw vertical lines from top to bottom.

3. Darken object lines.

Figure 3-2. Lines are sketched by locating their endpoints.

Sketching Plane Figures. Pictorial and orthographic drawings consist of lines, arcs, and plane figures in varying combinations. Common plane figures include circles, triangles, quadrilaterals, and polygons.

To sketch circles, locate the centerpoint and draw several intersecting diameter lines. Mark off the radius on these lines, and connect with a series of arcs. The diameter is commonly dimensioned. See Figure 3-3.

To sketch triangles, draw the base, determine the angles of the sides, and draw straight lines to complete. Generally, one or more of the sides is dimensioned and the angle is noted.

To sketch quadrilaterals, draw the base line and determine corner points. Connect the corner points with straight lines to complete. Dimensions of two sides and angles, as required, are often included.

To sketch polygons, locate the centerpoint, and draw a circle of the appropriate size. Mark off the length of each side on the circumference of the circle, and connect with a series of straight lines. Darken the lines to complete the polygon.

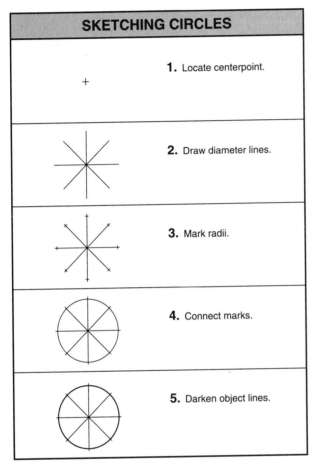

SKETCHING CIRCLES

1. Locate centerpoint.

2. Draw diameter lines.

3. Mark radii.

4. Connect marks.

5. Darken object lines.

Figure 3-3. Circles are sketched by marking off their radii.

PICTORIALS

Pictorial sketches or drawings look like a "picture" because they convey a sense of perspective and realism of the object being viewed. Height, length, and depth of the object are easily shown with various types of pictorial drawings.

While pictorial drawings are only occasionally used when preparing prints, an understanding of their basic concepts aids in the interpretation of prints. Additionally, the ability to quickly sketch a pictorial drawing of an object or detail aids in conveying technical information to others. Three

basic types of pictorial drawings are axonometric, oblique, and perspective. Perspective drawings are seldom used on welding prints. See Figure 3-4.

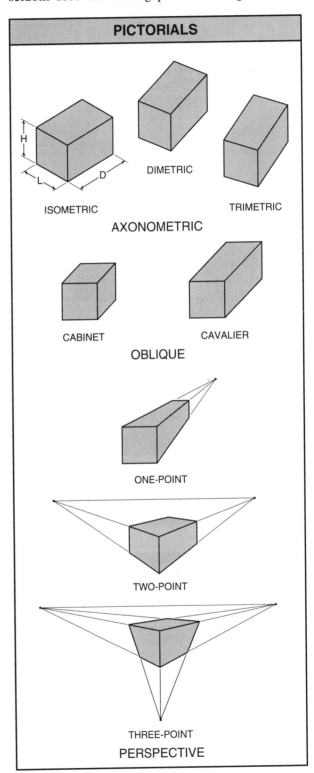

PICTORIALS

DIMETRIC

ISOMETRIC

TRIMETRIC

AXONOMETRIC

CABINET

CAVALIER

OBLIQUE

ONE-POINT

TWO-POINT

THREE-POINT

PERSPECTIVE

Figure 3-4. Three basic types of pictorial drawings are axonometric, oblique, and perspective.

Axonometric

An *axonometric* drawing is a pictorial drawing showing three sides of an object with horizontal and vertical dimensions drawn to scale and containing no true view of any side. The three basic types of axonometric drawings are isometric, dimetric, and trimetric. Of these, the isometric is the most commonly used. An *isometric* is an axonometric drawing with the axes drawn 120° apart. A *dimetric* is an axonometric drawing with two axes drawn on equal angles and one axis containing either fewer or more degrees. A *trimetric* is an axonometric drawing with all axes drawn at different angles. See Figure 3-5.

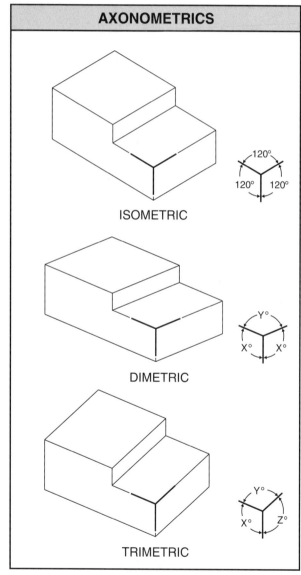

Figure 3-5. Isometrics are the most commonly drawn axonometric drawing.

Isometric. Isometric drawings contain three equal axes that are drawn 120° apart. Because of this 120° angle, no surface appears as a true view; however, the object has a natural appearance because three sides are seen. A *true view* is a view in which the line of sight is perpendicular to the surface.

Because of the skewed sides, circles (drilled holes, counterbores, etc.) appear as ellipses on isometric surfaces. Additionally, arcs appear as portions of ellipses. All surfaces not in one of the three principle isometric planes must be drawn by locating end points of the skewed surface. The end points are connected to complete the skewed surface. See Figure 3-6.

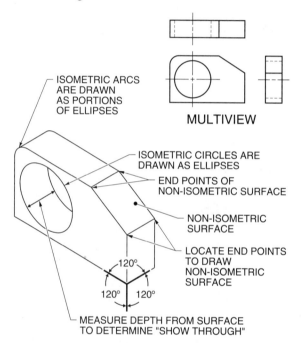

Figure 3-6. Isometrics contain three equal axes drawn 120° apart.

Sketching Isometrics. Isometric sketches are made by the following procedure. See Figure 3-7.

1. Locate the isometric axes and block in the front view using length and height measurements from the multiview. There are two front surfaces on this particular object. These two surfaces are parallel to one another. The depth dimension used to establish the location of the second front surface is taken from either the top or right side view of the multiview.

2. Sketch the outline shape of the front surfaces. Measurements are taken from the multiview. Notice that the arc on the second front surface is drawn as a portion of an ellipse.

3. Locate the centerpoint of the drilled hole on the second front surface. Construct an ellipse using measurements from the front view of the multiview.
4. Draw receding lines long enough to mark the depth of the object. Receding lines are parallel to the isometric axis.
5. Draw lines to establish the back surface. These lines are parallel to the isometric axis.

The skewed line on the back surface representing the V portion is drawn to its corresponding line in the front surface. Find the depth through the drilled hole to determine if the back portion of the drilled hole will show through on the front view. Draw a portion of the ellipse as required.
6. Darken all object lines to complete the isometric sketch.

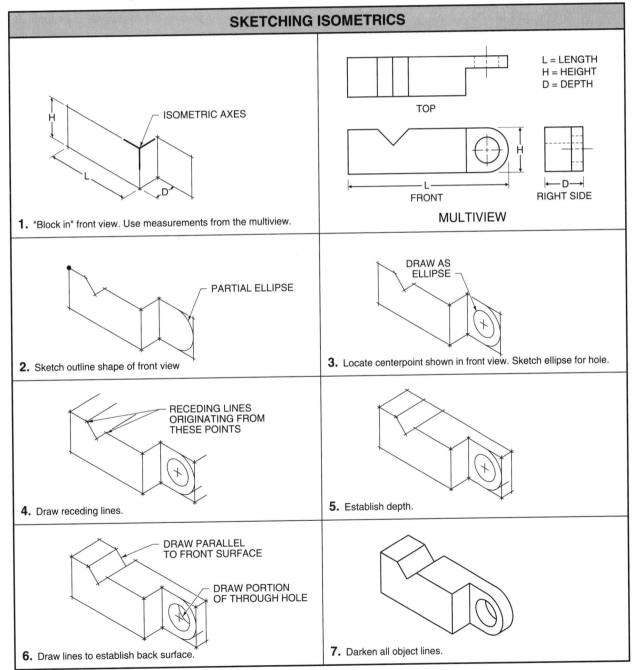

Figure 3-7. Isometrics can be quickly sketched using measurements from the multiview.

Circles on Isometrics. Circles on isometric surfaces are drawn as ellipses. An *ellipse* is a plane curve with two focal points. The sum of the distances from these two focal points to any point on the ellipse determines the shape of the ellipse. As the distance between the focal points decreases, the ellipse becomes more circular in shape.

For sketching, the parallelogram method of constructing ellipses is often used. In this method, dimensions from the multiview are used to determine the size of the parallelogram. Arcs are then drawn or sketched using the intersecting points as centerpoints. See Figure 3-8.

Oblique

Oblique drawings are pictorial drawings that show one surface of an object as a true view. All other surfaces of the object are distorted by the angle of the receding, oblique lines. All features shown on the face containing the true view are drawn as they appear. Additionally, right angles are shown at 90° on surfaces having a true view. For example, a drilled hole shown on the true view face of an oblique drawing is shown as a circle. Drilled holes on any other surface of the oblique drawing appear as ellipses. The angle at which an ellipse is drawn

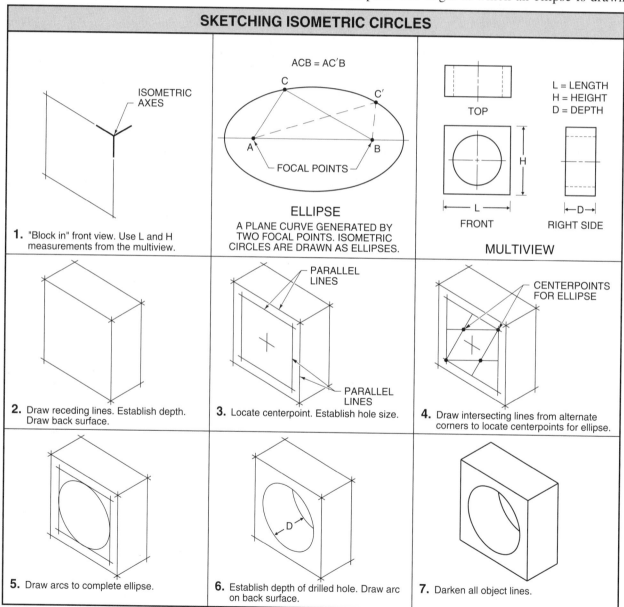

Figure 3-8. Circles on isometrics appear as ellipses.

on these surfaces is determined by the angle of the receding lines. Normally, these lines are drawn on a 30° or 45° angle.

The front view of an object is generally the view which shows the most shape, has the most detail, or is commonly thought of as the front view. The two types of oblique drawings are cabinet and cavalier.

Cabinet drawings are obliques with receding lines drawn to one-half the scale of lines in the true view. This is the most commonly used type of oblique drawing. *Cavalier* drawings are obliques with receding lines drawn to the same scale as the lines in the true view. The use of the same scale to draw all oblique lines of a cavalier drawing produces a distorted pictorial. Consequently, this type of drawing is seldom used.

Sketching Obliques. Oblique sketches are made by the following procedure. See Figure 3-9.

1. Determine which oblique will be sketched. (An oblique cabinet is shown.)

2. Block in the front view (usually the view with the most detail). Use length and height measurements of the multiview.
3. Complete the outline shape as shown in the multiview.
4. Locate all centerpoints of circles and arcs. Sketch circles and arcs.
5. Draw receding lines long enough to mark the depth of the object. Receding lines shown are drawn at a 45° angle. They could also be drawn at a 30° angle.
6. Establish depth dimension from the right side or top view of the multiview. Note that only one-half the dimension shown is drawn for the depth of the oblique cabinet.
7. Draw lines to represent the object's back surface. Draw receding lines. Find the depth through the drilled hole to determine if the back portion of the drilled hole will show through on the front view. Draw portion of hole as required.
8. Darken all object lines to complete the oblique sketch.

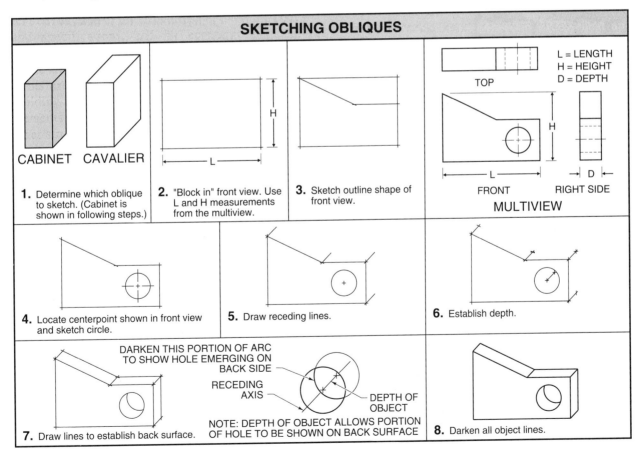

Figure 3-9. Obliques show one surface on an object as a true view.

ORTHOGRAPHICS

Orthographic projection (multiview drawing) is drawing at right angles. In multiview drawing, each view of an object is shown two-dimensionally. The six basic views of a multiview drawing are front, top, right side, back, left side, and bottom. The front, top, and right side are the most common views. Each view has a definite relationship of position with each other view. These views are projected from one another. See Figure 3-10.

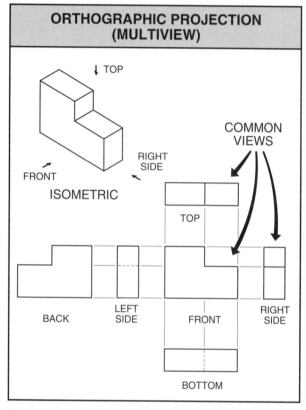

Figure 3-10. The front, top, and right side of a multiview are the most common views.

Multiviews

Welding prints utilize multiview (orthographic) drawings to show the various views of the object. Multiview drawings are based on the principle of showing each view of an object in detail. All objects have three basic dimensions: length, height, and depth. Any one view of a multiview drawing can only show two of these dimensions. For example, the front view shows length and height; the top view shows length and depth; and the right side view shows height and depth.

Only two views of any object are required to show the three basic dimensions. For example, two views of a cylindrical object show the three basic dimensions of length, height, and depth. The front view shows length and height, and the top view shows length and depth (which is the diameter of the cylindrical object). For most multiview drawings, however, the front, top, and right side views are required to clearly show all details. See Figure 3-11.

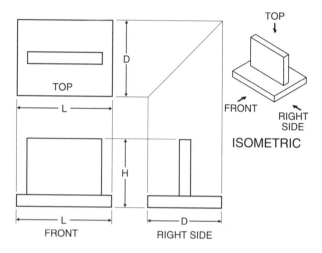

VIEW	SHOWS
Front	Length and height
Top	Length and depth
Right Side	Height and depth

Figure 3-11. Any one view of a multiview shows two dimensions.

Sketching Multiviews. Multiview sketches are made by the following procedure. See Figure 3-12.

1. Block in the length and height dimensions of the front view. Locate the front view to allow even spacing of views.
2. Project the length and height dimensions of the front view to the top and right side views.
3. Establish depth dimensions of the top and right side views. The depth dimension is shown with a 45° miter line on the right side of the top view to establish the same depth dimension in the right side view.
4. Sketch the drilled hole in the top view. Project hole dimensions to the front and right side views and draw hidden lines.
5. Draw center lines. Darken all object lines to complete the multiview sketch.

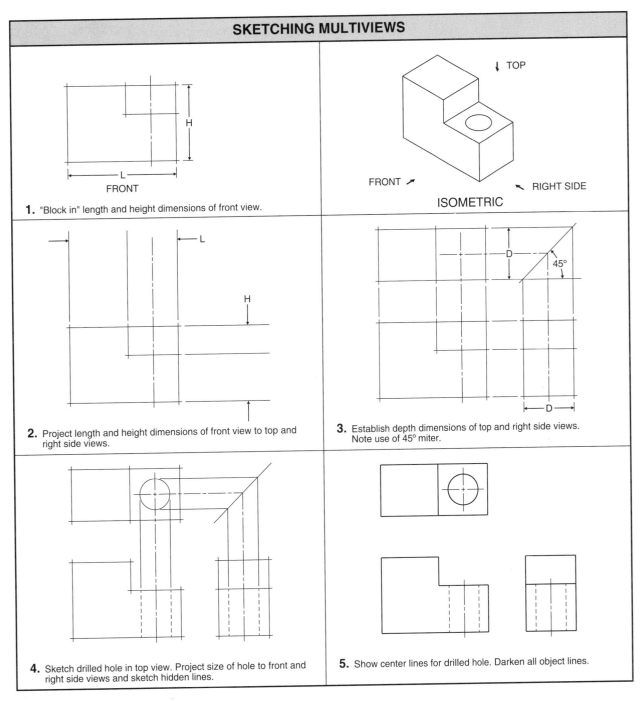

SKETCHING MULTIVIEWS

1. "Block in" length and height dimensions of front view.

ISOMETRIC

2. Project length and height dimensions of front view to top and right side views.

3. Establish depth dimensions of top and right side views. Note use of 45° miter.

4. Sketch drilled hole in top view. Project size of hole to front and right side views and sketch hidden lines.

5. Show center lines for drilled hole. Darken all object lines.

Figure 3-12. Dimensions of views in a multiview are projected from adjacent views.

Dimensions

Dimensions are numerical values that give size, form, or location of objects. All dimensions required to fabricate the part should be given. Dimensions are placed according to the function and mating relationship of a part. They should be placed so that they allow only one interpretation.

Drawings do not commonly specify manufacturing methods. The manufacturing method is commonly determined by the engineering department. Unless otherwise specified, all dimensions apply at 68°F (20°C).

Linear dimensions are dimensions that measure lines and are commonly expressed as decimal inches or millimeters. See Appendix. The decimal inch is predominately used in the United States. Decimal dimensions on prints are commonly expressed in hundredths of an inch (.01″), thousandths of an inch (.001″), and ten thousandths of an inch (.0001″). The use of decimal dimensions in hundredths of an inch to replace fractional dimensions is becoming increasingly popular.

Angular dimensions are dimensions that measure angles and are commonly expressed as degrees and decimal parts of a degree, or as degrees, minutes, and seconds. If the same dimensioning system is used throughout a drawing, inch marks (″) or millimeter (mm) designations may be omitted. A general note on the print, however, should specify the dimensioning system used. See Figure 3-13.

Fractional dimensions are dimensions that measure lines and are commonly expressed as inches and fractional parts of an inch. They are generally used on ordinary work not requiring a high degree of tolerance. *Tolerance* is the amount of variation allowed above or below a dimension. For example, a part with a specified dimension of $4\frac{1}{8}″ \pm \frac{1}{64}″$ is acceptable anywhere between $4\frac{7}{64}″$ ($4\frac{1}{8}″ - \frac{1}{64}″ = 4\frac{7}{64}″$) and $4\frac{9}{64}″$ ($4\frac{1}{8}″ + \frac{1}{64}″ = 4\frac{9}{64}″$).

Fractional dimensions are commonly based on fractional sizes having a denominator that is a multiple of 4. For example, $\frac{1}{2}$, $\frac{1}{4}$, $\frac{1}{8}$, $\frac{1}{16}$, $\frac{1}{32}$, and $\frac{1}{64}$ are the common denominators used with fractional parts of an inch.

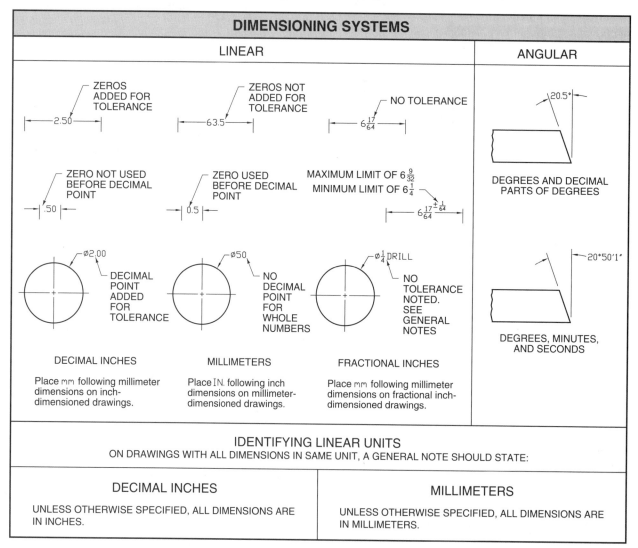

Figure 3-13. Dimensions are linear or angular.

Dimensions and notes are placed to read from the bottom of the sheet. They are generally placed outside the view of the object. Intermediate dimensions are placed nearest the object, and the overall dimension is placed farthest from the object. One intermediate dimension is omitted, or indicated as a reference dimension, if an overall dimension is used. A reference dimension is indicated by the placement of parenthetical brackets around the dimension. See Figure 3-14.

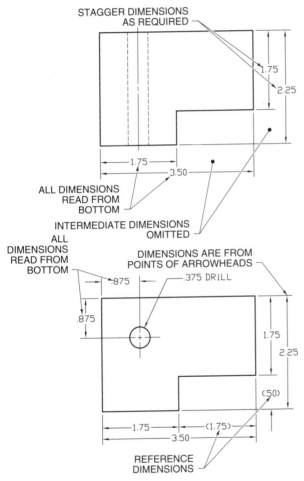

Figure 3-14. Dimensions read from the bottom of the sheet.

Lines

All drawings are composed of lines to show the shape of the drawn object. Lines may be drawn by the conventional method or by CAD. CAD lines are in linetype libraries. A *linetype library* is a CAD file that contains dashed, hidden, center, phantom, dot, dotdash, border, and divide lines.

Lines on prints have specific meanings. They are drawn based on conventions recommended by "Line Conventions and Lettering" (ANSI Y14.2M).

All lines should be sharp and dark. Lines are either thick or thin. Thick lines should be approximately .032″ (0.7 mm) wide. Thin lines should be approximately .016″ (0.35 mm) wide.

Common lines on drawings include object, hidden, center, dimension, extension, leader, cutting plane, section, and break lines. Each of these lines has a unique appearance that allows it to be easily recognized.

Dimension lines, extension lines, and leaders are used in conjunction with numerals to apply dimensions to the object. Dimensions are specified by "Dimensioning and Tolerance" (ANSI Y14.5M). See Figure 3-15.

Object Lines. *Object lines* are lines that define the visible shape of an object. They are used wherever there is a distinct change in the surface. Object lines are thick and dark. They are the most common lines shown on a print.

Hidden Lines. *Hidden lines* are lines that represent shapes which cannot be seen. Hidden lines are used wherever there is a distinct change in the surface. Hidden lines are thin and dark. They are drawn with a series of 1/8″ (3 mm) dashes separated by 1/32″ (0.75 mm) spaces.

Views to be drawn should be selected to minimize the use of hidden lines. A large number of hidden lines in one view makes the view difficult to read.

Center Lines. *Center lines* are lines that locate the centerpoints of arcs and circles. Center lines are thin and dark. They are drawn as a series of long and short dashes separated by spaces. The long dashes are from 3/4″ (19 mm) to 1 1/2″ (38 mm) long. The short dashes are 1/8″ (3 mm) long. The spaces are 1/16″ (1.5 mm) long.

Centerpoints for arcs and circles are represented by two short dashes crossing one another at right angles. Centerpoints are drawn in the centers of all circles and arcs.

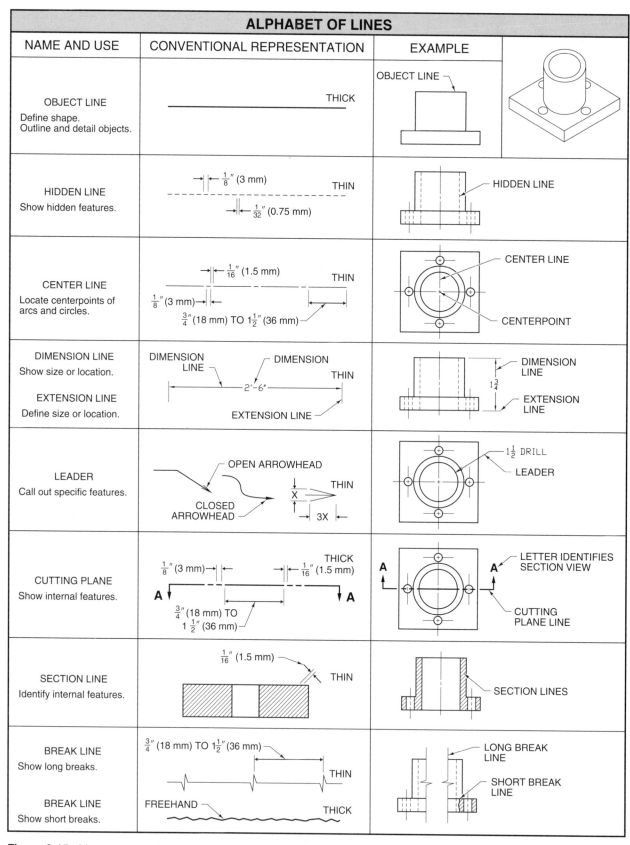

Figure 3-15. Lines on prints have specific meanings.

Dimension Lines. *Dimension lines* are lines that are used with dimensions to show size or location. They are thin and dark. Dimension lines are commonly broken for the placement of numerals giving the measurement. If a horizontal dimension line is not broken, the numeral is placed above the dimension line. See Figure 3-16.

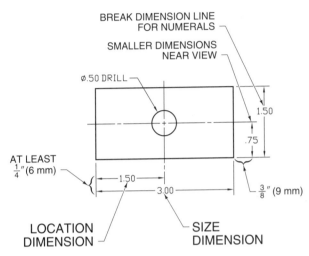

Figure 3-16. Dimension lines show size or location.

Dimension lines are terminated by arrowheads. An *arrowhead* is a symbol that indicates the extent of a dimension. Arrowheads are approximately $\frac{1}{8}''$ (3 mm) long and are three times as long as they are wide. Arrowheads may be open or filled-in.

Smaller dimensions are placed near the view. Intermediate dimensions are placed outside of smaller dimensions. Overall dimensions are placed the farthest from the view. The dimension line nearest the view should be at least $\frac{3}{8}''$ (9 mm) from the view. Succeeding dimension lines should be no closer than $\frac{1}{4}''$ (6 mm).

For parallel dimension lines, the dimensions are staggered to avoid crowding. Dimensions are aligned for appearance. Dimension lines should not cross other dimension lines.

Extension Lines. *Extension lines* are lines that extend from surface features and terminate dimension lines. They are thin and dark. Extension lines should not touch the feature from which they are extended. A $\frac{1}{16}''$ (1.5 mm) gap separates the extension line from the part.

Generally, extension lines should not cross other extension lines or dimension lines. If they must

cross, they are not broken. If extension lines must cross arrowheads, they are broken. Extension lines extend $\frac{1}{8}''$ (3 mm) beyond the dimension line. See Figure 3-17.

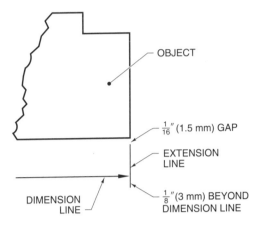

Figure 3-17. Extension lines terminate dimension lines.

Leaders. *Leaders* are lines that connect a dimension, note, or specification with a particular feature of the drawn object. They call out specific features such as drills, countersinks, counterdrills, etc. They are thin and dark.

One end of a leader terminates with an arrowhead or a dot. Arrowheads terminate on a line. Dots are placed within an area. The other end of the leader is generally connected to a shoulder although this is optional. The $\frac{1}{4}''$ (6 mm) shoulder is centered on the lettering at the beginning or ending of the note.

Leaders are inclined (slanted) lines. Leaders are commonly not drawn in a horizontal or vertical position. They are drawn at an angle so they will show up easily. See Figure 3-18.

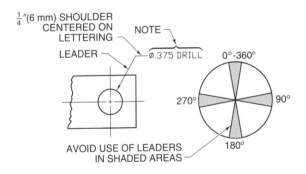

Figure 3-18. Leaders connect notes with a particular feature of the drawn object.

Cutting Plane Lines. *Cutting plane lines* are lines that show where an object is imagined to be cut in order to view internal features. They are thick and dark. Cutting plane lines are drawn as a series of long single and short double dashes. Long single dashes are from ¾″ (19 mm) to 1½″ (38 mm) long. Short double dashes are ⅛″ (3 mm) long. Spaces are 1/16″ (1.5 mm) long.

Arrowheads on the ends of the cutting plane line indicate the direction of sight. Letters near the arrowheads identify the section view.

Section Lines. *Section lines* are lines that identify the internal features cut by a cutting plane line. They are thin and dark. Section lines are drawn at an angle and are spaced 1/16″ (1.5 mm) apart for general section lines. Specific configurations of section lines identify the particular type of material.

Break Lines. *Break lines* are lines that can show internal features or avoid showing continuous features. Long break lines are drawn with a thin, dark line containing a zig-zag every ¾″ (19 mm) to 1½″ (38 mm). Short break lines are drawn with a thick, dark, freehand line.

SURFACE FEATURES

Metal may be cast or forged, and then machined as required. *Cast* metal parts are made by heating metal to its liquid state and pouring it into a mold where it cools and resolidifies. *Forged* metal parts are made by forming metal by a mechanical or hydraulic press with or without heat.

The surface of a metal part may be rough or finished. A rough surface appears as cast or forged. A finished surface is machined. Finished surfaces are required where parts mate or fit together. Surfaces, holes, fillets and rounds, and runouts produce surface features.

Surface features are any part of the surface where change occurs. A *normal surface* is a plane surface parallel to a plane of projection. It appears as a true view in the orthographic view to which it is parallel. It appears as a vertical or horizontal line on the remaining two orthographic views. An *oblique surface* is a plane surface not parallel to any plane of projection. It does not appear as a true view in any

orthographic view. An oblique surface appears foreshortened in all orthographic views.

An *inclined surface* is a plane surface perpendicular to one plane of projection and inclined to the remaining two orthographic views. *Intersecting surfaces* are created when one surface meets another surface. Distinct changes in surfaces are represented by object lines. Smooth changes in surfaces, such as when a curved surface becomes tangent to a flat surface, are not represented by object lines. See Figure 3-19.

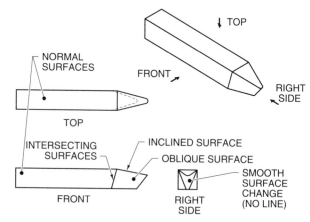

Figure 3-19. Surfaces are defined by lines.

Holes

Holes appear as circles in one view and in profile in adjacent views. Hole sizes are specified by their diameters. They are never specified by their radii. Notes give the size and type of hole. For example, a typical note that reads ∅ .750 DRILL-2.50 DEEP indicates that the diameter of the drill used to drill the hole is ¾″ and the hole is to be drilled 2½″ deep. The diameter of the hole (.750) is given first and is followed by the operation (drill).

A *through hole* is a drilled hole passing completely through the material. A *blind hole* is a drilled hole that does not pass through the material. The depth of blind holes is specified in the notes. The depth is measured only along the cylindrical portion of the drilled hole. The point of the drill leaves a cone shape in the bottom of the drilled hole. This cone shape is generally drawn as a 90° angle although it may be from approximately 80° to approximately 118°. Common operations to produce holes include drilling, reaming, counterboring, countersinking, and spotfacing. See Figure 3-20.

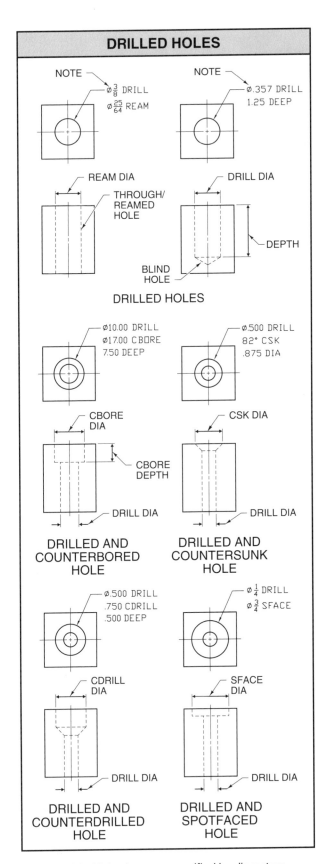

Figure 3-20. Hole sizes are specified by diameters.

Drilled Holes. A *drill* is a round hole in a material produced by a twist drill. Twist drills (also known as drills) are sized by their diameter. Sizes are designated by numbers 1 through 80, letters A through Z, and fractions from $\frac{1}{64}''$ to $3\frac{1}{2}''$. A No. 1 drill has a diameter of .2280″. A No. 80 drill has a diameter of .0135″. An A drill has a diameter of .234″. A Z drill has a diameter of .413″.

Fractional drills from $\frac{1}{64}''$ to $1\frac{3}{4}''$ are available in $\frac{1}{64}''$ increments. From $1\frac{3}{4}''$ to $2\frac{1}{4}''$, they are available in $\frac{1}{32}''$ increments. From $2\frac{1}{4}''$ to $3\frac{1}{2}''$, they are available in $\frac{1}{16}''$ increments.

Drills in metric sizes are also available. See Appendix. A drill appears as a circle in one view and as parallel hidden lines in the adjacent view.

The diameter of the drill is dimensioned in the circular view by the diameter extended to a note containing the dimension, extension lines terminating the diameter which contains the dimension, or a leader from a note containing the dimension. The symbol for diameter is $\varnothing$. The $\varnothing$ precedes the actual dimension.

The centerpoint is common on all circular views of drilled holes. The centerline is common on all side views of drilled holes. The centerpoint and center ine represent the axis of the concentric surfaces.

The depth of a drill is measured from the surface to the bottom of the cylindrical hole. The angle at the head of the drill is not included in the depth measurement. If it is not clear that a drill passes completely through the material, the dimension for the drill is followed by the abbreviation THRU. Drilled holes may be counterbored, countersunk, counterdrilled, or spotfaced.

Reaming is enlarging and improving the surface quality of a hole. Standard reamers are available in $\frac{1}{64}''$ (0.375 mm) increments.

Counterbored Holes. A *counterbored hole* is an enlarged and recessed hole with square shoulders. Counterbored holes permit screw heads or other mating parts to be recessed below the surface of the part.

The diameter of the smaller hole is given first. The diameter of the larger (counterbored) hole is given next, followed by the depth of the counterbore, if given. The abbreviation CBORE is commonly used on notes.

Countersunk Holes. A *countersunk hole* is a hole with a cone-shaped opening or recess at the outer surface. Countersunk holes permit the flush seating of parts with angled or tapered heads. For example, a flat-head screw can be positioned in a countersunk hole so that the head is flush (level) with the surface. A *countersink* is the tool that produces a countersunk hole. Countersinks are commercially available with standard included angles of 60° or 82°.

Countersunk holes are dimensioned by giving their drill diameter first. The diameter of the countersunk portion is given next, followed by the angle of the countersunk portion. The abbreviation CSK is commonly used on notes.

Counterdrilled Holes. A *counterdrilled hole* is a hole with a cone-shaped opening below the outer surface. Counterdrilled holes permit the recessed seating of parts with angled or tapered heads. For example, a flat-head screw can be positioned in a countersunk hole so that the head is recessed below the surface.

Counterdrilled holes are dimensioned by giving their drill diameter first. The diameter of the counterdrilled portion is given next, followed by the depth of the counterdrilled portion. The included angle of the counterdrill is optional. The abbreviation CDRILL is commonly used on notes.

Spotfaces. A *spotface* is a flat surface machined at a right angle to a drilled hole. Spotfaces may be recessed below the surrounding surfaces or may be machined slightly above the surrounding surfaces. They provide an area for tight fits of square-shouldered parts.

Spotfaces are dimensioned by giving their drill diameter first. The diameter of the spotface is given next. The depth of the spotface or the remaining thickness of the material is optional. The abbreviation SF is commonly used on notes.

Edges and Corners

Edges are the intersection of two surfaces. *Corners* are angular spaces at the intersection of surfaces. Dimensions are required to note the sizes of the surfaces and geometric characteristics of the edges and corners. Fillets, rounds, and runouts are examples of corners. Bevels and chamfers are examples of edges.

Fillets and Rounds. A *fillet* is a rounded interior corner. A *round* is a rounded exterior corner. Fillets and rounds are used to avoid sharp corners on objects. A rounded corner is produced by two intersecting rough surfaces. Sharp corners are produced by machining either or both intersecting surfaces. See Figure 3-21.

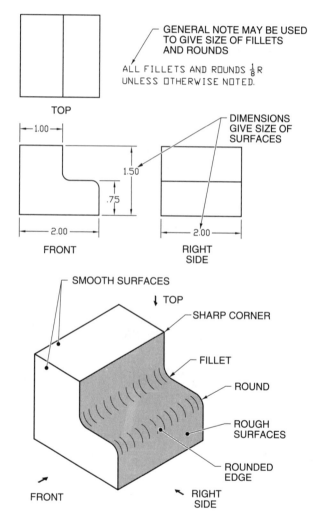

Figure 3-21. Fillets and rounds are used to avoid sharp corners on objects.

Runouts. A *runout* is the curve produced by a plane surface tangent to a cylindrical surface. The radius of a runout is commonly equal to that of a fillet. The direction of the runout is determined by the intersecting surfaces that form the runout. See Figure 3-22.

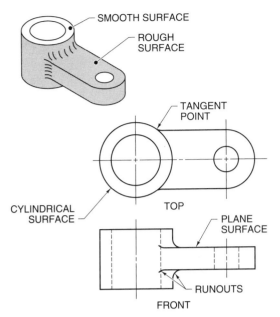

Figure 3-22. Runouts are curves produced by a plane surface tangent to a cylindrical surface.

Bevels. A *bevel* is a sloped edge of an object running from surface to surface. Neither intersection with the surfaces of the object is 90°. A bevel is commonly dimensioned by an angle and a linear dimension. See Figure 3-23.

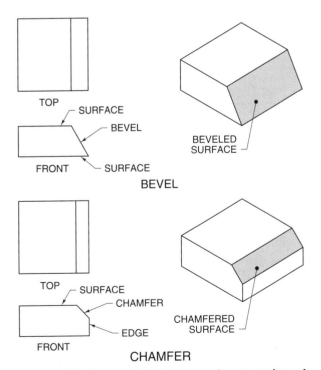

Figure 3-23. A bevel extends from surface to surface. A chamfer extends from surface to edge.

Chamfers. A *chamfer* is a sloped edge of an object running from surface to side. Chamfers are often used to reduce sharp corners. They are commonly dimensioned by an angle and a linear dimension or by two linear dimensions.

SECTION VIEWS

A *section view* is the interior view of an object through which a cutting plane has been passed. Section views show internal features more clearly than they could be shown with hidden lines. Generally, hidden lines are omitted in section views. The portion of the object through which the cutting plane passes is shown with section lines.

Symbols for section lines indicate the type of material. Arrows on of the cutting plane line indicate the direction of sight. Sections may be full, half, broken-out, removed, revolved, offset, or aligned. See Figure 3-24.

A *full section* is a section view created by passing the cutting plane line completely through the object. For a cylindrical object, the full section appears to be one-half of the object. A *half section* is a section view created by passing the cutting plane line halfway through the object. For a cylindrical object, the half section appears to be one-fourth of the object.

A *broken-out section* is a partial section view which appears to have been broken out of the object. A short break line is commonly used to show the broken-out section. A *revolved section* shows the cross-sectional shape of elongated objects.

An *offset section* is a section view created by an offset cutting plane line that shows internal features not in a straight line. An *aligned section* is a section view in which the cutting plane line is bent to pass through detailed features and the section view is revolved.

Conventional Breaks

A *conventional break* is a standard method of showing shortened views of elongated objects. Parts drawn with conventional breaks must be uniform in cross-sectional appearance. The drawing is dimensioned as if the complete part is shown. See Figure 3-25.

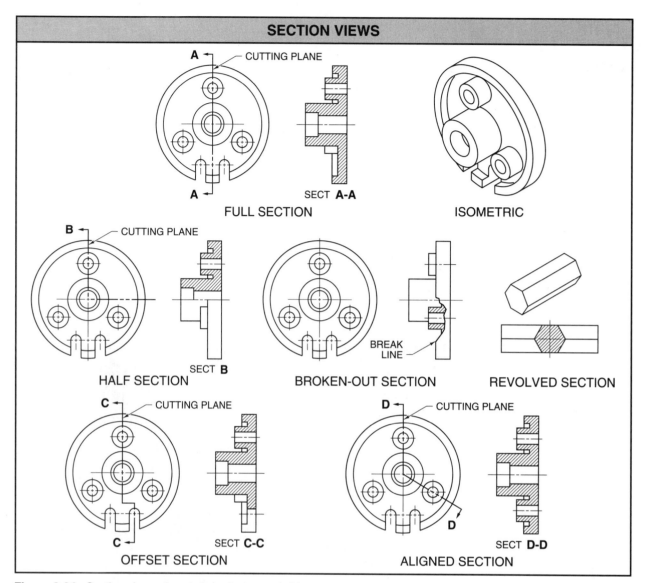

Figure 3-24. Section views show interior features of objects.

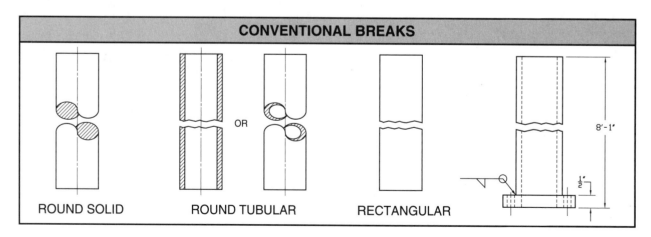

Figure 3-25. Conventional breaks allow details of elongated objects to be shown clearly.

Sketching

Name _____ Date _____

Sketching — Isometrics

Sketch isometrics of the weld joints.

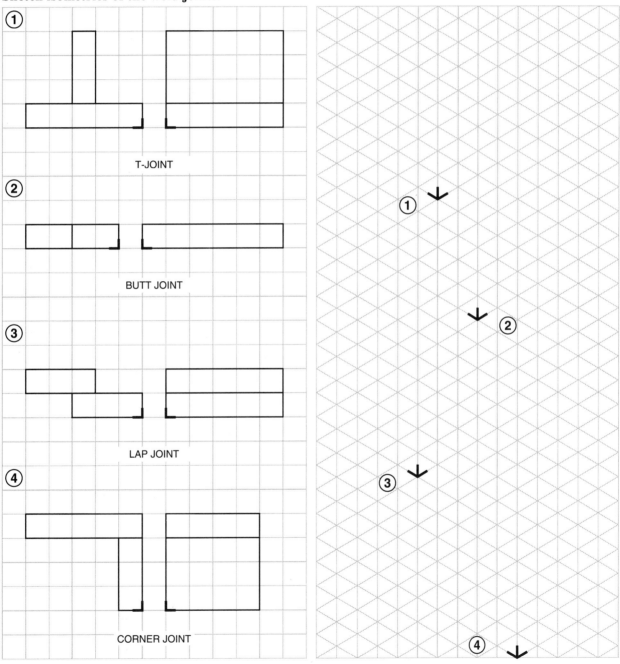

① T-JOINT

② BUTT JOINT

③ LAP JOINT

④ CORNER JOINT

Sketching — Multiviews (Missing Views)

Sketch the missing view of the multiviews.

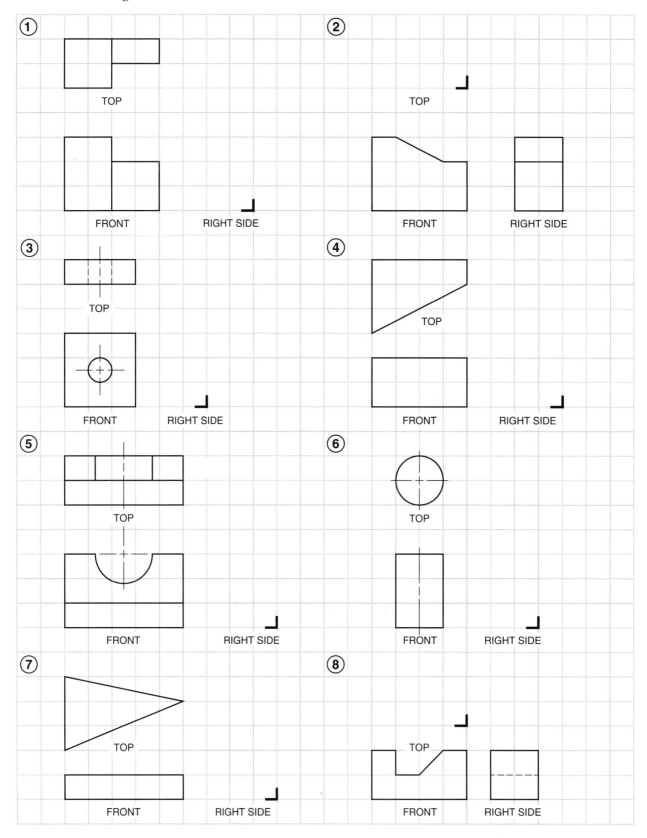

Sketching — Obliques

Sketch oblique cabinets of the multiviews.

Sketching — Multiviews (Front Views)

Sketch the front view (multiview) of the edge preparations for the groove welds.

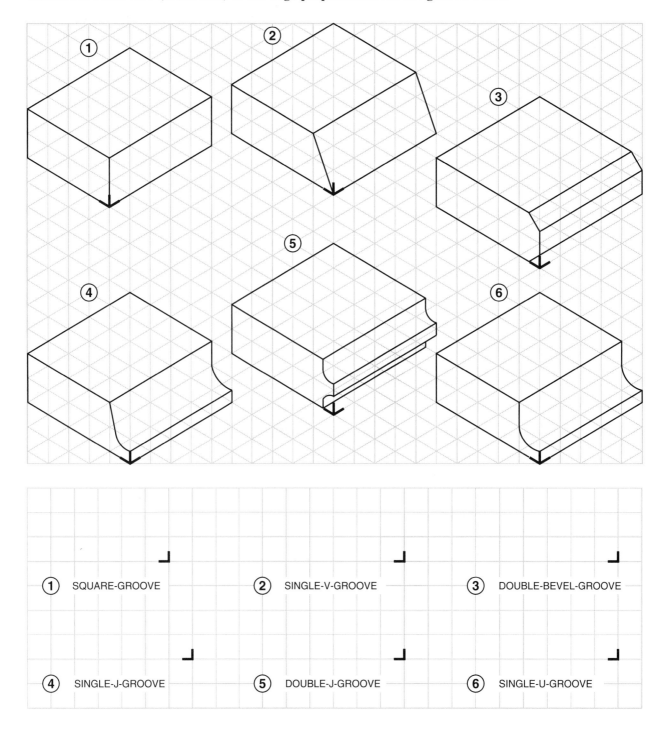

① SQUARE-GROOVE   ② SINGLE-V-GROOVE   ③ DOUBLE-BEVEL-GROOVE

④ SINGLE-J-GROOVE   ⑤ DOUBLE-J-GROOVE   ⑥ SINGLE-U-GROOVE

Review Questions

Name_____ Date _____

True-False

T   F   **1.** Sketching pencils may be wooden or mechanical.

T   F   **2.** Size B paper is $17'' \times 22''$.

T   F   **3.** The pencil point should be pulled across the paper while sketching.

T   F   **4.** Shading techniques are used with orthographic drawings.

T   F   **5.** Perspective drawings are seldom used on welding prints.

T   F   **6.** Oblique drawings show one surface of an object as a true view.

T   F   **7.** Object lines are thin and dark.

T   F   **8.** Break lines conserve space on drawings.

T   F   **9.** Hole sizes are specified by their diameters.

T   F   **10.** Spotfaces are commonly $1/8''$ deep.

T   F   **11.** A circle on an isometric drawing appears as an ellipse.

T   F   **12.** The front view of an object is normally the view that shows the most shape.

T   F   **13.** A cavalier drawing is a type of isometric drawing.

T   F   **14.** Each view of an object is shown two-dimensionally in a multiview drawing.

T   F   **15.** Dimensions of a drawing apply at $68°$.

Completion

_____   **1.** An axonometric drawing shows _____ sides of an object.

_____   **2.** Isometric drawings contain three equal axes that are drawn _____° apart.

_____   **3.** A(n) _____ is a plane curve with two focal points.

_____   **4.** A(n) _____ oblique drawing has receding lines drawn to one-half the scale of lines in the true view.

_____   5. _____ projection, or multiview drawing, is drawing at right angles.

_____   6. The _____ view of a multiview drawing shows length and height.

_____   7. _____ lines define the visible shape of an object.

_____   8. _____ lines are terminated by arrowheads on both ends.

_____   9. A(n) _____ surface is a plane surface parallel to a plane of projection.

_____   10. A(n) _____ hole is a drilled hole that does not pass through the material.

_____   11. A(n) _____ hole is an enlarged and recessed hole with square shoulders.

_____   12. A(n) _____ is a rounded interior corner.

_____   13. A(n) _____ is a rounded exterior corner.

_____   14. _____ views show internal features of an object more clearly than they could be shown with hidden lines.

_____   15. A(n) _____ is a standard method of showing shortened views of elongated objects.

Matching — Alphabet of Lines

_____   1. Object lines

_____   2. Cutting plane line

_____   3. Short break line

_____   4. Long break line

_____   5. Hidden line

_____   6. Center line

_____   7. Dimension line

_____   8. Section line

_____   9. Extension line

_____   10. Arrowhead

Matching — Orthographic Projection

_____ **1.** Front view - height

_____ **2.** Front view - length

_____ **3.** Top view - depth

_____ **4.** Right side view - depth

_____ **5.** Top view - length

_____ **6.** Right side view - height

Multiple Choice

_____ **1.** A true view is a view in which the _____.

    A. surface can be completely seen
    B. line of sight is parallel to the surface
    C. line of sight is perpendicular to the surface
    D. none of the above

_____ **2.** Arrows on the end(s) of the cutting plane line of a section view indicate the _____.

    A. size of the section view
    B. direction from the multiview to the section view
    C. direction of sight
    D. all of the above

_____ **3.** A spotface is _____.

    A. a flat surface
    B. either slightly above or below the surrounding surface
    C. abbreviated by the letters SF
    D. all of the above

_____ **4.** Twist drill sizes are designated by numbers _____.

    A. 1 through 40
    B. 10 through 40
    C. 1 through 80
    D. 10 through 80

_____ **5.** _____ surfaces are created when one surface meets another surface.

    A. Inclined
    B. Intersecting
    C. Oblique
    D. Normal

_____ **6.** Break lines may be drawn as _____.

    A. thin, dark lines with a zig-zag every $\frac{3}{4}''$ to $1\frac{1}{2}''$
    B. thick, dark freehand lines
    C. both A and B
    D. neither A nor B

_____ **7.** Angular dimensions may be expressed as _____.

    A. degrees and decimal parts of a degree
    B. inches and fractional parts of an inch
    C. both A and B
    D. neither A nor B

_____ **8.** Receding lines of oblique drawings are commonly drawn on a _____ angle.

    A. 30° or 45°
    B. 45° or 60°
    C. 60° or 90°
    D. 90° or 120°

_____ **9.** Thick lines on drawings are commonly drawn _____″ wide.

    A. .016
    B. .032
    C. .048
    D. .060

_____ **10.** Extension lines should extend _____″ beyond the dimension line.

    A. ⅛
    B. ³⁄₁₆
    C. ¼
    D. none of the above

Trade Competency Test

Name_____ Date _____

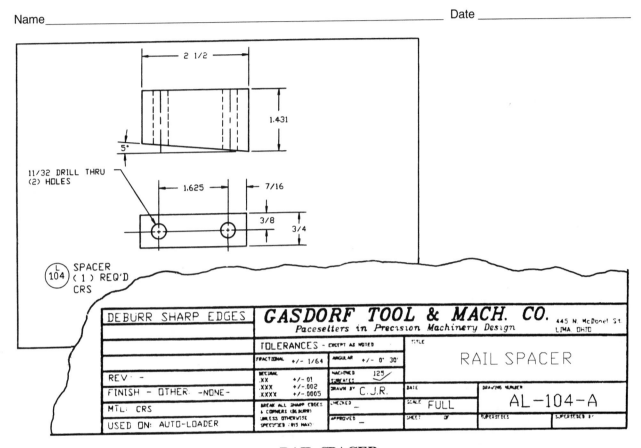

RAIL SPACER

Refer to the Rail Spacer print on page 69.

_____ 1. The tolerance for angular dimensions of the Rail Spacer are _____.

_____ 2. The size holes specified by notation is _____".

T   F   3. All sharp edges of the part are to be deburred.

_____ 4. The part is drawn to _____ scale.

_____ 5. The _____ and _____ views of the Rail Spacer are shown.
- A. top; front
- B. front; side
- C. top; side
- D. all of the above

_____ **6.** Dimensions shown on the print are _____ and _____.

    A. linear; angular
    B. fractional; decimal inch
    C. both A and B
    D. neither A nor B

_____ **7.** The thickness of the Rail Spacer is _____″.

_____ **8.** The Rail Spacer is used on a(n) _____.

_____ **9.** The overall depth of the Rail Spacer is _____″.

_____ **10.** The overall length of the Rail Spacer is _____″.

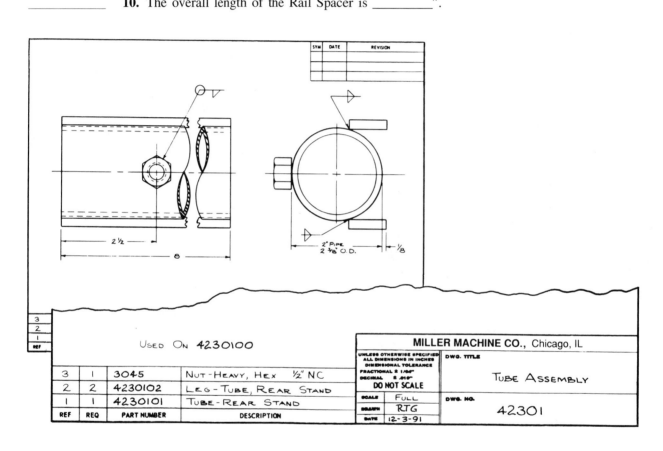

REF	REQ	PART NUMBER	DESCRIPTION
3	1	3045	Nut – Heavy, Hex ½″ NC
2	2	4230102	Leg – Tube, Rear Stand
1	1	4230101	Tube – Rear Stand

Used On 4230100

MILLER MACHINE CO., Chicago, IL

UNLESS OTHERWISE SPECIFIED ALL DIMENSIONS IN INCHES DIMENSIONAL TOLERANCE
FRACTIONAL ± 1/16″
DECIMAL ± .010″
DO NOT SCALE

DWG. TITLE
TUBE ASSEMBLY

SCALE FULL
DRAWN RTG
DATE 12-3-91

DWG. NO.
42301

TUBE ASSEMBLY

Refer to the Tube Assembly print on page 70.

_____ **1.** The weld to attach the hex nut to the tube is specified _____.

    A. to be welded-all-around
    B. as a fillet weld
    C. both A and B
    D. neither A nor B

_____ **2.** The overall length of the tube is _____″.

3. Two _____ are used to weld the legs to the tube.

    A. fillet welds, arrow side
    B. groove welds, arrow side
    C. fillet welds, both sides
    D. groove welds, both sides

4. Hidden lines in the front view show the _____.

    A. legs
    B. inside diameter of the tube
    C. both A and B
    D. neither A nor B

T   F   **5.** The wall thickness of the tube is $\frac{3}{16}''$.

6. The drawing number of the Tube Assembly is _____.

7. Freehand _____ lines are used to show the shortened legs in the front view.

8. _____ lines are used to indicate the tube wall in the front view.

9. A total of _____ parts are required for the Tube Assembly.

T   F   **10.** The Tube Assembly is manufactured by Miller Machine Co.

Refer to the Exit Tube Vise print on page 72.

1. The thickness of the plate used for the Exit Tube Vise is _____″.

2. _____ welds are used extensively on the Exit Tube Vise.

T   F   **3.** The decimal tolerance for this part is $\pm\frac{1}{32}''$.

T   F   **4.** All welds are specified as both sides.

5. The drawing number for this part is _____.

6. The _____ view gives location dimensions for the drilled holes.

7. The Exit Tube Vise is manufactured by _____.

8. The overall size of the Exit Tube Vise is _____.

    A. 5.000 × 5.000 × 13.000
    B. 5.000 × 5.500 × 13.000
    C. 5.000 × 5.500 × 17.000
    D. none of the above

9. The top view shows two corners notched _____″ along each edge.

10. The length between the centerpoints for drilled holes is _____″.

    A. 5.500
    B. 7.000
    C. both A and B
    D. neither A nor B

_____ 11. The drawing was completed by PAT on _____.

_____ 12. Drill holes in the top view are shown with _____ lines.

_____ 13. _____ pieces are welded together to produce the Exit Tube Vise.

T   F   14. The print for the Exit Tube Vise has been revised.

_____ 15. _____ lines are not required to show and size the Exit Tube Vise.
  A. Hidden
  B. Dimension
  C. Extension
  D. Cutting plane

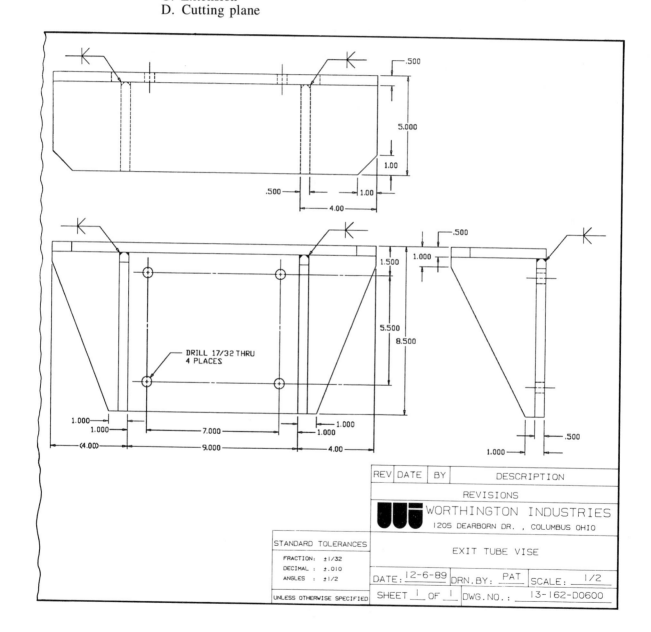

EXIT TUBE VISE

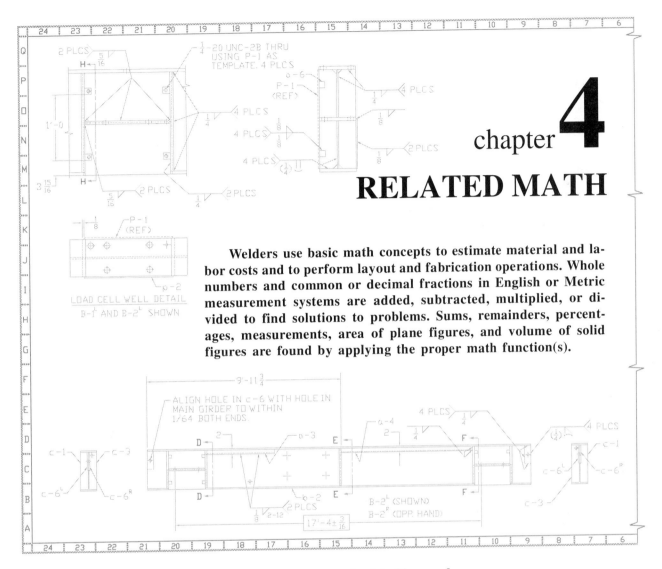

chapter 4

RELATED MATH

Welders use basic math concepts to estimate material and labor costs and to perform layout and fabrication operations. Whole numbers and common or decimal fractions in English or Metric measurement systems are added, subtracted, multiplied, or divided to find solutions to problems. Sums, remainders, percentages, measurements, area of plane figures, and volume of solid figures are found by applying the proper math function(s).

WHOLE NUMBERS

Whole numbers are all numbers that have no fractional or decimal parts. For example, numbers such as 1, 2, 19, 46, 67, 328, etc. are whole numbers. *Odd numbers* are any numbers that cannot be divided by 2 an exact number of times. For example, numbers such as 1, 3, 5, 57, 109, etc. are odd numbers. *Even numbers* are any numbers that can be divided by 2 an exact number of times. For example, numbers such as 2, 4, 6, 48, 432, etc. are even numbers.

Prime numbers are numbers that can be divided an exact number of times only by themselves and the number 1. For example, numbers such as 1, 2, 3, 5, 7, 11, 13, 17, 19, 23, etc. are prime numbers. Arabic and Roman numerals are the two common numeral systems used for calculations and notations. See Figure 4-1.

Arabic Numerals

Arabic numerals are expressed by the ten digits 0, 1, 2, 3, 4, 5, 6, 7, 8, and 9. These digits may be used alone or in combination to represent quantities indicating how much, how many, how far, how long, how hot, how expensive, etc. This is the numeral system most commonly used in the United States. See Figure 4-2.

The Arabic numeral system is the most commonly used numeral system. Large Arabic numerals are made easier to read by the use of periods. A *period* is a group of three digits separated from other periods by a comma. The *units period* (000 through 999) is the first period. The *thousands period* (1,000 through 999,999) is the second period. The *millions period* (1,000,000 through 999,999,999) is the third period, etc.

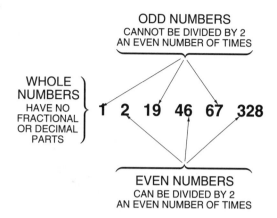

Figure 4-1. Whole numbers have no fractional or decimal parts.

ARABIC NUMERALS

0 ZERO
1 ONE
2 TWO
3 THREE
4 FOUR
5 FIVE
6 SIX
7 SEVEN
8 EIGHT
9 NINE

789,643,549,276,851

DIGITS

7 6 3

SEVEN HUNDREDS
SIX TENS
THREE ONES

Figure 4-2. Arabic numerals are expressed by digits.

Roman Numerals

Roman numerals are expressed by the letters I, X, L, C, D, and M. While not commonly used in the trades, this numeral system is occasionally seen as chapter numbers in a book, on clock faces, and on public buildings such as libraries, museums, etc. See Figure 4-3.

When a letter is followed by the same letter, or one lower in value, add the value of the letters. For example, XX = 20 and XV = 15. When a letter is followed by another letter greater in value, sub-

tract the smaller letter. For example, IV = 4, IX = 9, and XC = 90.

When a letter is placed between two letters of greater value, subtract the smaller letter from the sum of the other two. For example, XIV = 14. A superscript rule (line above) placed over a letter increases the value of the letter a thousand times. For example:

$$\overline{V} = 5,000$$
$$\overline{X} = 10,000$$

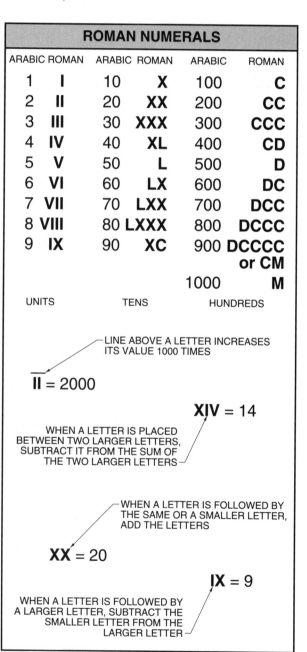

Figure 4-3. Roman numerals are expressed by letters.

Measurement Systems

Three common systems of measurement are the British (U.S.) System, Decimal Inch System, and SI Metric System (International System of Units). Arabic numerals are used with these three measurement systems. The British (U.S.) System is also known as the English System and is the system in primary use in the United States. This system uses the inch, foot, and pound units of measurement. The Decimal Inch System is based on tenths and hundredths to simplify measurements. The Decimal Inch System is used by surveyors, scientists, engineers, etc.

The Metric System is the most common measurement system used in most of the world. Prefixes are used in the Metric System to represent multipliers. For example, the distance of 3,000 meters is expressed as 3 kilometers. Metric measurements are converted to English measurements (and vice versa) by applying the appropriate conversion factors. See Figure 4-4. See Appendix.

BASE UNITS		
UNIT	SI SYMBOL	QUANTITY
Meter	m	Length
Gram	g	Mass
Second	s	Time
Ampere	A	Electric current

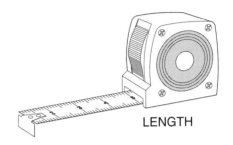

LENGTH

UNIT PREFIXES			
PREFIX	UNIT	SYMBOL	NUMBER
Other larger multiples			
Mega	Million	M	$1,000,000 = 10^6$
Kilo	Thousand	k	$1,000 = 10^3$
Hecto	Hundred	h	$100 = 10^2$
Deka	Ten	d	$10 = 10^1$
			Unit $1 = 10^0$
Deci	Tenth	d	$0.1 = 10^{-1}$
Centi	Hundreth	c	$0.01 = 10^{-2}$
Milli	Thousandth	m	$0.001 = 10^{-3}$
Micro	Millionth	μ	$0.000001 = 10^{-6}$
Other smaller multiples			

MASS

TIME

EXAMPLES

COMBINE UNIT PREFIX SYMBOL
AND BASE UNIT SYMBOL

mm = millimeter
kg = kilogram
mA = milliamp

UNIT SYMBOL ⎯ ⎯ BASE UNIT SYMBOL

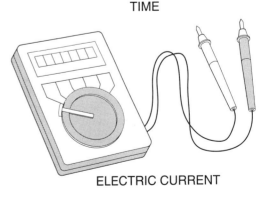

ELECTRIC CURRENT

Figure 4-4. The Metric System is the common measurement system used in most of the world.

Addition

Addition is the process of uniting two or more numbers to make one number. It is the most common operation in mathematics. The sign + (plus) indicates addition and is used when numbers are added horizontally or when two numbers are added vertically. When more than two numbers are added vertically, the operation is apparent, and no sign is required. The *sum* is the result obtained from adding two or more numbers.

To add whole numbers vertically, place all numbers in aligned columns. The units must be in the ones (units) column, tens in the tens column, hundreds in the hundreds column, etc. Add the columns from top to bottom, beginning with the ones column. When the sum of the numbers in the ones column is 0 – 9, record the sum and add the tens column. When the sum of the numbers in the ones column is 10 or more, record the last digit and carry the remaining digit(s) to the tens column. Follow this same procedure for remaining columns.

To add whole numbers horizontally is more difficult than adding them vertically. For example, 15 + 120 + 37 + 9 = 181 shows whole numbers added horizontally. This method is not as commonly used as the vertical alignment method because mistakes can occur more easily.

Check vertically aligned addition problems by adding the numbers from bottom to top. Check horizontally aligned addition problems by adding the numbers from right to left. The same sum will occur if both operations have been added correctly. See Figure 4-5.

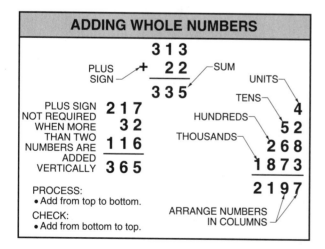

Figure 4-5. Addition is the process of uniting two or more numbers to make one number.

Subtraction

Subtraction is the process of taking one number away from another number. It is the opposite of addition. The sign − (minus) indicates subtraction. The *minuend* is the number from which the subtraction is made. The *subtrahend* is the number which is subtracted. The *remainder* is the difference between the two. Place the minuend above the subtrahend when vertically aligning numbers.

As in addition, the first column of numbers represents ones, the second column represents tens, etc. Whenever a subtrahend digit is larger than the corresponding minuend digit, borrow one unit (tens, hundreds, etc.) from the column immediately to the left and continue the operation. For example, when subtracting 8 from 22, borrow a 1 from the tens column, subtract 8 from 12, record the 4 in the units column, and record the remaining 1 in the tens column for a remainder of 14. See Figure 4-6.

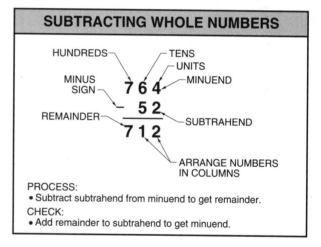

Figure 4-6. Subtraction is the process of taking one number away from another number.

Multiplication

Multiplication is the process of adding one number as many times as there are units in the other number. For example, $3 \times 4 = 12$ produces the same result as adding $4 + 4 + 4 = 12$. The sign × (times or multiplied by) indicates multiplication. The *multiplicand* is the number which is multiplied. The *multiplier* is the number by which multiplication is done. The *product* is the result of the multiplication.

The larger number is commonly used as the multiplicand when the units being multiplied are the same. For example, $6' \times 4' = 24'$. If a number to be multiplied represents a unit of measurement (inches, feet, pounds, etc.), identify the unit of measurement in the multiplicand, multiplier, and product. Numbers may be arranged vertically (preferred) or horizontally when multiplying. An effective method of checking the product is to reverse the multiplicand and the multiplier and perform the operation again. The same product will occur if both operations have been multiplied correctly.

Zeros have no value, therefore any number multiplied by a zero equals zero. For example, $18 \times 0 = 0$. To multiply a multiplicand by 10, add one zero. For example, to multiply 84 by 10, add one zero to the 84 to get 840 ($84 \times 10 = 840$). Add two zeros to multiply by 100, etc. See Figure 4-7.

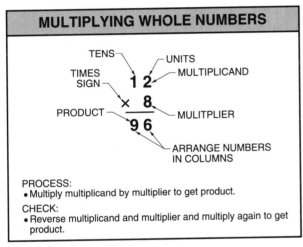

Figure 4-7. Multiplication is the process of adding one number as many times as there are units in the other number.

Division

Division is the process of finding how many times one number contains the other number. It is the reverse of multiplication. The sign ÷ (divided by) indicates division. The sign) also indicates division. The *dividend* is the number to be divided. The *divisor* is the number by which division is done. The *quotient* is the result of the division. The *remainder* is the part of the quotient left over whenever the quotient is not a whole number.

To divide a number by 10, 100, etc. remove as many places from the right of the dividend as their are zeros in the divisor. For example, $600 \div 10 = 60$. Notice that one zero was removed from the dividend (600) to yield the quotient of 60.

Any remainder is placed over the divisor and expressed as a fraction. For example, $27 \div 4 = 6\frac{3}{4}$. Notice that 4 goes into 27 six times with a remainder of 3. The 3 is placed over the 4 (divisor).

To check division, multiply the divisor by the quotient. For example, $48 \div 3 = 16$. To check this problem, multiply 3 (divisor) by 16 (quotient). For example, $3 \times 16 = 48$. See Figure 4-8.

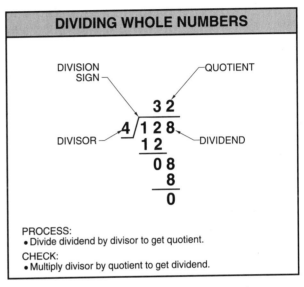

Figure 4-8. Division is the process of finding how many times one number contains the other number.

COMMON FRACTIONS

A *fraction* is one part of a whole number. The number 1 is the smallest whole number. Anything smaller than 1 is a fraction and can be divided into any number of fractional parts. Fractions are written above and below or on both sides of a fraction bar. Fraction bars may be horizontal or inclined.

The *denominator* shows how many parts the whole number has been divided into. The denominator is the lower (or right-hand) number of a fraction. The *numerator* shows the number of parts in the fraction. The numerator is the upper (or left-hand) number. For example, the fraction $\frac{3}{4}$ shows that a whole number is divided into four equal parts (denominator), and three of these parts (numerator) are present.

A *proper fraction* has a denominator larger than its numerator. An *improper fraction* has a numerator larger than its denominator. A *mixed number* is a combination of a whole number and a fraction.

For example, ⅜ is a proper fraction, ⁴⁄₃ is an improper fraction, and 1⁹⁄₁₆ is a mixed number.

Any type of units can be divided into fractional parts. For example, inches are commonly divided into fractional parts of an inch based upon halves, fourths, eighths, sixteenths, thirty-seconds, and sixty-fourths. Fractional parts of an inch are always expressed in their lowest common denominator.

The *lowest common denominator (LCD)* is found by dividing the highest number that will divide equally into the denominator and numerator. For example, the LCD of the fraction ¹²⁄₁₆ is ¾. This is obtained by dividing 4 into 12 and 4 into 16. Always reduce fractions to their lowest common denominators. See Figure 4-9.

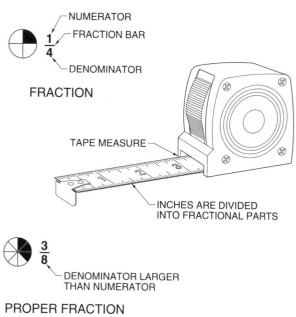

FRACTION

TAPE MEASURE

INCHES ARE DIVIDED INTO FRACTIONAL PARTS

$\frac{3}{8}$

DENOMINATOR LARGER THAN NUMERATOR

PROPER FRACTION

$\frac{5}{4}$

NUMERATOR LARGER THAN DENOMINATOR

IMPROPER FRACTION

Figure 4-9. A fraction is one part of a whole number.

Addition

Fractions may be added horizontally or vertically. Horizontal placement is the most common, as identification of numerators and denominators is easier. Fractions which may be added include

proper fractions, improper fractions, mixed numbers, and fractions with unlike denominators. There is a different rule for each of these four combinations. See Figure 4-10.

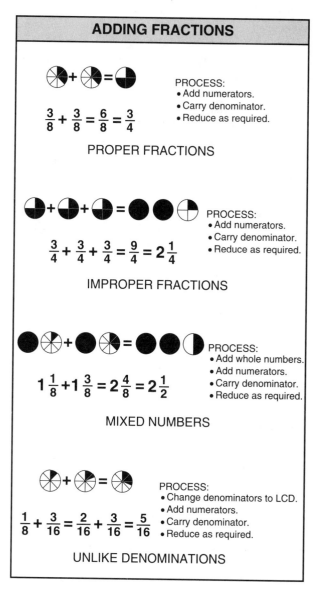

ADDING FRACTIONS

$\frac{3}{8} + \frac{3}{8} = \frac{6}{8} = \frac{3}{4}$

PROCESS:
• Add numerators.
• Carry denominator.
• Reduce as required.

PROPER FRACTIONS

$\frac{3}{4} + \frac{3}{4} + \frac{3}{4} = \frac{9}{4} = 2\frac{1}{4}$

PROCESS:
• Add numerators.
• Carry denominator.
• Reduce as required.

IMPROPER FRACTIONS

$1\frac{1}{8} + 1\frac{3}{8} = 2\frac{4}{8} = 2\frac{1}{2}$

PROCESS:
• Add whole numbers.
• Add numerators.
• Carry denominator.
• Reduce as required.

MIXED NUMBERS

$\frac{1}{8} + \frac{3}{16} = \frac{2}{16} + \frac{3}{16} = \frac{5}{16}$

PROCESS:
• Change denominators to LCD.
• Add numerators.
• Carry denominator.
• Reduce as required.

UNLIKE DENOMINATIONS

Figure 4-10. Fractions which may be added include proper fractions, improper fractions, mixed numbers, and fractions with unlike denominators.

Adding Proper Fractions. Fractions having the same denominator are added by adding the numerators and placing them over the denominator. For example, in the problem ⅓ + ⅓ = ⅔, the numerators (1 + 1) are added to produce 2. The denominator (3) remains constant.

Adding Improper Fractions. Fractions which produce a sum in which the numerator is larger than the denominator (improper fractions) are changed to a mixed number by dividing the numerator by the denominator, recording the quotient obtained, and treating the remainder as a numerator placed over the original denominator. For example, in the problem $\frac{3}{8} + \frac{3}{8} + \frac{3}{8} = \frac{9}{8}$, the improper fraction ($\frac{9}{8}$) is changed to $1\frac{1}{8}$ by dividing 9 by 8. The fraction is then written $\frac{3}{8} + \frac{3}{8} + \frac{3}{8} = \frac{9}{8} = 1\frac{1}{8}$.

Adding Mixed Numbers. To add fractions containing mixed numbers, add the whole numbers, add the numerators, and carry the denominator. For example, in the problem $1\frac{1}{4} + 3\frac{1}{4} + 4\frac{1}{4} = 8\frac{3}{4}$, the whole numbers (1 + 3 + 4) are added to produce 8. The numerators (1 + 1 + 1) are added to produce 3, which is placed over the denominator 4.

Adding Fractions with Unlike Denominators. To add fractions in which the denominators are not the same, change the denominators to the LCD, add the numerators, and carry the denominator. If an improper fraction occurs, change the improper fraction to a mixed number. For example, to add $\frac{3}{8} + \frac{1}{2} + \frac{3}{4}$, change the denominators to 8 and multiply the number of times the original denominators will go into 8 by the numerators to get $\frac{3}{8} + \frac{4}{8} + \frac{6}{8}$. Add the numerators $3 + 4 + 6 = 13$ and place over the denominator to get $\frac{13}{8}$. Change the improper fraction $\frac{13}{8}$ by dividing 13 by 8. Thirteen can be divided by 8 one time with a remainder of 5, which is placed over the 8 to produce $1\frac{5}{8}$ ($\frac{3}{8} + \frac{1}{2} + \frac{3}{4} = \frac{3}{8} + \frac{4}{8} + \frac{6}{8} = \frac{13}{8} = 1\frac{5}{8}$).

Subtraction

Subtraction of fractions is similar to addition of fractions. Fractions may be subtracted horizontally or vertically. Horizontal placement is the most common, as identification of numerators and denominators is easier. All fractions must have a common denominator before one can be subtracted from another. Fractions which may be subtracted include fractions with like denominators, fractions with unlike denominators, and mixed numbers. There is a

different rule for each of these three combinations. See Figure 4-11.

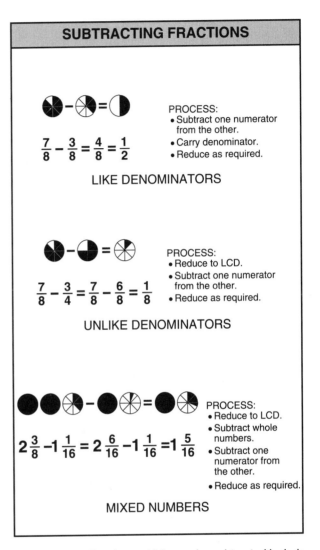

SUBTRACTING FRACTIONS

$$\frac{7}{8} - \frac{3}{8} = \frac{4}{8} = \frac{1}{2}$$

PROCESS:
• Subtract one numerator from the other.
• Carry denominator.
• Reduce as required.

LIKE DENOMINATORS

$$\frac{7}{8} - \frac{3}{4} = \frac{7}{8} - \frac{6}{8} = \frac{1}{8}$$

PROCESS:
• Reduce to LCD.
• Subtract one numerator from the other.
• Reduce as required.

UNLIKE DENOMINATORS

$$2\frac{3}{8} - 1\frac{1}{16} = 2\frac{6}{16} - 1\frac{1}{16} = 1\frac{5}{16}$$

PROCESS:
• Reduce to LCD.
• Subtract whole numbers.
• Subtract one numerator from the other.
• Reduce as required.

MIXED NUMBERS

Figure 4-11. Fractions which may be subtracted include fractions with like denominators, fractions with unlike denominators, and mixed numbers.

Subtracting Fractions with Like Denominators. To subtract fractions having the same denominators, subtract one numerator from the other numerator and place over the denominator. For example, to subtract $\frac{7}{16}$ from $\frac{11}{16}$, subtract the numerator 7 from the numerator 11 to get 4. Place the 4 over the denominator 16. Reduce $\frac{4}{16}$ by dividing by the largest number that will go into the numerator and denominator an even number of times. In this example, divide the numerator and denominator by 4 to get $\frac{1}{4}$ ($\frac{11}{16} - \frac{7}{16} = \frac{4}{16} = \frac{1}{4}$).

Subtracting Fractions with Unlike Denominators. To subtract fractions having unlike denominators, reduce the fractions to their LCD and subtract one numerator from the other. For example, to subtract $\frac{7}{16}$ from $\frac{3}{4}$, reduce $\frac{3}{4}$ to $\frac{12}{16}$ and subtract $\frac{7}{16}$ to get $\frac{5}{16}$ ($\frac{3}{4} - \frac{7}{16} = \frac{12}{16} - \frac{7}{16} = \frac{5}{16}$).

Subtracting Mixed Numbers. To subtract fractions having mixed numbers, follow the applicable procedure for denominators, subtract the numerators, subtract the whole numbers, and if necessary, reduce the fraction to its lowest common denominator. For example, to subtract $1\frac{1}{4}$ from $3\frac{1}{2}$, reduce the fractions to their LCD and subtract one numerator from another to get $\frac{2}{4} - \frac{1}{4} = \frac{1}{4}$. Subtract the whole numbers to get $3 - 1 = 2$. Add the whole number and the fraction to get $2 + \frac{1}{4} = 2\frac{1}{4}$.

Multiplication

Fractions may be multiplied horizontally or vertically. Horizontal placement of fractions is the most common, as identification of numerators and denominators is easier. Fractions which may be multiplied include two fractions, fractions and a whole number, a mixed number and a whole number, and two mixed numbers. There is a different rule for each of these four combinations. See Figure 4-12.

Multiplying Two Fractions. To multiply two fractions, multiply the numerator of one fraction by the numerator of the other fraction. Do the same with the denominators. Reduce the answer as required. For example, to multiply $\frac{3}{8}$ by $\frac{1}{8}$, multiply the numerators to get 3 ($3 \times 1 = 3$) and multiply the denominators to get 64 ($8 \times 8 = 64$). Thus, $\frac{3}{8} \times \frac{1}{8} = \frac{3}{64}$.

Multiplying Fractions and a Whole Number. To multiply a fraction and a whole number, multiply the numerator of the fraction by the whole number and place over the denominator. Reduce the answer as required. For example, to multiply $\frac{1}{8} \times 3$, multiply the numerator 1×3 (whole number) to get 3 ($1 \times 3 = 3$) and place the 3 over the denominator 8 to get $\frac{3}{8}$. Thus, $\frac{1}{8} \times 3 = \frac{3}{8}$.

Multiplying a Mixed Number and a Whole Number. To multiply a mixed number and a whole number, multiply the fraction of the mixed number by the whole number, multiply the whole numbers, and add the two products. For example, to multiply $4\frac{7}{8} \times 3$, multiply $\frac{7}{8}$ (fraction of the mixed number) by 3 (whole number) to get $2\frac{5}{8}$ ($\frac{7}{8} \times 3 = \frac{21}{8} = 2\frac{5}{8}$). Multiply the whole numbers to get 12 ($4 \times 3 = 12$) and add the two products to get $14\frac{5}{8}$ ($2\frac{5}{8} + 12 = 14\frac{5}{8}$). Thus, $4\frac{7}{8} \times 3 = 14\frac{5}{8}$.

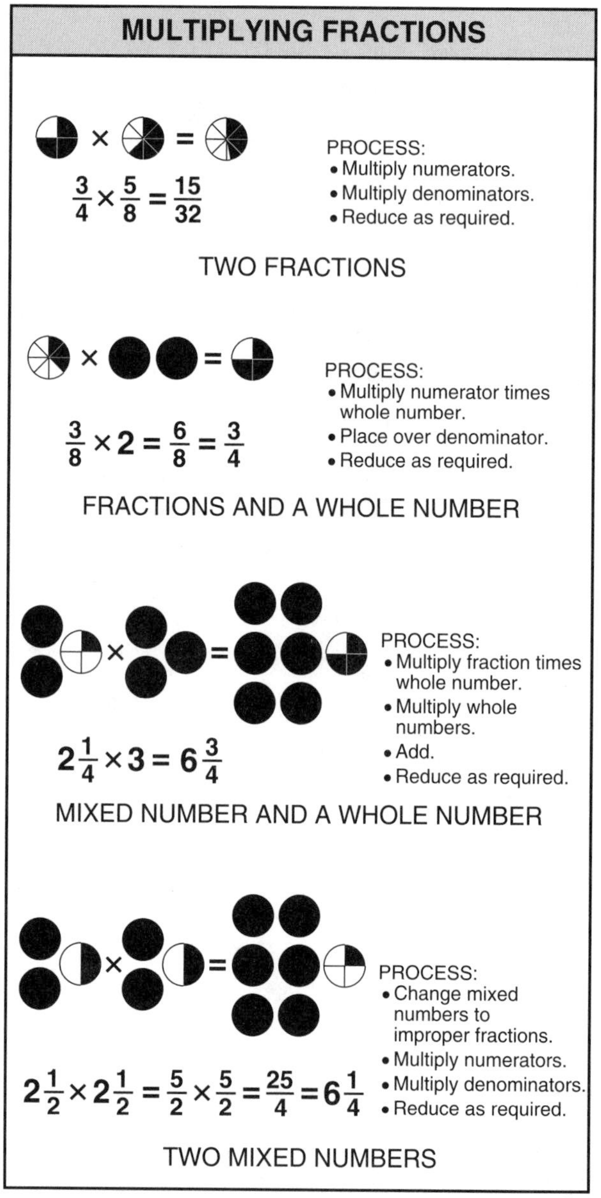

MULTIPLYING FRACTIONS

$\frac{3}{4} \times \frac{5}{8} = \frac{15}{32}$

PROCESS:
- Multiply numerators.
- Multiply denominators.
- Reduce as required.

TWO FRACTIONS

$\frac{3}{8} \times 2 = \frac{6}{8} = \frac{3}{4}$

PROCESS:
- Multiply numerator times whole number.
- Place over denominator.
- Reduce as required.

FRACTIONS AND A WHOLE NUMBER

$2\frac{1}{4} \times 3 = 6\frac{3}{4}$

PROCESS:
- Multiply fraction times whole number.
- Multiply whole numbers.
- Add.
- Reduce as required.

MIXED NUMBER AND A WHOLE NUMBER

$2\frac{1}{2} \times 2\frac{1}{2} = \frac{5}{2} \times \frac{5}{2} = \frac{25}{4} = 6\frac{1}{4}$

PROCESS:
- Change mixed numbers to improper fractions.
- Multiply numerators.
- Multiply denominators.
- Reduce as required.

TWO MIXED NUMBERS

Figure 4-12. Fractions which may be multiplied include two fractions, fractions and a whole number, a mixed number and a whole number, and two mixed numbers.

Multiplying Two Mixed Numbers. To multiply two mixed numbers, change both mixed numbers to improper fractions and multiply. For example, to multiply $3\frac{1}{4}$ by $4\frac{1}{2}$, change the mixed number $3\frac{1}{4}$ to the improper fraction $\frac{13}{4}$ by multiplying the whole number 3 by the denominator 4 and adding the 1 ($3 \times 4 = 12 + 1 = \frac{13}{4}$). Change the mixed number $4\frac{1}{2}$ to the improper fraction $\frac{9}{2}$ by multiplying the whole number 4 by the denominator 2 and adding the 1 ($4 \times 2 = 8 + 1 = \frac{9}{2}$). Multiply the improper fractions to get $12\frac{1}{8}$ ($\frac{13}{4} \times \frac{9}{2} = \frac{117}{8} = 14\frac{5}{8}$). Thus, $3\frac{1}{4} \times 4\frac{1}{2} = 14\frac{5}{8}$.

Division

Fractions are divided horizontally. Fractions which may be divided include a fraction by a whole number, a mixed number by a whole number, two fractions, a whole number by a fraction, and two mixed numbers. There is a different rule for each of these five combinations. See Figure 4-13.

Dividing a Fraction by a Whole Number. To divide a fraction by a whole number, multiply the denominator of the fraction by the whole number. For example, to divide $\frac{3}{8}$ by 4, multiply the denominator 8 by the whole number 4 to get 32 ($8 \times 4 = 32$). Place the numerator 3 over the 32 to get $\frac{3}{32}$. Thus, $\frac{3}{8} \div 4 = \frac{3}{32}$.

Dividing a Mixed Number by a Whole Number. To divide a mixed number by a whole number, change the mixed number to an improper fraction and multiply the denominator of the improper fraction by the whole number. For example, to divide $2\frac{7}{8}$ by 3, change the mixed number $2\frac{7}{8}$ to $\frac{23}{8}$. Multiply the denominator of the improper fraction $\frac{23}{8}$ by the whole number 3 to get $\frac{23}{24}$. Thus, $2\frac{7}{8} \div 3 = \frac{23}{24}$.

Dividing Two Fractions. To divide two fractions, invert the divisor fraction and multiply the numerator by the numerator and the denominator by the denominator. For example, to divide $\frac{3}{8}$ by $\frac{1}{4}$, invert the divisor fraction $\frac{1}{4}$ and multiply by

$\frac{3}{8}$ to get $1\frac{1}{2}$ ($\frac{3}{8} \times \frac{4}{1} = \frac{12}{8} = 1\frac{4}{8} = 1\frac{1}{2}$). Thus, $\frac{3}{8} \div \frac{1}{4} = 1\frac{1}{2}$.

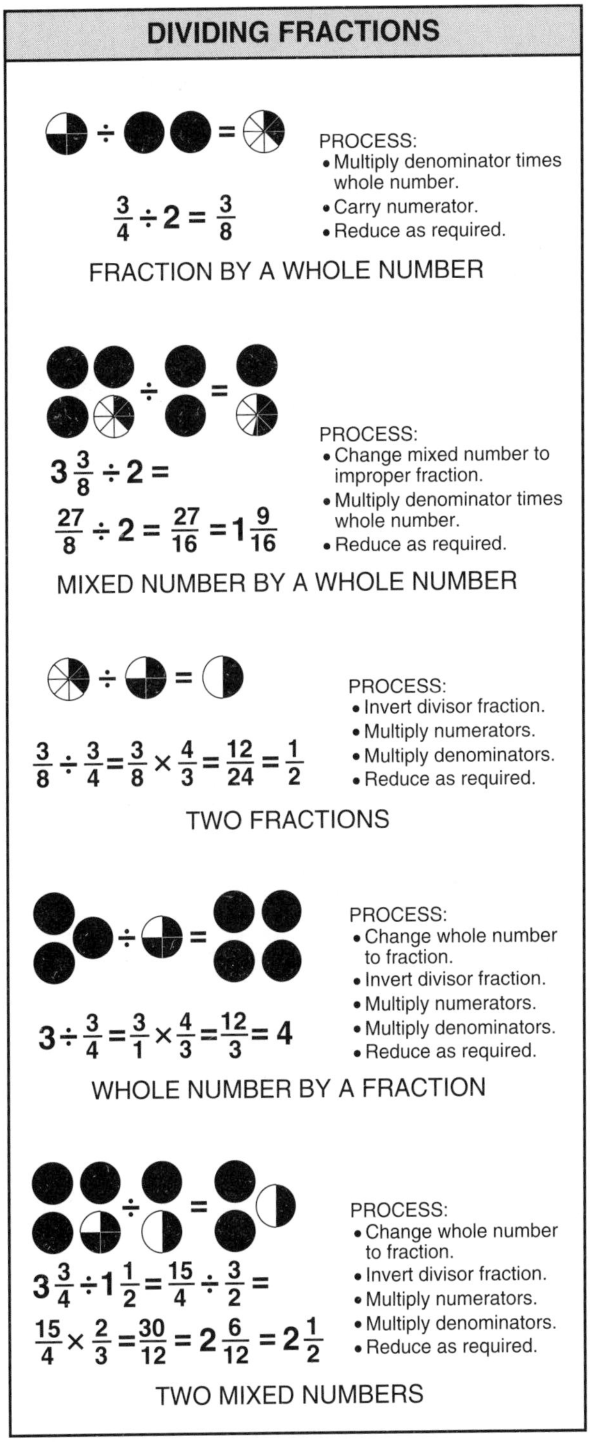

DIVIDING FRACTIONS

$$\frac{3}{4} \div 2 = \frac{3}{8}$$

FRACTION BY A WHOLE NUMBER

PROCESS:
- Multiply denominator times whole number.
- Carry numerator.
- Reduce as required.

$$3\frac{3}{8} \div 2 =$$
$$\frac{27}{8} \div 2 = \frac{27}{16} = 1\frac{9}{16}$$

MIXED NUMBER BY A WHOLE NUMBER

PROCESS:
- Change mixed number to improper fraction.
- Multiply denominator times whole number.
- Reduce as required.

$$\frac{3}{8} \div \frac{3}{4} = \frac{3}{8} \times \frac{4}{3} = \frac{12}{24} = \frac{1}{2}$$

TWO FRACTIONS

PROCESS:
- Invert divisor fraction.
- Multiply numerators.
- Multiply denominators.
- Reduce as required.

$$3 \div \frac{3}{4} = \frac{3}{1} \times \frac{4}{3} = \frac{12}{3} = 4$$

WHOLE NUMBER BY A FRACTION

PROCESS:
- Change whole number to fraction.
- Invert divisor fraction.
- Multiply numerators.
- Multiply denominators.
- Reduce as required.

$$3\frac{3}{4} \div 1\frac{1}{2} = \frac{15}{4} \div \frac{3}{2} =$$
$$\frac{15}{4} \times \frac{2}{3} = \frac{30}{12} = 2\frac{6}{12} = 2\frac{1}{2}$$

TWO MIXED NUMBERS

PROCESS:
- Change whole number to fraction.
- Invert divisor fraction.
- Multiply numerators.
- Multiply denominators.
- Reduce as required.

Figure 4-13. Fractions which may be divided include a fraction by a whole number, a mixed number by a whole number, two fractions, a whole number by a fraction, and two mixed numbers.

Dividing a Whole Number by a Fraction. To divide a whole number by a fraction, change the whole number into fraction form, invert the divisor fraction, and multiply the numerator by the numerator and the denominator by the denominator. For example, to divide 12 by $\frac{3}{4}$, change the whole number 12 to fraction form $\frac{12}{1}$. Invert the divisor fraction $\frac{3}{4}$ and multiply to get 16 ($\frac{12}{1} \times \frac{4}{3} = \frac{48}{3} = 16$). Thus, $12 \div \frac{3}{4} = 16$.

Dividing Two Mixed Numbers. To divide two mixed numbers, change both mixed numbers to improper fractions, invert the divisor fraction, and multiply the numerator by the numerator and the denominator by the denominator. For example, to divide $12\frac{1}{2}$ by $3\frac{1}{8}$, change the mixed number $12\frac{1}{2}$ to $\frac{25}{2}$ and the mixed number $3\frac{1}{8}$ to $\frac{25}{8}$. Invert the divisor fraction $\frac{25}{8}$ and multiply to get 4 ($\frac{25}{2} \times \frac{8}{25} = \frac{200}{50} = 4$). Thus, $12\frac{1}{2} \div 3\frac{1}{8} = 4$.

DECIMALS

A *decimal* is a fraction with a denominator of 10, 100, 1000, etc. The number 1 is the smallest whole number. Anything smaller than 1 is a decimal and can be divided into any number of decimal parts. For example, the decimal .75 shows that the whole number 1 is divided into 100 equal parts, and 75 of these parts are present. Any fraction with 10, 100, 1000, or other multiple of ten for the denominator, may be written as a decimal. For example, the fraction $\frac{1}{10}$ is .1 in decimals, $\frac{1}{100}$ is .01, and $\frac{1}{1000}$ is .001. See Figure 4-14.

A *decimal point* is the period in a decimal number. Shop workers and others who use decimals in their work often say *point* at the decimal point. For example, to denote 3.22, the worker says "three point twenty-two." This denotes $3\frac{22}{100}$. Others may say *and* at the decimal point. For example, to denote 4.37, they may say "four and thirty-seven hundredths." Both methods of expressing decimals are acceptable.

The United States monetary system is based on decimals. The dollar ($1.00) is valued at 100 cents. Each cent is $\frac{1}{100}$ of a dollar or $.01. Each nickel is $\frac{5}{100}$ of a dollar or $.05. Each dime is $\frac{10}{100}$ of a

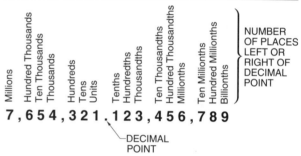

DECIMALS			
CURRENCY	VALUE	DECIMAL	FRACTION
DOLLAR BILL	$1.00	1.00	$\frac{100}{100}$
HALF-DOLLAR	50¢	.50	$\frac{50}{100}$
QUARTER	25¢	.25	$\frac{25}{100}$
DIME	10¢	.10	$\frac{10}{100}$
NICKEL	5¢	.05	$\frac{5}{100}$
PENNY	1¢	.01	$\frac{1}{100}$

Millions, Hundred Thousands, Ten Thousands, Thousands, Hundreds, Tens, Units, Tenths, Hundredths, Thousandths, Ten Thousandths, Hundred Thousandths, Millionths, Ten Millionths, Hundred Millionths, Billionths — NUMBER OF PLACES LEFT OR RIGHT OF DECIMAL POINT

7,654,321.123,456,789 — DECIMAL POINT

```
.25
.03
.17      12.55
1.08   - 3.32
1.53     9.23
```
PROCESS:
• Align decimal points.
• Add or subtract same as whole numbers.

ADDING OR SUBTRACTING DECIMALS

```
   2.36
  ×.7
 1.652
```
POINT OFF DECIMAL PLACE

PROCESS:
• Multiply same as whole numbers.
• Point off decimal place.

MULTIPLYING DECIMALS

```
          3.
3.25. /9.75.
       9 75
```
PROCESS:
• Divide same as whole numbers.
• Point off decimal place.

DIVIDING DECIMALS

Figure 4-14. A decimal is a fraction with a denominator of 10, 100, 1000, etc.

dollar or $.10. Each quarter is $^{25}/_{100}$ of a dollar or $.25. Each half-dollar is $^{50}/_{100}$ of a dollar or $.50.

More places written in a decimal indicate a higher degree of accuracy. For example, while .2 and .20 represent the same value, .20 is measured in hundredths and .2 is measured in tenths.

Adding or Subtracting Decimals

To add or subtract decimals, align the numbers vertically on the decimal points. Thus units will be added or subtracted to units, tenths to tenths, hundredths to hundredths, etc. Add or subtract as in whole numbers and place the decimal point of the sum or remainder directly below the other decimal points. For example, to add 27.08 and 9.127, align the numbers vertically on the decimal points and add to get 36.207.

Multiplying Decimals

To multiply decimals, multiply as in whole numbers. Then begin at the right of the product and point off to the left the same number of decimal places in the quantities multiplied. Prefix zeros when necessary. For example, to multiply 20.45 by 3.15, align the numbers vertically on the decimal points, multiply as in whole numbers, and point off four places from the right to get 64.4175.

Dividing Decimals

To divide decimals, divide as though dividend and divisor are whole numbers. Then point off from right to left as many decimal places as the difference between the number of decimal places in the dividend and divisor. If the dividend has less decimal places than the divisor, add zeros to the dividend. There must be at least as many decimal places in the dividend as in the divisor. For example, to divide 2.5 into 16.75, divide as in whole numbers and point off one decimal place from right to left to get 6.7.

Changing Decimals to Common Fractions

To change a decimal to a common fraction, use the figures in the quantity as a numerator. For the denominator, place the figure one followed by as many zeros as there are figures to the right of the decimal point in the quantity. For example, to change the decimal .4 to a common fraction, place the 4 as a numerator and place a 1 followed by a zero as the denominator to get $^{4}/_{10}$.

To change the decimal .47 to a common fraction, place the 47 as a numerator and place a 1 followed by two zeros as the denominator to get $^{47}/_{100}$. To change the decimal .479 to a common fraction, place the 479 as a numerator and place a 1 followed by three zeros as the denominator to get $^{479}/_{1000}$.

PLANE FIGURES

A *plane figure* is a flat figure. It has no depth. Plane figures are the basis for sketching and drawing. All plane figures are composed of lines drawn at various angles or with arcs. Plane figures include circles, triangles, quadrilaterals, and polygons. A *regular plane figure* has equal angles and equal sides. An *irregular plane figure* does not have equal angles and equal sides.

Lines

A *straight line* is the shortest distance between two points. It is commonly referred to as a line. A *horizontal line* is parallel to the horizon. It is a level line. A *vertical line* is perpendicular to the horizon. Vertical, perpendicular, and plumb mean being at right angles to a base line. *Vertical* is a line in a straight upward position. *Perpendicular* stresses the straightness of a line making a right angle with another (not necessarily horizontal) line. The symbol for perpendicular is ⊥. *Plumb* is an exact verticality (determined by a plumb bob and line) with Earth's gravity.

Lines may be drawn in any position. An *inclined (slanted) line* is neither horizontal nor vertical. *Parallel lines* remain the same distance apart. The symbol for parallel lines is ‖. See Figure 4-15.

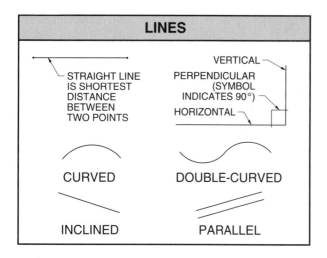

Figure 4-15. A straight line is the shortest distance between two points.

Angles. An *angle* is the intersection of two lines. The symbol for angle is ∠. Angles are measured in degrees, minutes, and seconds. The symbol for degrees is °. The symbol for minutes is ′. The symbol for seconds is ″. There are 360° in a circle. There are 60 minutes in one degree and 60 seconds in one minute. For example, an angle might contain 112°-30′-12″. See Figure 4-16.

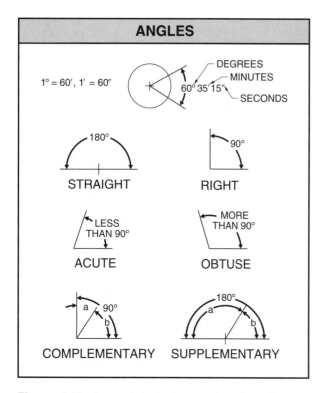

Figure 4-16. An angle is the intersection of two lines.

A *straight angle* contains 180°. A *right angle* contains 90°. An *acute angle* contains less than 90°. An *obtuse angle* contains more than 90°. *Complementary angles* equal 90°. *Supplementary angles* equal 180°.

Circles

A *circle* is a plane figure generated around a centerpoint. All circles contain 360°. The *diameter* is the distance from circumference (outside) to circumference through the centerpoint. The *circumference* is 3.1416 times the diameter of a circle. The *radius* is one-half the length of the diameter. See Figure 4-17.

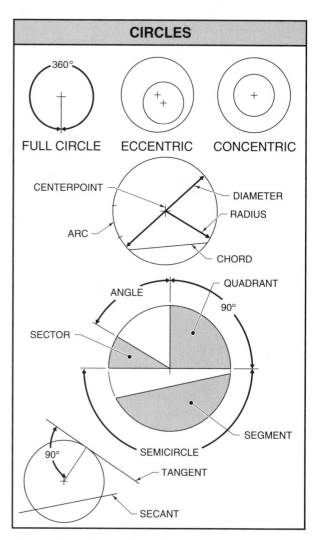

Figure 4-17. A circle is a plane figure generated around a centerpoint.

A *chord* is a line from circumference to circumference not through the centerpoint. An *arc* is a portion of the circumference. A *quadrant* is one-fourth of a circle. Quadrants have a right angle. A *sector* is a pie-shaped piece of a circle. A *segment* is the portion of a circle set off by a chord. A *semicircle* is one-half of a circle. Semicircles always contain 180°. *Concentric circles* have different diameters and the same centerpoint. *Eccentric circles* have different diameters and different centerpoints. A *tangent* is a straight line touching the circumference at only one point. It is 90° to the radius. A *secant* is a straight line touching the circumference at two points.

Triangles

A *triangle* is a three-sided plane figure. All triangles contain 180°. A *right triangle* contains one 90° angle. An *isosceles triangle* has two equal sides and two equal angles. An *equilateral triangle* has three equal sides and three equal angles. A *scalene triangle* has no equal sides or equal angles. A scalene triangle may be acute or obtuse. See Figure 4-18.

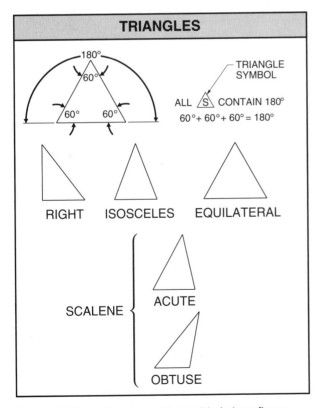

Figure 4-18. A triangle is a three-sided plane figure.

Quadrilaterals

A *quadrilateral* is a four-sided plane figure. All quadrilaterals contain 360°. A *square* has four equal sides and four 90° angles. A *rectangle* has opposite sides equal and four 90° angles. A *rhombus* has four equal sides with opposite angles equal and no 90° angles. A *rhomboid* has opposite sides equal with opposite angles equal and no 90° angles. A *trapezoid* has two sides parallel. A *trapezium* has no sides parallel.

Squares, rectangles, rhombuses, and rhomboids are classified as parallelograms in mathematics. A *parallelogram* is a four-sided plane figure with opposite sides parallel and equal. See Figure 4-19.

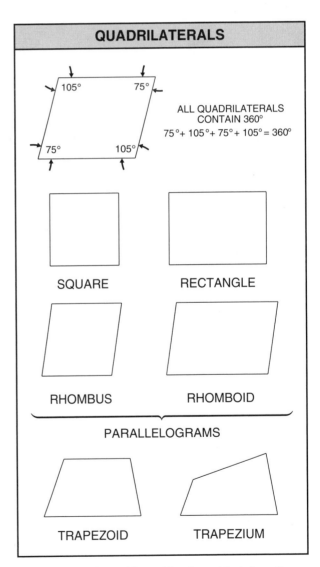

Figure 4-19. A quadrilateral is a four-sided plane figure.

Polygons

A *polygon* is a many-sided plane figure. All polygons contain 360°. A *regular polygon* has equal sides and equal angles. An *irregular polygon* has unequal sides and unequal angles. Polygons are named according to their number of sides. For example, a *pentagon* has five sides; a *hexagon* has six sides; a *heptagon* has seven sides; an *octagon* has eight sides; etc. See Figure 4-20.

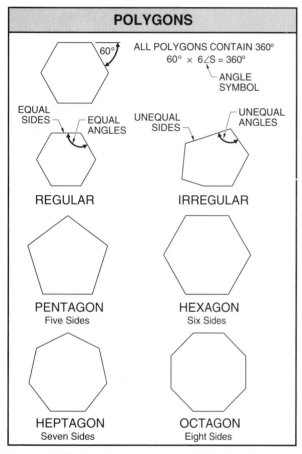

Figure 4-20. A polygon is a many-sided plane figure.

Conic Sections

A *conic section* is a curve produced by a plane intersecting a right circular cone. A *right circular cone* is a cone with the axis at a 90° angle to the circular base. The four types of conic sections are the circle, ellipse, parabola, and hyperbola.

A *circle* has the plane perpendicular to the axis. An *ellipse* has the plane oblique to the axis but making a greater angle with the axis than with the elements of the cone. A *parabola* has the plane oblique to the

axis and making the same angle with the axis as with the elements of the cone. A *hyperbola* has the plane making a smaller angle with the axis than with the elements of the cone. An *oblique* plane is not perpendicular to the base. See Figure 4-21.

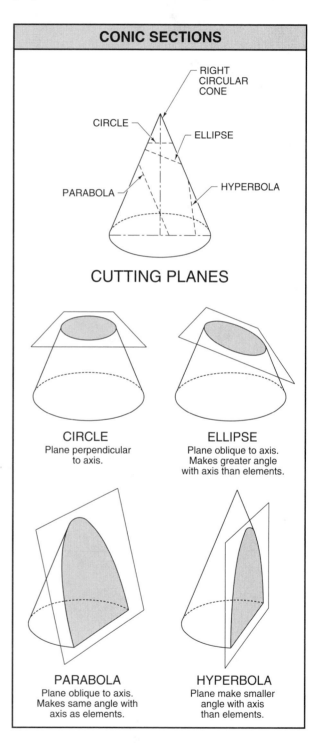

Figure 4-21. A conic section is a curve produced by a plane intersecting a right circular cone.

BASIC MATH FORMULAS

An *equation* is a means of showing that two numbers or two groups of numbers are equal to the same amount. For example, a baker has two pies of the same size. Pie A is cut into six equal pieces, and Pie B is cut into eight equal pieces. One customer buys three pieces of Pie A, and another customer buys four pieces of Pie B. The customers have bought the same amount of pie because $\frac{3}{6} = \frac{1}{2}$ and $\frac{4}{8} = \frac{1}{2}$. All equations must balance, and as $\frac{3}{6}$ and $\frac{4}{8}$ each equal $\frac{1}{2}$, $\frac{3}{6} = \frac{4}{8}$ is an equation.

A *formula* is a mathematical equation which contains a fact, rule, or principle. Letters are used in formulas to represent values (amount). In the common electrical formula, $I = VA/V$, the I represents ampacity (denoted A), the VA represents wattage or volt amps (denoted VA), and the V represents voltage (denoted V). If any two of these values are known, the other value can be found by rearranging the formula. See Figure 4-22.

Ampacity is found by applying the formula:

$$I = \frac{VA}{V}$$

where

I = ampacity (in A)

VA = wattage or volt amps (in VA)

V = voltage (in V)

For example, what is the ampacity of a 120 V electric circuit with 2400 watts?

$$I = \frac{VA}{V}$$

$$I = \frac{2400}{120}$$

$$I = \textbf{20 A}$$

Voltage is found by applying the formula:

$$V = \frac{VA}{I}$$

For example, what is the voltage of a 15 A circuit with 1800 VA?

$$V = \frac{VA}{I}$$

$$V = \frac{1800}{15}$$

$$E = \textbf{120 V}$$

Wattage is found by applying the formula:

$$VA = I \times V$$

For example, what is the volt amps of a 120 V, 30 A circuit?

$$VA = I \times V$$

$$VA = 30 \times 120$$

$$VA = \textbf{3600 VA}$$

I = AMPACITY (A)
VA = WATTAGE OR VOLT AMPS (VA)
V = VOLTAGE (V)

ELECTRICAL FORMULA

How many amps will flow in a 120 V circuit servicing a 1500 VA sign load?

$$I = \frac{VA}{V}$$
$$I = \frac{1500}{120}$$
$$I = \textbf{12.5 A}$$

AMPACITY

How many volt amps does a 6 A grinder use under full load on a 120 V circuit?

$$VA = I \times V$$
$$VA = 6 \times 120$$
$$VA = \textbf{720 VA}$$

VOLT AMPS

What voltage is required for a 2400 VA, 20 A circuit?

$$V = \frac{VA}{I}$$
$$V = \frac{2400}{20}$$
$$V = \textbf{120 V}$$

VOLTAGE

Figure 4-22. When two values of a formula are known, the third value can be found.

Common Formulas

In the metal trades, common formulas related to plane and solid figures are used when laying out jobs. For example, a welder may be required to lay out and build a cylindrical tank to hold a specified number of gallons of liquid. By applying the volume formula for cylinders, the welder can determine the size of the cylindrical tank.

Area. *Area* is the number of unit squares equal to the surface of an object. For example, a standard size piece of plywood contains 32 sq ft (4 × 8 = 32 sq ft). Area is expressed in square inches, square feet, and other units of measure. A *square inch* measures 1″ × 1″ or its equivalent. A *square foot* contains 144 sq in. (12″ × 12″ = 144 sq in.). The area of any plane figure can be determined by applying the proper formula. See Figure 4-23.

Circumference of a Circle (Diameter). When the diameter is known, the circumference of a circle is found by applying the formula:

$$C = \pi D$$

where

C = circumference

π = 3.1416

D = diameter

For example, what is the circumference of a 20″ diameter circle?

$$C = \pi D$$

$$C = 3.1416 \times 20$$

$$C = \textbf{62.832″}$$

Circumference of a Circle (Radius). When the radius is known, the circumference of a circle is found by applying the formula:

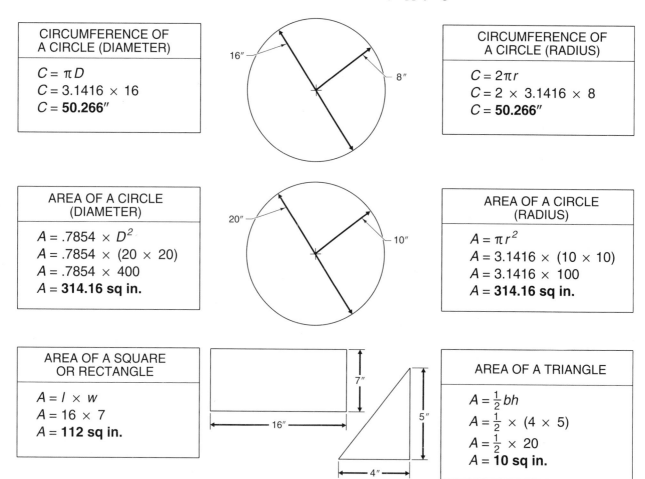

CIRCUMFERENCE OF A CIRCLE (DIAMETER)
$C = \pi D$ $C = 3.1416 \times 16$ $C = \textbf{50.266″}$

CIRCUMFERENCE OF A CIRCLE (RADIUS)
$C = 2\pi r$ $C = 2 \times 3.1416 \times 8$ $C = \textbf{50.266″}$

AREA OF A CIRCLE (DIAMETER)
$A = .7854 \times D^2$ $A = .7854 \times (20 \times 20)$ $A = .7854 \times 400$ $A = \textbf{314.16 sq in.}$

AREA OF A CIRCLE (RADIUS)
$A = \pi r^2$ $A = 3.1416 \times (10 \times 10)$ $A = 3.1416 \times 100$ $A = \textbf{314.16 sq in.}$

AREA OF A SQUARE OR RECTANGLE
$A = l \times w$ $A = 16 \times 7$ $A = \textbf{112 sq in.}$

AREA OF A TRIANGLE
$A = \frac{1}{2}bh$ $A = \frac{1}{2} \times (4 \times 5)$ $A = \frac{1}{2} \times 20$ $A = \textbf{10 sq in.}$

Figure 4-23. Area is the number of unit squares equal to the surface of an object.

$C = 2\pi r$

where

C = circumference

2 = constant

π = 3.1416

r = radius

For example, what is the circumference of a 10″ radius circle?

$C = 2\pi r$

$C = 2 \times 3.1416 \times 10$

$C = \mathbf{62.832''}$

Area of a Circle (Diameter). When the diameter is known, the area of a circle is found by applying the formula:

$A = .7854 \times D^2$

where

A = area

.7854 = constant

D^2 = diameter squared

For example, what is the area of a 28″ diameter circle?

$A = .7854 \times D^2$

$A = .7854 \times (28 \times 28)$

$A = .7854 \times 784$

$A = \mathbf{615.754}$ **sq in.**

Area of a Circle (Radius). When the radius is known, the area of a circle is found by applying the formula:

$A = \pi r^2$

where

A = area

π = 3.1416

r^2 = radius squared

For example, what is the area of a 14″ radius circle?

$A = \pi r^2$

$A = 3.1416 \times (14 \times 14)$

$A = 3.1416 \times 196$

$A = \mathbf{615.754}$ **sq in.**

Area of a Square or Rectangle. The area of a square or the area of a rectangle is found by applying the formula:

$A = l \times w$

where

A = area

l = length

w = width

For example, what is the area of a 22′-0″ × 16′-0″ storage room?

$A = l \times w$

$A = 22 \times 16$

$A = \mathbf{352}$ **sq ft**

Area of a Triangle. The area of a triangle is found by applying the formula:

$A = \frac{1}{2}bh$

where

A = area

$\frac{1}{2}$ = constant

b = base

h = height

For example, what is the area of a triangle with a 10″ base and a 12″ height?

$A = \frac{1}{2}bh$

$A = \frac{1}{2} \times (10 \times 12)$

$A = \frac{1}{2} \times 120$

$A = \mathbf{60}$ **sq in.**

Pythagorean Theorem. The *Pythagorean Theorem* states that the square of the hypotenuse of a right triangle is equal to the sum of the squares of the other two sides. The *hypotenuse* is the side of a right triangle opposite the right angle. Because a right triangle can have a 3-4-5 relationship, it is often used in laying out right angles and checking corners for squareness. See Figure 4-24.

The length of the hypotenuse of a right triangle is found by applying the formula:

$c = \sqrt{a^2 + b^2}$

where

c = length of hypotenuse

a = length of one side

b = length of other side

For example, what is the length of the hypotenuse of a triangle having sides of 3′ and 4′?

$c = \sqrt{a^2 + b^2}$

$c = \sqrt{(3 \times 3) + (4 \times 4)}$

$c = \sqrt{9 + 16}$

$c = \sqrt{25}$

$c = \mathbf{5′}$

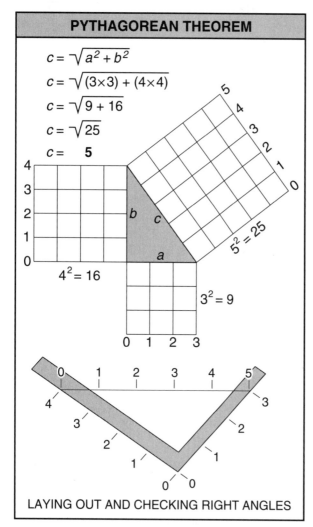

PYTHAGOREAN THEOREM

$c = \sqrt{a^2 + b^2}$

$c = \sqrt{(3 \times 3) + (4 \times 4)}$

$c = \sqrt{9 + 16}$

$c = \sqrt{25}$

$c = \mathbf{5}$

$4^2 = 16$

$3^2 = 9$

$5^2 = 25$

LAYING OUT AND CHECKING RIGHT ANGLES

Figure 4-24. The square of the hypotenuse of a right triangle is equal to the sum of the squares of the other two sides of the triangle.

Volume. *Volume* is the three-dimensional size of an object measured in cubic units. For example, the volume of a standard size concrete block is 1024 cu in. ($8 \times 8 \times 16 = 1024$ cu in.). Volume is expressed in cubic inches, cubic feet, cubic yards and other units of measure. A *cubic inch* measures 1″ × 1″ × 1″ or its equivalent. A *cubic foot* contains 1728 cu in. (12″ × 12″ × 12″ = 1728 cu in.). A cubic yard contains 27 cu ft (3′ × 3′ × 3′ = 27 cu ft). The volume of a solid figure can be determined by applying the proper formula. See Figure 4-25.

Volume of a Rectangular Solid. The volume of a rectangular solid is found by applying the formula:

$V = l \times w \times h$

where

V = volume

l = length

w = width

h = height

For example, what is the volume of a 24″ × 12″ × 8″ rectangular solid?

$V = l \times w \times h$

$V = 24 \times 12 \times 8$

$V = \mathbf{2304}$ **cu in.**

Volume of a Cylinder (Diameter). When the diameter is known, the volume of a cylinder is found by applying the formula:

$V = .7854 \times D^2 \times h$

where

V = volume

.7854 = constant

D^2 = diameter squared

h = height

For example, what is the volume of a tank that is 4′-0″ in diameter and 12′-0″ long?

$V = .7854 \times D^2 \times h$

$V = .7854 \times (4 \times 4) \times 12$

$V = .7854 \times 16 \times 12$

$V = \mathbf{150.797}$ **cu ft**

$V = \pi r^2 \times l$

where

V = volume

π = 3.1416

r^2 = radius squared

l = length

For example, what is the volume of a tank that has a 2'-0″ radius and is 12'-0″ long?

$V = \pi r^2 \times l$

$V = 3.1416 \times (2 \times 2) \times 12$

$V =$ **150.797 cu ft**

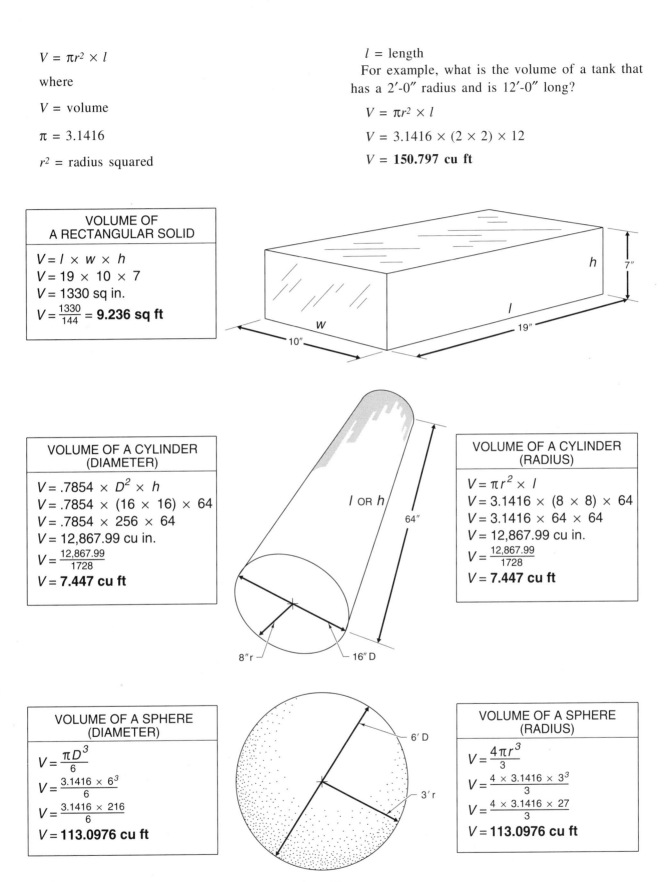

VOLUME OF A RECTANGULAR SOLID

$V = l \times w \times h$
$V = 19 \times 10 \times 7$
$V = 1330$ sq in.
$V = \frac{1330}{144} =$ **9.236 sq ft**

VOLUME OF A CYLINDER (DIAMETER)

$V = .7854 \times D^2 \times h$
$V = .7854 \times (16 \times 16) \times 64$
$V = .7854 \times 256 \times 64$
$V = 12{,}867.99$ cu in.
$V = \frac{12{,}867.99}{1728}$
$V =$ **7.447 cu ft**

VOLUME OF A CYLINDER (RADIUS)

$V = \pi r^2 \times l$
$V = 3.1416 \times (8 \times 8) \times 64$
$V = 3.1416 \times 64 \times 64$
$V = 12{,}867.99$ cu in.
$V = \frac{12{,}867.99}{1728}$
$V =$ **7.447 cu ft**

VOLUME OF A SPHERE (DIAMETER)

$V = \frac{\pi D^3}{6}$
$V = \frac{3.1416 \times 6^3}{6}$
$V = \frac{3.1416 \times 216}{6}$
$V =$ **113.0976 cu ft**

VOLUME OF A SPHERE (RADIUS)

$V = \frac{4\pi r^3}{3}$
$V = \frac{4 \times 3.1416 \times 3^3}{3}$
$V = \frac{4 \times 3.1416 \times 27}{3}$
$V =$ **113.0976 cu ft**

Figure 4-25. Volume is the three-dimensional size of an object.

Volume of a Sphere (Diameter). When the diameter is known, the volume of a sphere is found by applying the formula:

$$V = \frac{\pi D^3}{6}$$

where

V = volume

π = 3.1416

D^3 = diameter cubed

6 = constant

For example, what is the volume of a sphere that is 4'-0" in diameter?

$$V = \frac{\pi D^3}{6}$$

$$V = \frac{3.1416 \times 4^3}{6}$$

$$V = \frac{3.1416 \times 64}{6}$$

$$V = \textbf{33.5104 cu ft}$$

Volume of a Sphere (Radius). When the radius is known, the volume of a sphere is found by applying the following formula:

$$V = \frac{4\pi r^3}{3}$$

where

V = volume

4 = constant

π = 3.1416

r^3 = radius cubed

3 = constant

For example, what is the volume of a sphere that is 4'-0" in diameter?

$$V = \frac{4\pi r^3}{3}$$

$$V = \frac{4 \times 3.1416 \times 8}{3}$$

$$V = \textbf{33.5104 cu ft}$$

Review Questions

Name_____ Date _____

True-False

T   F   **1.** Arabic numerals are expressed by ten digits.

T   F   **2.** Odd numbers cannot be divided by 2 an even number of times.

T   F   **3.** A mixed number has a numerator larger than the denominator.

T   F   **4.** An improper fraction can be changed to a mixed number.

T   F   **5.** Fractions are divided horizontally.

T   F   **6.** The United States monetary system is based on Roman numerals.

T   F   **7.** More places written in a decimal number indicate a higher degree of accuracy.

T   F   **8.** Any number multiplied by a zero equals that number.

T   F   **9.** To check division, multiply the sum by the quotient.

T   F   **10.** The decimal point is the period in a decimal number.

T   F   **11.** A circle is a plane figure.

T   F   **12.** All plane figures are composed of lines drawn at various angles or with arcs.

T   F   **13.** A vertical line is the shortest distance between two points.

T   F   **14.** Parallel lines may vary in length and distance apart.

T   F   **15.** Angles are measured in degrees, minutes, and seconds.

T   F   **16.** A trapezoid is also a parallelogram.

T   F   **17.** A scalene triangle has two equal sides and two equal angles.

T   F   **18.** An octagon is a plane figure with six sides.

T   F   **19.** The hypotenuse of a right triangle is the side opposite the right angle.

T   F   **20.** A cubic foot contains 1278 cu in.

Matching — Roman Numerals

_____   **1.** 4   **A.** XX

_____   **2.** 90   **B.** D

_____   **3.** 1000   **C.** I

_____   **4.** 20   **D.** X

_____   **5.** 1   **E.** V

_____   **6.** 10   **F.** VII

_____   **7.** 40   **G.** IV

_____   **8.** 7   **H.** XL

_____   **9.** 500   **I.** M

_____   **10.** 5   **J.** XC

Completion

_____   **1.** A(n) _____ number is any number that has no fractional or decimal parts.

_____   **2.** A(n) _____ number is any number that can be divided by 2 an exact number of times.

_____   **3.** The _____ numeral system is the most commonly used numeral system in the United States.

_____   **4.** _____ are groups of three digits separated by a comma.

_____   **5.** _____ is the process of uniting two or more numbers to make one number.

_____   **6.** _____ is the opposite of addition.

_____   **7.** A(n) _____ or decimal is one part of a whole number.

_____   **8.** The _____ shows into how many parts a whole number has been divided.

_____   **9.** A(n) _____ is a fraction with a denominator of 10, 100, 1000, etc.

_____   **10.** The smallest whole number is _____.

_____   **11.** A(n) _____ figure is a flat figure.

_____   **12.** A(n) _____ line is parallel to the horizon.

_____   **13.** A straight angle always contains _____°.

_____ **14.** A circle always contains _____°.

_____ **15.** The _____ of a circle is one-half the length of the diameter.

_____ **16.** A right triangle contains one _____° angle.

_____ **17.** A(n) _____ is a many-sided plane figure generated around a centerpoint.

_____ **18.** _____ circles have different diameters and different centerpoints.

_____ **19.** A(n) _____ is a mathematical equation which contains a fact, rule, or principle.

_____ **20.** A square foot contains _____ sq in.

Matching — Math

_____ **1.** Minuend

_____ **2.** Remainder

_____ **3.** Multiplier

_____ **4.** Divisor

_____ **5.** Sum

_____ **6.** Dividend

_____ **7.** Quotient

_____ **8.** Subtrahend

_____ **9.** Product

_____ **10.** Multiplicand

$$217 + 31 = 248$$

$$853 - 142 = 711$$

$$6\overline{)180}$$

$$15 \times 7 = 105$$

Multiple Choice

_____ **1.** _____ numbers are numbers that can be divided an exact number of times only by themselves and the number 1.
A. Proper
B. Period
C. Product
D. Prime

_____ **2.** The most common operation in mathematics is _____.
A. addition
B. subtraction
C. multiplication
D. division

_____ **3.** A(n) _____ fraction has a denominator larger than its numerator.

    A. proper
    B. improper
    C. mixed
    D. none of the above

_____ **4.** The numerator is the _____-hand number of a fraction.

    A. lower right
    B. lower left
    C. upper right
    D. upper left

_____ **5.** Fractional parts of an inch are always expressed in their _____ common denominator.

    A. highest
    B. most
    C. lowest
    D. none of the above

_____ **6.** Two numbers to the right of the decimal point indicate _____.

    A. tenths
    B. hundredths
    C. thousandths
    D. none of the above

_____ **7.** A superscript rule placed over a Roman numeral letter increases the letter's value _____ times.

    A. 10
    B. 100
    C. 1000
    D. 10,000

_____ **8.** The larger number is commonly used as the _____ when the units being multiplied are the same.

    A. multiplier
    B. multiplicand
    C. product
    D. sum

_____ **9.** All fractions must have a common _____ before one can be subtracted from the other.

    A. sum
    B. remainder
    C. numerator
    D. denominator

_____ **10.** To multiply fractions, multiply the _____ and reduce as required.

    A. numerator times the numerator
    B. denominator times the denominator
    C. both A and B
    D. neither A nor B

_____ **11.** A regular plane figure has _____ angles and _____ sides.

    A. equal; equal
    B. equal; unequal
    C. unequal; unequal
    D. unequal; equal

12. An angle is _____.

    A. always at 90° to the horizon
    B. the intersection of two lines
    C. composed of lines and arcs
    D. none of the above

13. A _____ is one-half of a circle.

    A. sector
    B. segment
    C. quadrant
    D. semicircle

14. An oblique plane of a right circular cone is _____.

    A. parallel to the base
    B. parallel to the axis
    C. perpendicular to the axis
    D. not perpendicular to the base

15. A triangle _____.

    A. contains 180°
    B. has three sides
    C. both A and B
    D. neither A nor B

16. The volume of a cylinder may be found by applying the formula: _____.

    A. $V = .7854 \times r^2 \times h$
    B. $V = \pi d^2 \times l$
    C. either A or B
    D. neither A nor B

17. A(n) _____ triangle can have a 3-4-5 relationship.

    A. oblique
    B. equilateral
    C. right
    D. none of the above

18. A circle is produced by passing a plane through a right circular cone at _____° to the axis.

    A. 60
    B. 90
    C. 120
    D. 180

19. A rhomboid is a plane figure with _____.

    A. opposite sides equal
    B. opposite angles equal
    C. no 90° angles
    D. all of the above

20. Supplementary angles always contain _____°.

    A. 60
    B. 90
    C. 120
    D. 180

Matching — Circle

_____ **1.** Diameter

_____ **2.** Sector

_____ **3.** Arc

_____ **4.** Segment

_____ **5.** Radius

_____ **6.** Tangent

_____ **7.** Centerpoint

_____ **8.** Circle

_____ **9.** Chord

_____ **10.** Quadrant

Matching — Metrics

_____ **1.** Hecto

_____ **2.** m

_____ **3.** Gram

_____ **4.** A

_____ **5.** mm

_____ **6.** M

_____ **7.** Hundreth

_____ **8.** Thousand

_____ **9.** Tenth

_____ **10.** s

A. Thousand

B. Mega

C. Time symbol

D. Kilo

E. Centi

F. Length symbol

G. Deci

H. Ampere

I. Hundred

J. Mass

Trade Competency Test

Name_____ Date _____

Word Problems

_____ **1.** Four pieces (A, B, C, and D) were cut from the piece of cold rolled steel. What is the length of the remaining piece? (Disregard the cutting waste).

_____ **2.** A welder spends $3\frac{1}{4}$ hours cutting and grinding steel to shape and $1\frac{1}{2}$ hours positioning, welding, and finishing parts to complete a job. How many hours did the job require?

_____ **3.** Four fillet welds and 32 spot welds are required to complete the welding on a parts bin. How many spot welds are required to produce 124 parts bins?

_____ **4.** How many $2\frac{1}{4}'' \times 16''$ strips can be cut from the piece of sheet metal (assuming there is no cutting waste)?

_____ **5.** What is the volume of the tank in cu ft?

_____ **6.** What is the length of Side c of Triangle A?

_____ **7.** How many VA will a 6.8 A, 120 V grinder pull under a full load?

_____ **8.** What is the area of Triangle B?

_____ **9.** What is the circumference of Circle A?

_____ **10.** What is the volume of the rectangular solid?

_____ **11.** What is the area of Circle A?

_____ **12.** What is the area of Triangle A?

_____ **13.** What is the volume of Sphere A?

_____ **14.** What is the total length of the sides of the square?

_____ **15.** What is the area of the square?

SPHERE A

4'-0"

$7\frac{1}{4}''$

$7\frac{1}{4}''$

Refer to the Dowel Positioner print on page 101.

_____ **1.** The overall length of Part B is _____".

_____ **2.** Part _____ has rounded corners.

_____ **3.** A _____ is used to weld the two parts together.
   A. butt weld, arrow side
   B. butt weld, both sides
   C. fillet weld, other side
   D. fillet weld, both sides

_____ **4.** Parts A and B are the same _____.
   A. size
   B. shape
   C. both A and B
   D. neither A nor B

T   F   **5.** Four $\frac{3}{8}''$ diameter holes are drilled in Part A.

_____ **6.** The horizontal center-to-center distance between the holes in Part B is _____".

_____ **7.** The Dowel Positioner was manufactured by _____.

_____ **8.** Part B has an area of _____ sq in. (disregarding the drilled holes).

_____ **9.** Part B has a volume of _____ cu in. (disregarding the drilled holes).

_____ **10.** The centerpoints of the two upper holes in Part B are _____" from the top edge of Part A.

_____ **11.** The drawing of the Dowel Positioner was completed on _____.

_____ **12.** The holes in Part A are _____" farther apart than the holes in Part B.

T   F   **13.** The drawing of the Dowel Positioner has been revised.

T   F   **14.** The overall height of the Dowel Positioner is $4\frac{1}{4}''$.

T   F   **15.** The Dowel Positioner is made from $\frac{1}{2}''$ thick stock.

T   F   **16.** The holes in Part B are larger than the holes in Part A.

T   F   **17.** Part B is welded perpendicular to Part A.

_____ **18.** The radius of all rounded corners is _____″.

_____ **19.** All drilled holes are shown in the right side view with _____ lines.

_____ **20.** The overall depth of the Dowel Positioner is _____″.

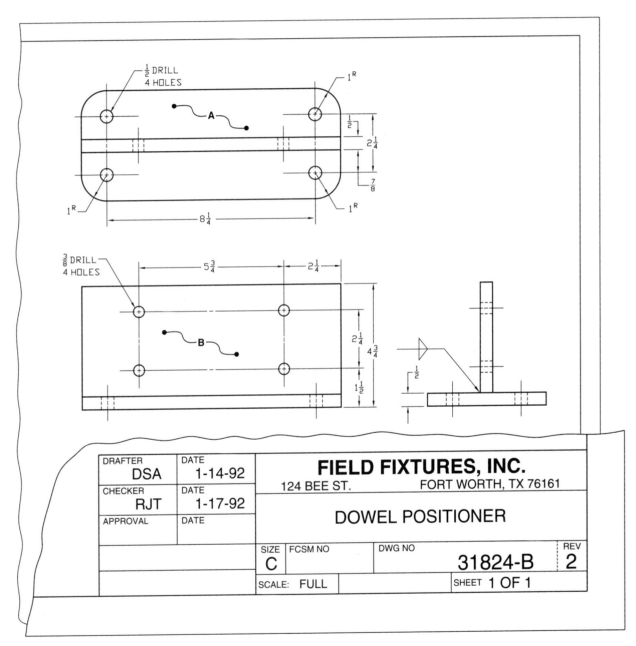

DRAFTER	DATE	**FIELD FIXTURES, INC.**		
DSA	1-14-92	124 BEE ST. FORT WORTH, TX 76161		
CHECKER	DATE			
RJT	1-17-92	**DOWEL POSITIONER**		
APPROVAL	DATE			
		SIZE	FCSM NO	DWG NO	REV
		C		31824-B	2
		SCALE: FULL		SHEET 1 OF 1	

DOWEL POSITIONER

Matching — Plane Figures

_____ 1. Trapezoid

_____ 2. Isosceles Triangle

_____ 3. Pentagon

_____ 4. Equilateral Triangle

_____ 5. Rhombus

_____ 6. Square

_____ 7. Right Triangle

_____ 8. Trapezium

_____ 9. Hexagon

_____ 10. Rectangle

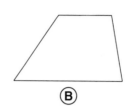

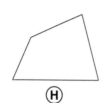
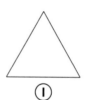
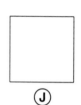

Conversions

Convert the following fractions to decimals. Convert the following decimals to fractions.

1. $\frac{1}{4} =$

2. $\frac{3}{8} =$

3. $\frac{5}{8} =$

4. $\frac{1}{2} =$

5. $\frac{9}{16} =$

6. $\frac{1}{8} =$

7. $\frac{3}{4} =$

8. $\frac{7}{8} =$

9. $\frac{7}{16} =$

10. $\frac{1}{16} =$

11. $\frac{3}{16} =$

12. $\frac{5}{16} =$

13. $\frac{9}{32} =$

14. $\frac{11}{32} =$

15. $\frac{11}{16} =$

16. $.1875 =$

17. $1.000 =$

18. $.9375 =$

19. $1.250 =$

20. $2.5 =$

21. $3.750 =$

22. $9.1875 =$

23. $6.50 =$

24. $.3125 =$

25. $8.375 =$

26. $8.625 =$

27. $2.875 =$

28. $4.125 =$

29. $3.3125 =$

30. $.750 =$

Math Problems

Use a separate sheet of paper to perform the operation indicated. Reduce all fractions.

1. Adding Fractions

A. $\frac{1}{4} + \frac{3}{4} =$

B. $\frac{3}{8} + \frac{1}{8} + 1\frac{1}{8} =$

C. $\frac{9}{16} + \frac{3}{16} =$

D. $\frac{5}{8} + 4\frac{3}{8} + 1\frac{1}{8} =$

E. $\frac{3}{16} + \frac{5}{8} + 11 =$

F. $1\frac{1}{2} + 3\frac{1}{4} + \frac{9}{16} =$

G. $3\frac{1}{8} + 13\frac{3}{4} + 11\frac{5}{8} =$

H. $27\frac{9}{16} + 15\frac{3}{8} =$

I. $3\frac{3}{8} + 3\frac{3}{8} + 3\frac{1}{8} =$

J. $9\frac{7}{16} + 2\frac{5}{16} + 1\frac{1}{16} =$

2. Subtracting Fractions

A. $\frac{13}{16} - \frac{9}{16} =$

B. $\frac{7}{8} - \frac{3}{4} =$

C. $2\frac{7}{16} - \frac{3}{8} =$

D. $11\frac{7}{8} - 3\frac{1}{2} =$

E. $13\frac{1}{2} - \frac{3}{4} =$

F. $12\frac{5}{6} - 3\frac{9}{16} =$

G. $3\frac{7}{8} - 2\frac{3}{8} =$

H. $24\frac{1}{2} - 12\frac{1}{2} =$

I. $3\frac{5}{8} - \frac{7}{8} =$

J. $12\frac{3}{4} - 3\frac{1}{16} =$

3. Multiplying Fractions

A. $\frac{1}{2} \times \frac{1}{2} =$

B. $\frac{1}{4} \times \frac{1}{8} =$

C. $\frac{3}{8} \times \frac{1}{2} =$

D. $\frac{7}{8} \times \frac{3}{4} =$

E. $1\frac{1}{2} \times \frac{1}{4} =$

F. $1\frac{3}{4} \times 2 =$

G. $12\frac{1}{4} \times \frac{1}{8} =$

H. $5\frac{3}{4} \times 2 =$

I. $6\frac{1}{8} \times 8 =$

J. $3\frac{7}{8} \times 2 =$

4. Dividing Fractions

A. $\frac{3}{8} \div 2 =$

B. $\frac{3}{4} \div \frac{1}{2} =$

C. $4\frac{9}{16} \div 2 =$

D. $11\frac{3}{8} \div 3 =$

E. $5\frac{7}{16} \div 2 =$

F. $3\frac{3}{8} \div 3 =$

G. $23\frac{1}{2} \div 2 =$

H. $50\frac{1}{2} \div 5 =$

I. $9\frac{9}{16} \div 3\frac{3}{16} =$

J. $3\frac{5}{8} \div 2 =$

5. Adding Whole Numbers

A.
```
  31
 +16
```

B.
```
  79
 +17
```

C.
```
 138
 +32
```

D.
```
 169
+312
```

E.
```
 398
+426
```

F.
```
 319
+672
```

G.
```
4821
 362
 467
```

H.
```
3834
2431
 379
 165
```

I.
```
3365
2240
3089
```

J.
```
4162
 360
 475
 388
```

6. Subtracting Whole Numbers

A.
```
  38
 -22
```

B.
```
  96
 -27
```

C.
```
 235
 -68
```

D.
```
 383
-139
```

E.
```
 796
-294
```

F.
```
2089
-137
```

G.
```
3642
-1388
```

H.   6847
     −3998

I.   4360
      −890

J.   8876
     −8431

7. Multiplying Whole Numbers

A.   24
        ×8

B.   92
       ×12

C.   138
       ×28

D.   144
       ×12

E.   830
       ×20

F.   312
       ×64

G.   665
       ×75

H.   1388
      ×241

I.   2428
      ×324

J.   3206
      ×412

8. Dividing Whole Numbers

A. 8 $\overline{)192}$

B. 12 $\overline{)5472}$

C. 32 $\overline{)15,360}$

D. 78 $\overline{)2496}$

E. 24 $\overline{)7896}$

F. 25 $\overline{)453}$

G. 16 $\overline{)78,854}$

H. 32 $\overline{)848}$

I. 24 $\overline{)1998}$

J. 12 $\overline{)2472}$

9. Adding and Subtracting Decimal Numbers

A.   .31
       +.82

B.   .69
       +.33

C.   1.55
      +2.08

D.   3.665
     +2.501

E.   3.157
      2.358
      3.159

F.   13.54
      −2.28

G.   24.68
      −6.05

H.   .705
      −.038

I.   .0667
     −.0389

J.   3.80
       −.72

10. Multiplying and Dividing Decimal Numbers

A.   .25
       ×.30

B.   .36
       ×.03

C.   .128
       ×.04

D.   3.14
        ×4

E.   98.60
       ×.55

F. 14 ÷ .5 =

G. 27.5 ÷ .25 =

H. 3.30 ÷ 1.1 =

I. 37.50 ÷ .20 =

J. 45.45 ÷ .90 =

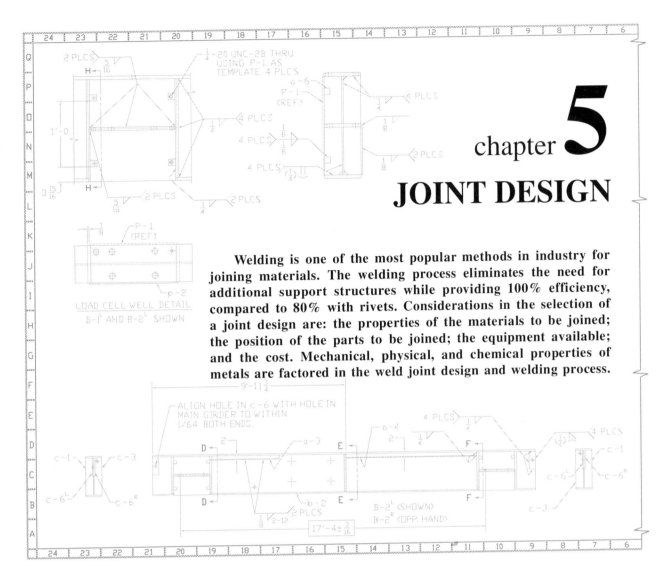

chapter 5

JOINT DESIGN

Welding is one of the most popular methods in industry for joining materials. The welding process eliminates the need for additional support structures while providing 100% efficiency, compared to 80% with rivets. Considerations in the selection of a joint design are: the properties of the materials to be joined; the position of the parts to be joined; the equipment available; and the cost. Mechanical, physical, and chemical properties of metals are factored in the weld joint design and welding process.

MECHANICAL PROPERTIES

Mechanical properties are the properties that describe the behavior of metals under applied loads. *Load* is an external force applied to an elastic body that causes stress in a material. Mechanical properties are largely influenced by the composition and treatment of the metal. Mechanical properties of metals are classified using standards established by the American Society of Testing and Materials (ASTM).

The designer must consider the mechanical properties of metals when specifying the welds required. Standard terms commonly used to describe the mechanical properties of solid metals include stress and strain.

Stress

Stress is the effect of an external force applied upon a solid material. To maintain equilibrium, the solid material has an internal resistance that absorbs the external force. This internal resistance is expressed in pounds per square inch (psi or lb/sq in.). *Equilibrium* is a state of balance between opposing forces. For example, a spring at rest is in a state of equilibrium. The forces (tensile) pulling on it equal the forces pushing (compressive) on it.

Every machine part or structural member is designed to safely withstand a certain amount of stress. The type of stress is determined by position and direction. The five types of stress are tensile, compressive, shearing, bending, and torsional. See Figure 5-1.

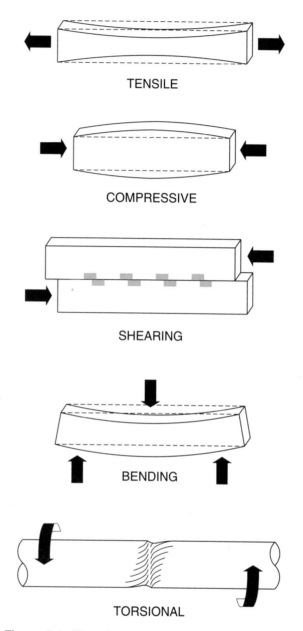

Figure 5-1. The five types of stress are tensile, compressive, shearing, bending, and torsional.

Tensile Stress. *Tensile stress* is stress caused by two equal forces acting on the same axial line to pull an object apart. Tensile stress tends to stretch an object. If the amount of tensile stress exceeds a certain value beyond the proportional limit of the material, the part is deformed permanently. This deformation reduces the cross-sectional area of the object and correspondingly increases its length. *Proportional limit* is the maximum stress a material can withstand without permanent deformation.

The magnitude of stress to which an object is subjected depends on the amount of external force placed on the object and the cross-sectional area of the object. The same external force causes greater stress to an object with a smaller cross-sectional area than to an object with a larger cross-sectional area. See Figure 5-2.

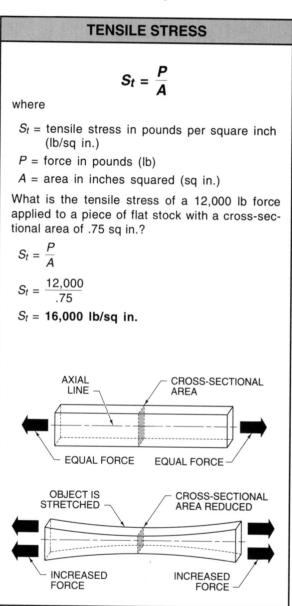

TENSILE STRESS

$$S_t = \frac{P}{A}$$

where

S_t = tensile stress in pounds per square inch (lb/sq in.)

P = force in pounds (lb)

A = area in inches squared (sq in.)

What is the tensile stress of a 12,000 lb force applied to a piece of flat stock with a cross-sectional area of .75 sq in.?

$$S_t = \frac{P}{A}$$

$$S_t = \frac{12,000}{.75}$$

$$S_t = \textbf{16,000 lb/sq in.}$$

Figure 5-2. Tensile stress is stress caused by two equal forces acting on the same axial line to pull an object apart.

To find tensile stress, apply the following formula:

$$S_t = \frac{P}{A}$$

where

S_t = tensile stress in pounds per square inch (lb/sq in.)

P = force in pounds (lb)

A = area in inches squared (sq in.)

For example, what is the tensile stress of an 8000 lb force applied to a square steel rod with a cross-sectional area of .50 sq in.?

$$S_t = \frac{P}{A}$$

$$S_t = \frac{8000}{.50}$$

$$S_t = \textbf{16,000 lb/sq in.}$$

Compressive Stress. *Compressive stress* is stress caused by two equal forces acting on the same axial line to crush the object. Compressive stress tends to squeeze an object. The deformation caused by compressive force consists of an increase in the cross-sectional area and a decrease in the original length of the object. See Figure 5-3.

To find compressive stress, apply the following formula:

$$S_c = \frac{P}{A}$$

where

S_c = compressive stress in pounds per square inch (lb/sq in.)

P = force in pounds (lb)

A = area in inches squared (sq in.)

For example, what is the compressive stress of a 120,000 lb force applied to a rectangular cast iron bar with a cross-sectional area of 6 sq in.?

$$S_c = \frac{P}{A}$$

$$S_c = \frac{120,000}{6}$$

$$S_c = \textbf{20,000 lb/sq in.}$$

Shearing Stress. *Shearing stress* is stress caused by two equal and parallel forces acting upon an object from opposite directions. Shearing stress tends to cause one side of the object to slide in relation to the other side. Shearing stress placed on

COMPRESSIVE STRESS

$$S_c = \frac{P}{A}$$

where

S_c = compressive stress in pounds per square inch (lb/sq in.)

P = force in pounds (lb)

A = area in inches squared (sq in.)

What is the compressive stress of a 10,000 lb force applied to a piece of flat stock with a cross-sectional area of .75 sq in.?

$$S_c = \frac{P}{A}$$

$$S_c = \frac{10,000}{.75}$$

$$S_c = \textbf{13,333.33 lb/sq in.}$$

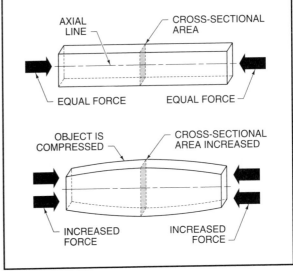

Figure 5-3. Compressive stress is stress caused by two equal forces acting on the same axial line to crush an object.

the cross-sectional area of an object is parallel to the force. The strength of materials in shearing stress is less than in tensile stress and compressive stress. See Figure 5-4.

To find shearing stress, apply the following formula:

$$S_s = \frac{P}{A} \text{ or } P = S_s \times A$$

where

S_s = shearing stress in pounds per square inch (lb/sq in.)

P = force in pounds (lb)

A = area in inches squared (sq in.)

For example, a .750″ hole is to be punched in a steel plate 0.5″ thick. What is the required force of the press if the ultimate strength of the steel plate in shearing is 42,000 lb/sq in.?

The shearing cross-sectional area (A) is equal to the circumference of the hole times the thickness of the plate (3.14 × .750 × 0.5 = 1.1775).

$P = S_s \times A$

$P = 42,000 \times 1.1775$

$P = \textbf{49,455 lb}$

Bending Stress. *Bending stress* is stress caused by equal forces acting perpendicular to the horizontal axis of an object. Bending stress tends to bend an object as the perpendicular force overcomes the reaction force. Bending stress is a combination of tensile stress and compressive stress.

Bending stress is commonly associated with beams and columns. The deformation caused by bending stress changes the shape of the object and creates a deflection. See Figure 5-5.

To find bending stress, apply the following formula:

$$S_b = \frac{Mc}{Z}$$

where

S_b = bending stress in pounds per square inch (lb/sq in.)

M = maximum bending moment in inch-pounds (in-lb)

c = distance from neutral axis to farthest point in cross section in inches (in.)

Z = section modulus in cubic inches (cu in.)

For example, what is the bending stress of a 1″ solid shaft subjected to a bending moment of 1400 in-lb? The distance from the neutral axis to the cross-sectional area c = 0.5″, and the section modulus (taken from another formula) is 0.049.

$$S_b = \frac{Mc}{Z}$$

$$S_b = \frac{1400 \times 0.5}{0.049}$$

S_b = 14,285.7143 = **14,286 lb/sq in.**

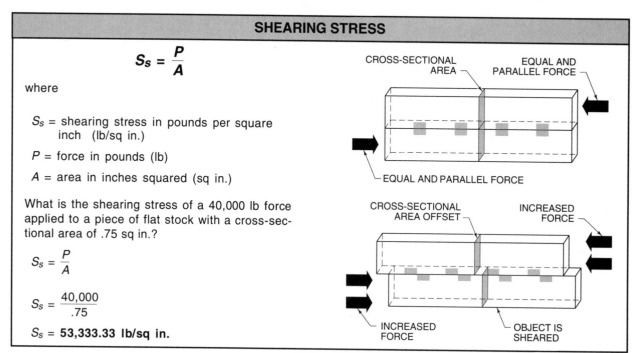

SHEARING STRESS

$$S_s = \frac{P}{A}$$

where

S_s = shearing stress in pounds per square inch (lb/sq in.)

P = force in pounds (lb)

A = area in inches squared (sq in.)

What is the shearing stress of a 40,000 lb force applied to a piece of flat stock with a cross-sectional area of .75 sq in.?

$$S_s = \frac{P}{A}$$

$$S_s = \frac{40,000}{.75}$$

S_s = **53,333.33 lb/sq in.**

CROSS-SECTIONAL AREA

EQUAL AND PARALLEL FORCE

EQUAL AND PARALLEL FORCE

CROSS-SECTIONAL AREA OFFSET

INCREASED FORCE

INCREASED FORCE

OBJECT IS SHEARED

Figure 5-4. Shearing stress is stress caused by two equal and parallel forces acting upon an object in opposite directions.

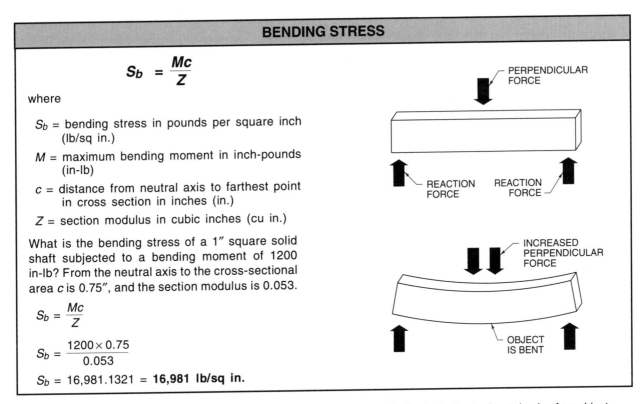

BENDING STRESS

$$S_b = \frac{Mc}{Z}$$

where

S_b = bending stress in pounds per square inch (lb/sq in.)

M = maximum bending moment in inch-pounds (in-lb)

c = distance from neutral axis to farthest point in cross section in inches (in.)

Z = section modulus in cubic inches (cu in.)

What is the bending stress of a 1″ square solid shaft subjected to a bending moment of 1200 in-lb? From the neutral axis to the cross-sectional area c is 0.75″, and the section modulus is 0.053.

$$S_b = \frac{Mc}{Z}$$

$$S_b = \frac{1200 \times 0.75}{0.053}$$

S_b = 16,981.1321 = **16,981 lb/sq in.**

Figure 5-5. Bending stress is stress caused by equal forces acting perpendicular to the horizontal axis of an object.

Torsional Stress. *Torsional stress* is stress caused by two forces acting in opposite twisting motions. Torsional stress tends to twist an object. Shafts used to transfer rotary motion are subject to torsional stress. The shafts are twisted by the excessive torque expressed in inch-pounds (in-lb). *Torque* is the product of the applied force (P) times the distance (L) from the center of application. See Figure 5-6.

To find torque, apply the following formula:

$$T = P \times L$$

where

T = torque in inch-pounds (in-lb)

P = force in pounds (lb)

L = distance in inches (in.)

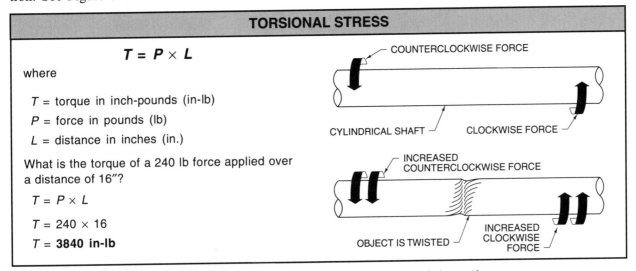

TORSIONAL STRESS

$$T = P \times L$$

where

T = torque in inch-pounds (in-lb)

P = force in pounds (lb)

L = distance in inches (in.)

What is the torque of a 240 lb force applied over a distance of 16″?

$$T = P \times L$$

$$T = 240 \times 16$$

T = **3840 in-lb**

Figure 5-6. Torsional stress is stress caused by two forces acting in opposite twisting motions.

For example, what is the torque of a 160 lb force applied over a distance of 12″?

$$T = P \times L$$

$$T = 160 \times 12$$

$$T = \mathbf{1920\ in\text{-}lb}$$

Strain

Strain is the deformation per unit length of a solid under stress. The magnitude of strain is equal to the total amount of deformation divided by the original size of the solid. Strain is usually expressed as a percentage of linear units. For example, strain may be expressed as percent of inch per inch (% in./in.).

Elastic deformation is the ability of a metal to return to its original size and shape after being loaded and unloaded. Elastic deformation occurs along the straight-line portion of the typical stress-strain curve. Most metal parts are designed to deform elastically. *Plastic deformation* is the failure of a metal to return to its original size and shape after being loaded and unloaded. *Elastic limit (yield)* is the last point at which a material can be deformed and still return to its original shape. It is the transitional area from elastic deformation to plastic deformation. See Figure 5-7.

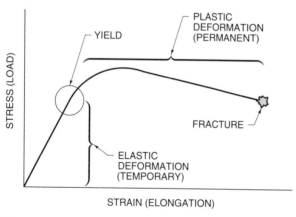

Figure 5-7. Strain is the deformation per unit length of a solid under stress.

Modulus of elasticity is the ratio of stress to strain within the elastic limit. It is an index of the ability of a solid material to deform when an external force is applied and then return to its original size and shape after removing the external force. The less a material deforms under a given stress,

the higher the modulus of elasticity. Modulus of elasticity does not measure the amount of stretch a particular metal can take before breaking or deforming. It indicates how much stress is required to deform metal a given amount. See Figure 5-8.

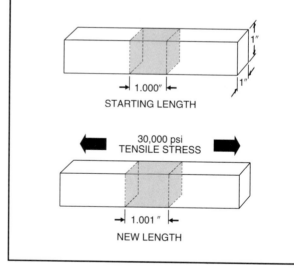

Figure 5-8. Modulus of elasticity is the ratio of stress to strain within the elastic limit.

To find modulus of elasticity, apply the following formula:

$$E = \frac{S_s}{S_n}$$

where

E = modulus of elasticity in millions of pounds per square inch (10^6 psi)

S_s = stress in psi

S_n = strain in inch per inch (in./in.)

For example, what is the modulus of elasticity of a 1″ square piece of metal subjected to 40,000 lb of tension and exhibiting 0.001 in./in. strain?

$$E = \frac{S_s}{S_n}$$

$$E = \frac{40,000}{0.001}$$

E = 40,000,000 psi = **40 × 10⁶ psi**

Ductility is the ability of a metal to stretch, bend, or twist without breaking or cracking. High ductility metals such as copper resist stress from loads and fail gradually. Low ductility metals deform only slightly before failure occurs.

Hardness is the ability of a metal to resist indentation. *Brittleness* is the lack of ductility in a given metal. Brittle metals will fracture quickly under low stress conditions. *Toughness* is a combination of strength and ductility. *Malleability* is the ability of the metal to be deformed by compressive forces without developing defects. *Creep* is a slow progressive strain that causes metal to fail. *Cryogenic properties* are the abilities of metals to resist failure when subjected to very low temperatures.

PHYSICAL PROPERTIES

Physical properties are the thermal, electrical, optical, magnetic, and general properties of metal. *Thermal properties* include melting point, thermal conductivity, and thermal expansion and contraction. *Melting point* is the amount of heat required to melt a given amount of metal. *Thermal conductivity* is the rate at which the metal transmits heat. Metals that have high thermal conductivity, such as copper and aluminum, require less heat than other metals in the welding process.

Thermal expansion is the expansion of a metal when subjected to heat. The amount of thermal expansion is expressed as the coefficient of thermal expansion. The *coefficient of thermal expansion* is the unit change in the length of a material caused by changing the temperature 1°F. See Figure 5-9.

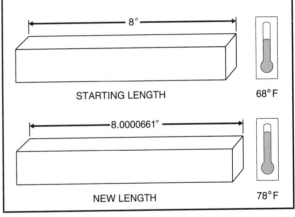

COEFFICIENT OF THERMAL EXPANSION

$$C = \frac{L_d}{\Delta T}$$

where

C = coefficient of thermal expansion

L_d = length differential

ΔT = temperature difference in °F

What is the coefficient of thermal expansion of an 8″ wrought iron bar increased to 8.0000661″ when the temperature is increased from 68°F to 78°F?

$$C = \frac{L_d}{\Delta T}$$

$$C = \frac{.0000661}{10}$$

C = **.00000661″/°F**

8″

STARTING LENGTH   68°F

8.0000661″

NEW LENGTH   78°F

Figure 5-9. The coefficient of thermal expansion is the unit change in the length of a material caused by changing the temperature 1°F.

Thermal expansion occurs in all dimensions of metal when exposed to heat. Different metals have different coefficients of thermal expansion. To find coefficient of thermal expansion, apply the following formula:

$$C = \frac{L_d}{\Delta T}$$

where

C = coefficient of thermal expansion

L_d = length differential

ΔT = temperature difference in °F

For example, what is the coefficient of thermal expansion of a 9″ copper bar increased to 9.000063″ when the temperature is increased from 72°F to 79°F?

$$C = \frac{L_d}{\Delta T}$$

$$C = \frac{.000063}{7}$$

$C = .00000900″/°F$

Electrical properties are the abilities of a metal to conduct or resist electricity or the flow of electrons. Weld temperature is affected by the ability of electrons to flow through metal. Metals with high conductivity heat more quickly. As metals are heated, conductivity decreases.

Optical properties of metals are the color of the metal and how it reflects light. Optical properties of metals allow visual identification of the content of a metal by analysis of color and luster.

Light falling on a metal surface is either absorbed, reflected, or transmitted. *Absortivity* is the fraction of light absorbed. *Reflectivity* is the fraction of light reflected. *Transmissivity* is the fraction of light transmitted. Absortivity and reflectivity levels of metals are fairly high, while transmissivity is very low, except in very thin layers of metal.

Metals which can be easily identified by their colors include:

gold.....yellow
copper.....red-orange
silver.....metallic white
chromium.....blue-white
stainless steel.....yellow-white

CHEMICAL PROPERTIES

Chemical properties of metals are directly related to molecular composition and pertain to the chemical reactivity of metals and the surrounding environment. Chemical reactivity includes corrosion, oxidation, and reduction of metal. *Corrosion* is the combining of metals with elements in the surrounding environment that leads to the deterioration or wasting away of the metal. *Oxidation* is the combining of metal and oxygen into metal oxides. *Reduction* is the loss or removal of oxygen during the welding process. Oxygen in the atmosphere has the greatest effect on welds and must be controlled to prevent contamination of the weld area.

LOADS

Load is an external force applied to an elastic body that causes stress in a material. A load, when applied to a structure or part, stresses the part and causes movement (strain) within the material. Structures are designed for strength and rigidity by considering the external load, the internal stress caused from the external load, and the selection of the material required to carry the load within the allowable stress or strain. Loads are static, impact, or variable. See Figure 5-10.

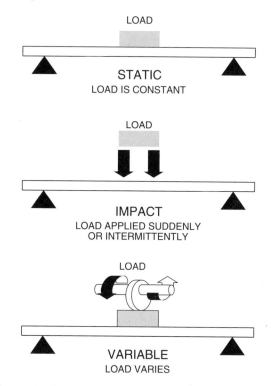

Figure 5-10. A load is an external force applied to an elastic body that causes stress in a material.

A *static load* is a load that remains constant. Static loads do not vary. An example of a static load is a constant amount of water stored in a storage tank.

An *impact load* is a load that is applied suddenly or intermittently. An example of an impact load is the action of a pile driver placing a pile.

A *variable load* is a load that varies with time and rate, but without the sudden change that occurs with an impact load. An example of a variable load is a revolving camshaft with a varying compressive and tensile load applied.

In a structure, welds are classified as primary or secondary. A *primary weld* is a weld that is an integral part of a structure and that directly transfers the load. A primary weld must possess or exceed the strength of the members in the structure. A *secondary weld* is a weld used to hold joint members and subassemblies together. Secondary welds are subjected to less stress than primary welds and have less of a load than primary welds.

WELD JOINT SELECTION

Welded joints provide strength and efficiency, and can be made more quickly than other joining methods. Welded joints have replaced many parts and structures which used fasteners or the casting process. All weld designs are determined by the intended function of the structure, by governing codes and specifications, and by economic considerations. Engineers and designers consider all factors of the welded part for safety and efficiency. Welded joints are used in virtually every industry. In the building industry, welds are used to join structural elements such as columns, trusses, girders, and other structural components.

Weld Joints and Type

Most weld joints are subjected to loads that require strength and rigidity to prevent failure of members. Loads in a structure are transferred from member to member through the welds. Weld joints subjected to minimum loads, such as their own weight, are considered to be "no-load" welds. For example, access covers and panels, safety guards, and protective covers require "no-load" welds.

Weld joint designs are governed by AWS Codes and other appropriate codes. For example, in building construction, AWS codes govern structural and welding materials, weld details, processes and techniques, weld quality, and inspection. The design of the structural elements is governed by American Institute of Steel Construction (AISC)

Specifications. The design selected also factors wind forces, loads, seismic conditions, and other conditions which may cause fatigue. Weld joint designs selected for use in buildings are governed by AISC Specifications. Additional codes, such as the Uniform Building Code, and other appropriate state and local codes, may also apply.

The five basic weld joints are the butt, lap, T, edge, and corner weld joints. Each of these may be used with applicable weld types to meet load requirements. See Figure 5-11. Load requirements dictate strength of the required welds. To maximize weld strength, basic rules are observed. These include:

Minimize edge preparation. Minimizing edge preparation reduces the cutting and the machining costs.

Provide weld access. Allow for access to the weld by welding equipment. Available welding equipment must be considered.

Minimize filler metal required. Minimizing filler metal reduces cost of filler metal. For example, a $\frac{3}{8}''$ fillet requires 44% more filler metal than a $\frac{5}{16}''$ fillet, but only provides 20% more strength.

Reduce excess heat. Reducing excess heat applied to the weld area during the welding process minimizes molecular changes of the base metal and filler metal.

Minimize number of welds. Minimizing the number of welds reduces filler metal required. Additionally, distortion of joint members from heat applied is reduced.

Size weld for thinnest joint member. The size of the weld should not exceed the strength of the thinnest joint member.

Welding Positions

The weld joint selected is also affected by the welding position. The four basic welding positions are flat, horizontal, vertical, and overhead. The flat welding position allows the greatest control of the welding process.

In horizontal, vertical, and overhead positions, gravity reduces penetration and filler metal control. This can cause weld defects resulting in weak welds. Smaller parts can be placed in the flat position for efficiency. However, some larger parts cannot be positioned for flat position welding

WELD JOINTS AND TYPES						
APPLICABLE WELDS	WELD SYMBOL	BUTT	LAP	T	EDGE	CORNER
SQUARE-GROOVE			—			
BEVEL-GROOVE						
V-GROOVE			—	—		
U-GROOVE			—	—		
J-GROOVE						
FLARE-BEVEL-GROOVE						
FLARE-V-GROOVE			—	—		
FILLET		—			—	
PLUG		—			—	
SLOT		—			—	
EDGE-FLANGE			—	—		—
CORNER-FLANGE		—	—	—		
SPOT		—			—	
PROJECTION		—			—	
SEAM		—				
BRAZE	BRAZE				—	

Figure 5-11. The five basic weld joints are used with applicable weld types to meet load requirements.

They must be welded in positions other than the flat position. See Figure 5-12.

In pipe welding, complete penetration of the joint members is crucial. In some instances, the pipe can be rotated so that all welding is completed in the flat position. This minimizes problems caused by welding in the horizontal, vertical, and overhead positions.

Weld Joint Access

Adequate access is required for the welding equipment and the penetration and deposition of the filler metal. For example, use of a large gun in flux cored arc welding limits the access to welds in tight areas. Proper fit-up also assures that the joint members are in correct alignment, have the correct edge preparation, and have the required root opening for proper penetration and sufficient weld reinforcement. See Figure 5-13. Subassemblies can eliminate some access problems.

Alignment of the joint members assures that the joint members are true to print specifications. Print specifications require joint members to be in the same or in a different plane with the proper edge-to-edge or end-to-end alignment. Edge preparation requires the correct angle or shape on the specified joint members. The root opening is

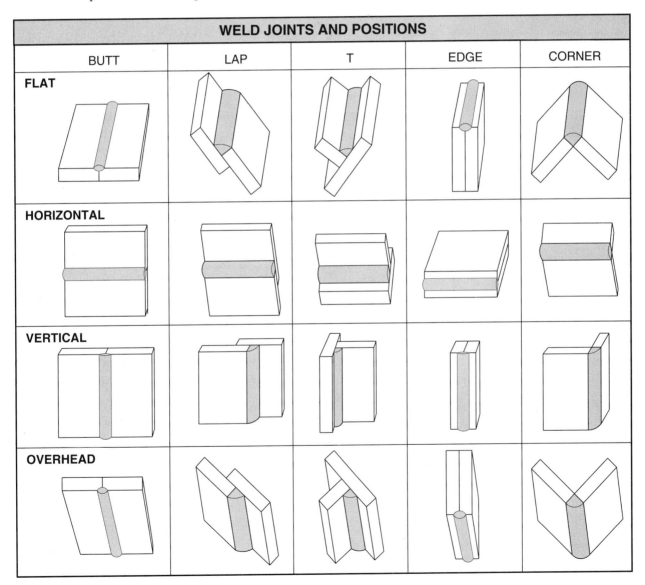

Figure 5-12. The four basic welding positions are flat, horizontal, vertical, and overhead.

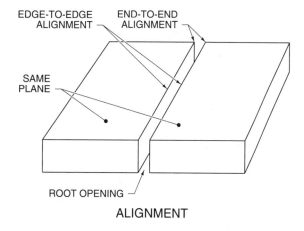

ALIGNMENT

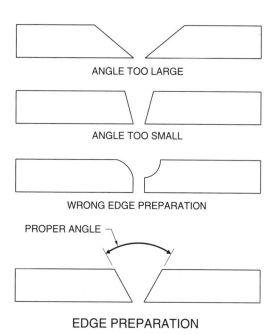

EDGE PREPARATION

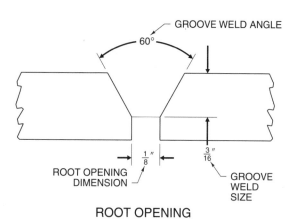

ROOT OPENING

Figure 5-13. Alignment, edge preparation, and root opening are necessary for good fit-up.

controlled using jigs, tack welds, spacers, or consumable inserts.

A *tack weld* is a weld that joins the joint members at random points to keep the joint members from moving out of their required positions. Subsequent welding on the weld joint melts through the tack welds. Spacers provide a gap between the joint members to be tack welded. After tack welding, the spacers are removed before continuing the welding process. Consumable inserts are melted during the welding process and become part of the filler metal added to the weld joint.

Welding Locations

Welding is performed in the shop or in the field, depending upon the structure's size and fabrication requirements. Welds on small parts, structures, and subassemblies are often completed in the shop. The shop provides a controlled environment in which fixtures and positioners can be used to move the part for improved welding productivity.

A *fixture* is a device used to maintain the correct positional relationship between joint members required by print specifications. A *positioner* is a mechanical device that supports and moves joint members for maximum loading, welding, and unloading efficiency.

Positioners can be used with hand- and machine-controlled welding equipment. In production settings, positioners and welding equipment are used together for maximum welding efficiency.

Welding is performed in the field when the structure's size or fabrication requirements prohibit assembly in the shop. Welding in the field often results in a decrease in welding productivity.

Ambient temperature, weather, welding conditions, and welder efficiency in the field affect welding productivity. For example, pipe welding of sections of pipe used for the transmission of water, chemicals, petroleum products, and other materials across great distances, must be done in the field. The pipeline right-of-way must be cut and graded for access of pipelaying and welding equipment. The ditch is dug and pipe joints are strung for fit-up and welding. The sections of pipe are aligned and welded in place to the required specifications. This requires highly-skilled welders to work quickly in various conditions without compromising weld quality.

Sketching

Name _____ Date _____

Sketching — Stress

Sketch the front view (multiview) of each isometric. Show force arrows.

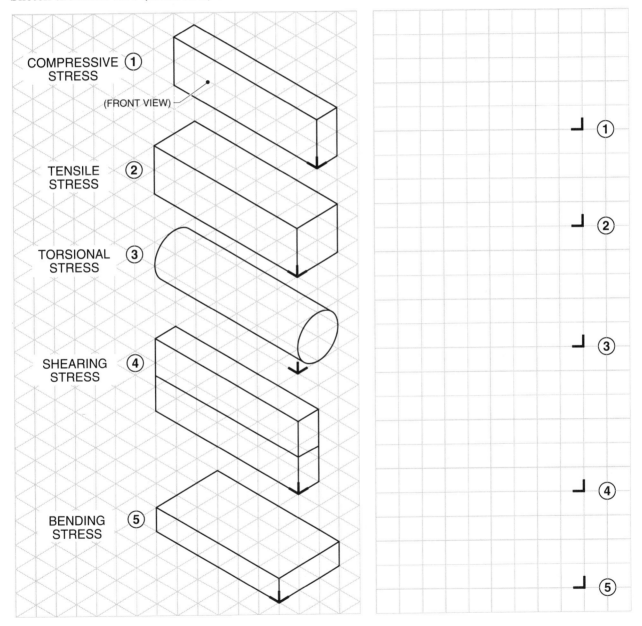

Sketching — Weld Joints and Positions — Isometrics

Sketch the isometric of each weld joint. The overall dimensions of each completed weld are given.

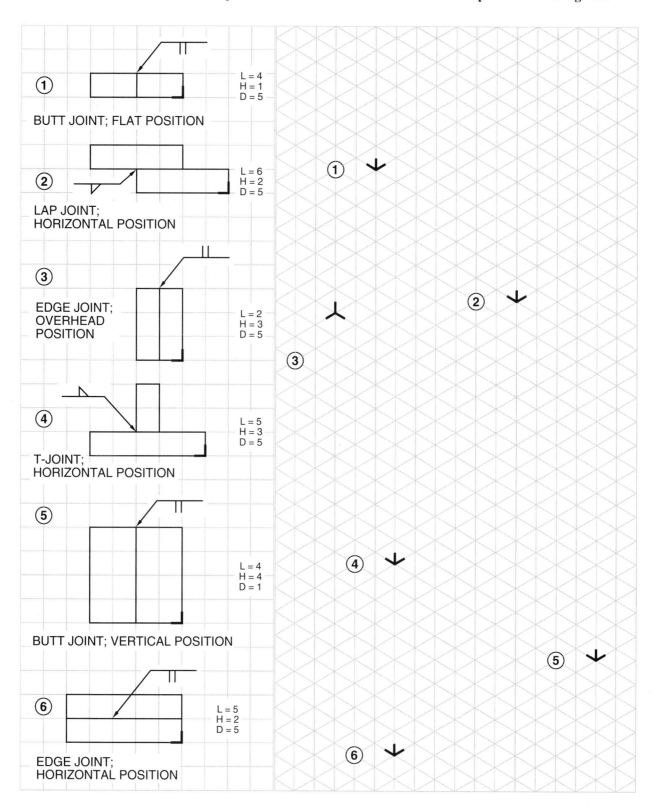

① BUTT JOINT; FLAT POSITION
L = 4
H = 1
D = 5

② LAP JOINT; HORIZONTAL POSITION
L = 6
H = 2
D = 5

③ EDGE JOINT; OVERHEAD POSITION
L = 2
H = 3
D = 5

④ T-JOINT; HORIZONTAL POSITION
L = 5
H = 3
D = 5

⑤ BUTT JOINT; VERTICAL POSITION
L = 4
H = 4
D = 1

⑥ EDGE JOINT; HORIZONTAL POSITION
L = 5
H = 2
D = 5

Sketching — Weld Joints and Positions — Multiviews

Sketch the front view (multiview) of each oblique. Show welding symbols as specified.

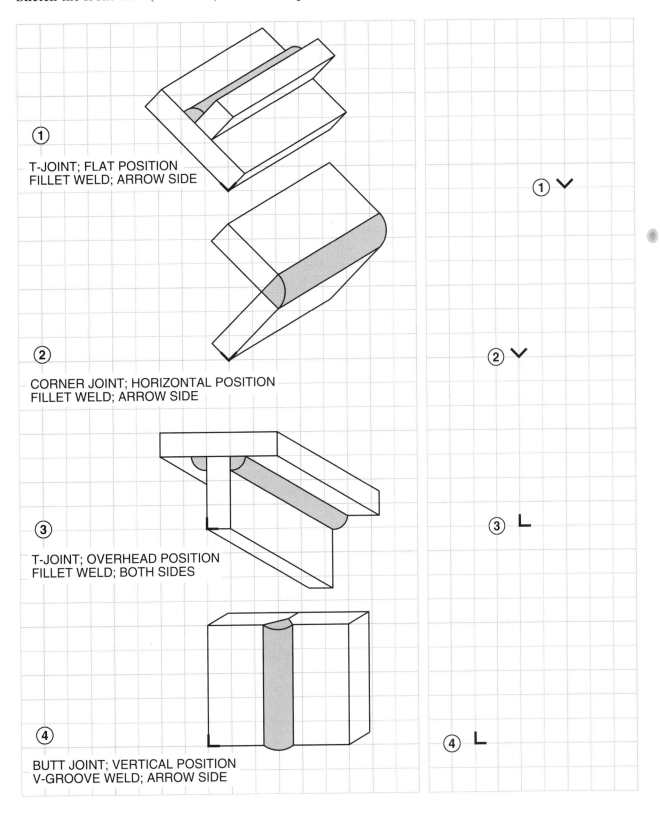

① T-JOINT; FLAT POSITION
FILLET WELD; ARROW SIDE

② CORNER JOINT; HORIZONTAL POSITION
FILLET WELD; ARROW SIDE

③ T-JOINT; OVERHEAD POSITION
FILLET WELD; BOTH SIDES

④ BUTT JOINT; VERTICAL POSITION
V-GROOVE WELD; ARROW SIDE

① ∨

② ∨

③ ∟

④ ∟

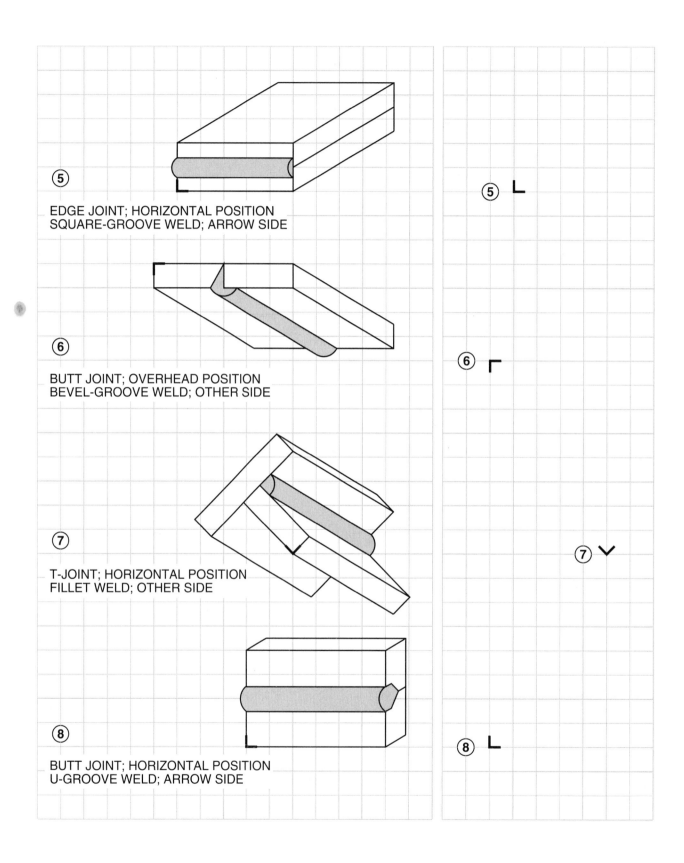

(5) EDGE JOINT; HORIZONTAL POSITION
SQUARE-GROOVE WELD; ARROW SIDE

(6) BUTT JOINT; OVERHEAD POSITION
BEVEL-GROOVE WELD; OTHER SIDE

(7) T-JOINT; HORIZONTAL POSITION
FILLET WELD; OTHER SIDE

(8) BUTT JOINT; HORIZONTAL POSITION
U-GROOVE WELD; ARROW SIDE

Review Questions

Name_____ Date _____

Matching — Stress

_____ **1.** Tensile stress

_____ **2.** Compressive stress

_____ **3.** Shearing stress

_____ **4.** Bending stress

_____ **5.** Torsional stress

Matching — Joint Design

_____ **1.** Elastic limit

_____ **2.** Thermal conductivity

_____ **3.** Impact load

_____ **4.** Static load

_____ **5.** Basic welding positions

_____ **6.** Basic weld joints

A. Applied suddenly or intermittently

B. Remains constant

C. Yield

D. Flat, horizontal, vertical, and overhead

E. Rate at which metal transmits heat

F. Butt, lap, T, edge, and corner

True-False

T   F   **1.** Equilibrium is a state of balance between opposing forces.

T   F   **2.** The type of stress is determined by its position and direction.

T   F   **3.** Stress is expressed in psi or lb sq/in.

T   F   **4.** Proportional limit is the minimum stress a material can withstand without permanent deformation.

T   F   **5.** Compressive stress tends to stretch an object.

T   F   **6.** The forces of shearing stress are applied in opposite directions.

T   F   **7.** Bending stress is a combination of tensile stress and shearing stress.

T   F   **8.** Torsional stress tends to twist an object.

T   F   **9.** The strength of materials in shearing stress is more than in tensile stress and compressive stress.

T   F   **10.** Compressive stress tends to increase the cross-sectional area of an object.

T   F   **11.** A primary weld is used to hold joint members and subassemblies together.

T   F   **12.** Oxygen must be controlled to prevent contamination of the weld area.

T   F   **13.** Strain is the deformation per unit length of a solid under stress.

T   F   **14.** Welding may be performed in the shop or in the field, depending upon the structure's size and fabrication requirements.

T   F   **15.** The size of the weld should not exceed the strength of the thinnest joint member.

Matching — Weld Joints and Positions

_____   **1.** Horizontal butt joint

_____   **2.** Flat lap joint

_____   **3.** Flat butt joint

_____   **4.** Vertical edge joint

_____   **5.** Overhead corner joint

_____   **6.** Vertical butt joint

_____   **7.** Vertical T-joint

_____   **8.** Overhead butt joint

_____   **9.** Flat edge joint

_____   **10.** Horizontal lap joint

Completion

_____   **1.** _____ properties describe the behavior of metals under applied loads.

_____   **2.** Shearing stress placed on the cross-sectional area of an object is _____ to the force.

_____   **3.** Bending stress is stress caused by equal forces acting _____ to the horizontal axis of an object.

4. _____ deformation is the failure of a metal to return to its original size and shape after being loaded and unloaded.

5. _____ is the ability of a metal to stretch, bend, or twist without breaking or cracking.

6. _____ properties of metals are the color of the metal and how it reflects light.

7. _____ is wasting away of metals caused by combination of metals with elements in the surrounding environment.

8. A(n) _____ is an external force applied to an elastic body that causes stress in a material.

9. The _____ welding position allows the greatest control of the welding process.

10. A(n) _____ is a device used to maintain the correct positional relationship between joint members while welding.

11. _____ is a slow progressive strain that causes metal to fail.

12. A(n) _____ weld is an integral part of a structure and directly transfers the load.

13. _____ reduces filler metal control in horizontal, vertical, and overhead positions.

14. A(n) _____ weld is a weld that joins the joint members at random points to keep the joint members from moving out of their required positions.

15. _____ deformation is the ability of a metal to return to its original size and shape after being loaded and unloaded.

Matching — Weld Joints and Types

1. Bevel-groove; butt joint

2. Flare-V-groove; edge joint

3. Flare-V-groove; corner joint

4. Edge-flange; butt joint

5. V-groove; butt joint

6. Square-groove; edge joint

7. U-groove; corner joint

8. J-groove; lap joint

9. U-groove; butt joint

10. Bevel-groove; edge joint

Matching — Weld Symbols

_____   1.

_____   2.

_____   3.

_____   4.

_____   5.

_____   6.

_____   7.

_____   8.

_____   9.

_____   10.

Ⓐ

Ⓑ

Ⓒ

Ⓓ

Ⓔ

Ⓕ

Ⓖ

Ⓗ

Ⓘ

Ⓙ

Trade Competency Test

Name_____ Date _____

Math — Stress

Answer the following questions in the spaces provided. Show all work.

_____ **1.** What is the tensile stress of a 12,500 lb force applied to a square steel rod with a cross-sectional area of 1.25 sq in.?

_____ **2.** What is the compressive stress of a 22,400 lb force applied to a rectangular steel bar with a cross-sectional area of 4 sq in.?

_____ **3.** What is the torque of a 2000 lb force applied over a distance of $3\frac{1}{2}$″?

_____ **4.** Find the compressive stress.

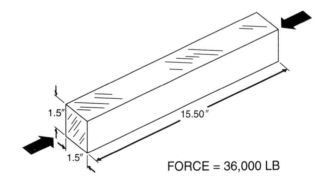

1.5″   15.50″   1.5″   FORCE = 36,000 LB

_____ **5.** Find the tensile stress.

1.25″   12″   1.25″   FORCE = 21,000 LB

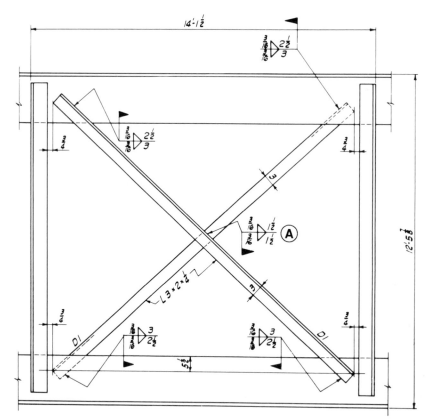

American Institute of Steel Construction, Inc.

BRACING PANEL

Refer to the Bracing Panel print on page 126.

_____ 1. The print specifies _____ welds to join braces to supports.

_____ 2. All welds are to be completed in the _____.

_____ 3. Weld size specified for all weld joints is _____″.

_____ 4. Structural steel used for braces is _____″ wide on the largest leg.

_____ 5. The vertical supports are _____″ from the brace angles.

_____ 6. The length of weld (A) is _____″.

T   F   7. The leg size of weld (A) is $^3/_{16}$″.

T   F   8. The angle iron braces are welded together on their longest legs.

T   F   9. The height of the bracing panel is 14′-1½″.

T   F   10. Short break lines are used to show breaks in the upper and lower horizontal supports.

Identification — Primary/Secondary Welds

Indicate whether each weld is a primary or secondary weld.

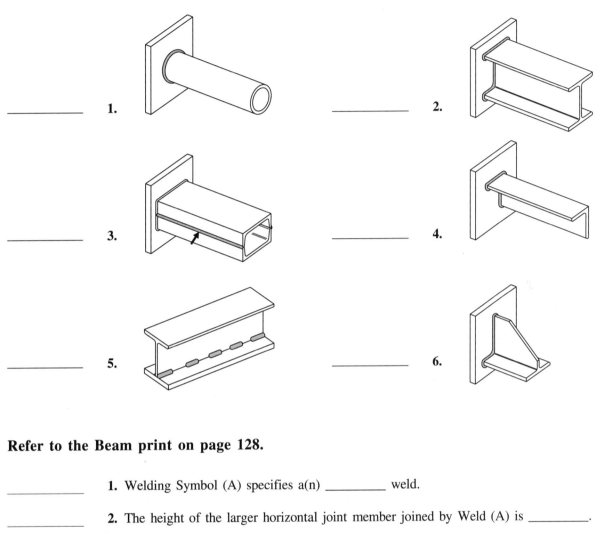

_____ 1.

_____ 2.

_____ 3.

_____ 4.

_____ 5.

_____ 6.

Refer to the Beam print on page 128.

_____ **1.** Welding Symbol (A) specifies a(n) _____ weld.

_____ **2.** The height of the larger horizontal joint member joined by Weld (A) is _____.

_____ **3.** Welding Symbol (B) specifies a(n) _____ weld on both sides.

_____ **4.** The weld size of Weld (B) is _____″.

_____ **5.** The joint member with holes at Weld (C) is _____″ thick.

T   F   **6.** The gusset plate angle for center holes at (D) is 45°.

_____ **7.** The thickness of the plate at (B) is _____″.
    A. $5/16$
    B. $1/2$
    C. $5/8$
    D. 1

T   F   **8.** The weight per running foot of the larger horizontal joint member is 27 lb.

T F **9.** Section A-A specifies $1\frac{3}{4}''$ center-to-center, horizontal bolt hole placement.

_____ **10.** Section A-A is centered _____ from mounting flange (E).

_____ **11.** The length of the larger I beam is _____.

T F **12.** Section B-B is taken in a vertical plane.

_____ **13.** Gusset plates are _____″ thick.

_____ **14.** The $\frac{5}{16}''$ radius notch at (B) is _____ long.

T F **15.** All gusset plates are welded with fillet welds.

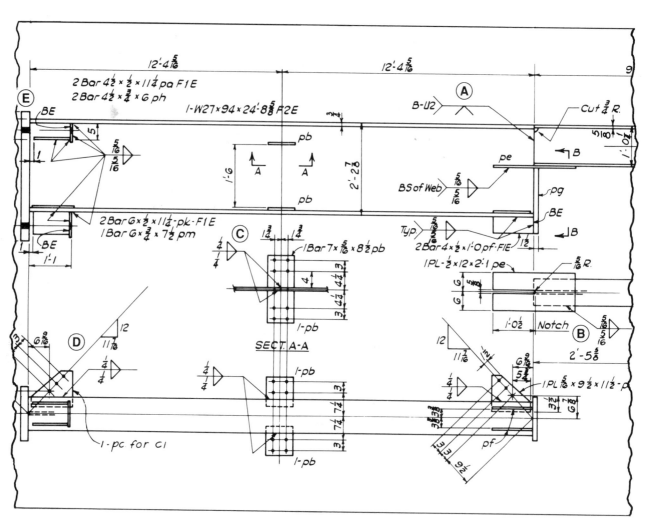

BEAM

Matching — Stress

_____ **1.** Shearing

_____ **2.** Bending

_____ **3.** Torsional

_____ **4.** Tensile

_____ **5.** Compressive

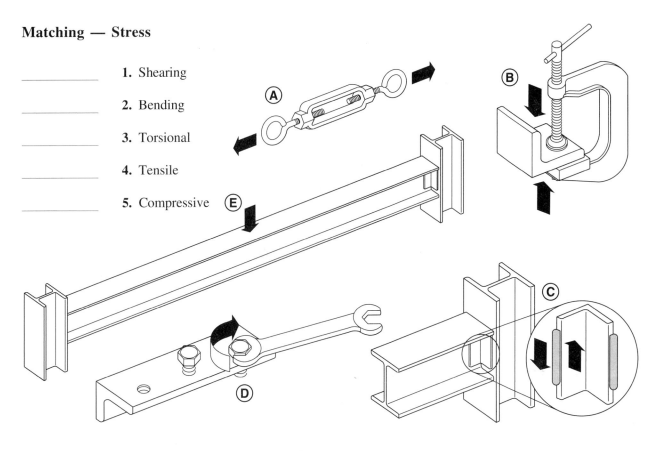

Refer to the Lift Mount print on page 130.

_____ **1.** The overall length of the Lift Mount is _____″.

_____ **2.** The angle iron is welded to the base of the unit with a 60° _____ weld.

T   F **3.** All holes are drilled to ³⁄₄″.

T   F **4.** The hex nut is welded-all-around.

T   F **5.** Weld (A) is specified as a field weld.

_____ **6.** The ¹⁄₂″ × 2″ guide is welded to the unit with a ¹⁄₄″ _____ weld.

T   F **7.** Weld (B) is welded-all-around.

_____ **8.** Weld (C) is a(n) _____-groove weld.

_____ **9.** The base of the unit is _____″ thick.

_____ **10.** The angle iron has a web thickness of _____″.

T   F **11.** The hex nut is centered vertically and horizontally on the Lift Mount.

_____ **12.** The length of the angle iron is _____″.

T   F   **13.** Weld (D) is a $\frac{3}{16}''$ V-groove weld

_____ **14.** The overall height of the Lift Mount is _____$''$.

_____ **15.** The two drilled holes in the base are centered _____$''$ from the ends.

T   F   **16.** The hex nut has coarse threads.

_____ **17.** The right side of the guide is _____$''$ from the right side of the Lift Mount.

T   F   **18.** The drawing of the Lift Mount is not to scale.

T   F   **19.** The guide is centered $2\frac{3}{4}''$ above the bottom of the base.

_____ **20.** Each leg of the angle iron is _____$''$ wide.

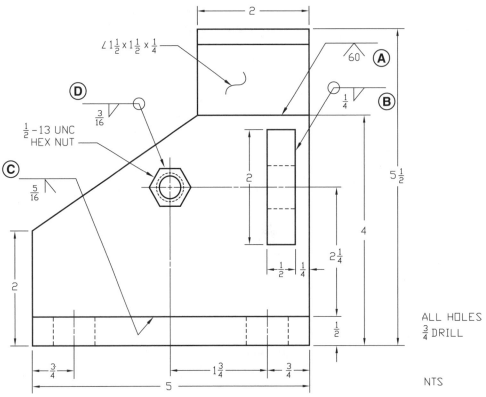

LIFT MOUNT

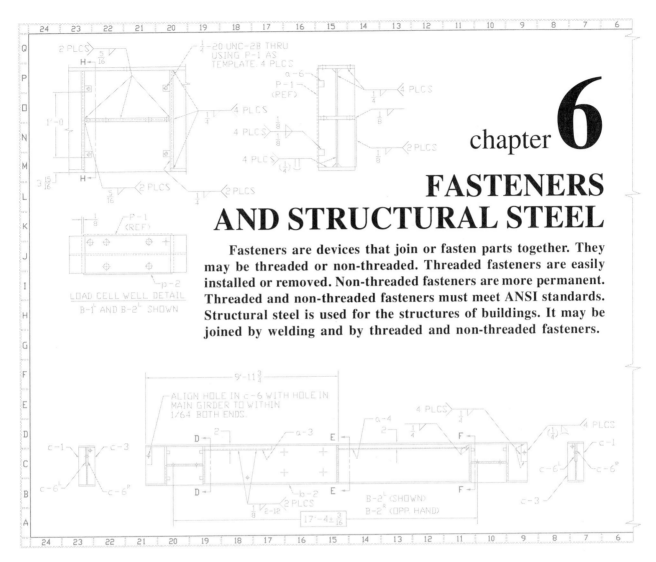

chapter 6

FASTENERS AND STRUCTURAL STEEL

Fasteners are devices that join or fasten parts together. They may be threaded or non-threaded. Threaded fasteners are easily installed or removed. Non-threaded fasteners are more permanent. Threaded and non-threaded fasteners must meet ANSI standards. Structural steel is used for the structures of buildings. It may be joined by welding and by threaded and non-threaded fasteners.

THREADED FASTENERS

Threaded fasteners are devices such as nuts and bolts that join or fasten parts together with threads. Threaded fasteners have several advantages for joining parts. For example, threaded fasteners are commercially available in a variety of sizes, styles, strengths, and materials and are capable of joining similar or dissimilar materials. They are easily installed in the shop or the field with hand or power tools and are easily removed and replaced.

Threaded fasteners are based upon the principle of an inclined plane wrapped around a cylinder. Early threaded fasteners were not uniform in size or thread profile and consequently were not interchangeable. The need for interchangeable parts prompted the work of Sir Joseph Whitworth of Great Britain and William Sellers of the United States. In the nineteenth century, Whitworth developed the Whitworth Thread, which had a standard thread angle of approximately 55°, and Sellers developed the basis for the American National thread, which had a standard thread angle of 60°.

The Unified Screw Thread Standard was developed by committees from the United States, Great Britain, and Canada, based on the 60° standard thread angle. Subsequent developments led to the Unified Inch Screw Thread (UN and UNR Thread Form). Profiles of UN (Unified National) and UNR (Unified National Rounded) threads are the same, except that roots and crests of UNR threads may be rounded. Additionally, the basic profile of UN and UNR threads is the same as ISO (International Organization for Standardization) metric threads, except for the diameter and number of threads per inch. See Figure 6-1. See Appendix.

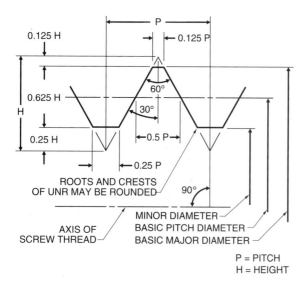

P = PITCH
H = HEIGHT

Figure 6-1. The basic profiles of UN and UNR threads are the same.

Screw Thread Series

Screw thread series are groups of diameter-pitch combinations. Screw thread series are distinguished from one another by the number of threads per inch for a series of specific diameters. For example, the *standard series* is a screw thread series of coarse (UNC/UNRC), fine (UNF/UNRF), and extra-fine (UNEF/UNREF) graded pitches and eight series with constant pitches. See Figure 6-2.

Graded pitch is a standard screw thread series with a different number of threads per inch based on the diameter. Generally, the smaller the diameter of the graded pitch series, the larger the number of threads per inch. For example, a 2 (0.0860″) UNC thread has 56 threads per inch while a ¾ (0.7500″) UNC thread has 10 threads per inch.

Constant pitch is a standard screw thread series with a set number of threads per inch regardless of diameter. Constant pitch series may have 4, 6, 8, 12, 16, 20, 28, or 32 threads per inch. The larger the diameter, the smaller the number of threads per inch. For example, a 2½ diameter bolt has 4, 6, 8, 12, 16, or 20 threads per inch, while a ¼ diameter bolt has 20, 28, or 32 threads per inch.

The *special series* is a screw thread series with combinations of diameter and pitch not in the standard screw thread series. Preference is given to standard series coarse and fine graded pitch threads.

STANDARD SERIES THREADS — GRADED PITCHES						
NOMINAL DIAMETER	UNC		UNF		UNEF	
	TPI	TAP DRILL	TPI	TAP DRILL	TPI	TAP DRILL
0 (.0600)			80	³⁄₆₄		
1 (.0730)	64	No. 53	72	No. 53		
2 (.0860)	56	No. 50	64	No. 50		
3 (.0990)	48	No. 47	56	No. 45		
4 (.1120)	40	No. 43	48	No. 42		
5 (.1250)	40	No. 38	44	No. 37		
6 (.1380)	32	No. 36	40	No. 33		
8 (.1640)	32	No. 29	36	No. 29		
10 (.1900)	24	No. 25	32	No. 21		
12 (.2160)	24	No. 16	28	No. 14	32	No. 13
¼ (.2500)	20	No. 7	28	No. 3	32	⁷⁄₃₂
⁵⁄₁₆ (.3125)	18	F	24	I	32	⁹⁄₃₂
⅜ (.3750)	16	⁵⁄₁₆	24	Q	32	¹¹⁄₃₂
⁷⁄₁₆ (.4375)	14	U	20	²⁵⁄₆₄	28	¹³⁄₃₂
½ (.5000)	13	²⁷⁄₆₄	20	²⁹⁄₆₄	28	¹⁵⁄₃₂
⁹⁄₁₆ (.5625)	12	³¹⁄₆₄	18	³³⁄₆₄	24	³³⁄₆₄
⅝ (.6250)	11	¹⁷⁄₃₂	18	³⁷⁄₆₄	24	³⁷⁄₆₄
¹¹⁄₁₆ (.6875)					24	⁴¹⁄₆₄
¾ (.7500)	10	²¹⁄₃₂	16	¹¹⁄₁₆	20	⁴⁵⁄₆₄
¹³⁄₁₆ (.8125)					20	⁴⁹⁄₆₄
⅞ (.8750)	9	⁴⁹⁄₆₄	14	¹³⁄₁₆	20	⁵³⁄₆₄
¹⁵⁄₁₆ (.9375)					20	⁵⁷⁄₆₄
1 (1.000)	8	⅞	12	⁵⁹⁄₆₄	20	⁶¹⁄₆₄

Figure 6-2. Screw threads series are groups of diameter-pitch combinations.

Threads are grouped into series by their pitch. *Pitch* is the distance between corresponding points on adjacent thread forms. Pitch is always measured parallel to the axis. To find pitch, apply the following formula:

$$P = \frac{1''}{N/in.}$$

where

P = pitch

$1''$ = constant

$N/in.$ = threads per inch

For example, what is the pitch of a thread form having 16 threads per inch?

$$P = \frac{1''}{N/in.}$$

$$P = \frac{1}{16}$$

$$P = \frac{1}{16}''$$

Metric threads are measured in millimeters (25.4 mm = 1″). The same basic formula can be used to determine the approximate pitch for metric threads. Threads may be either graded pitch or constant pitch.

Screw Thread Classes

The class of thread indicates its tolerance and allowance. *Tolerance* is amount of variation allowed above or below a dimension. *Allowance* is the difference between the design size and the basic size of a thread.

Classes of threads are 1A, 1B, 2A, 2B, 3A, and 3B. The A designates external threads. The B designates internal threads. The tolerance of threads decreases as the class number increases. For example, the tolerance for a Class 2A thread is less than the tolerance for a Class 1A thread. Class 2A and Class 2B are the most commonly used thread classes for bolts, nuts, screws, etc.

Screw Thread Designation

Threads are designated by thread notes. The thread note specifies in sequence the nominal size, number of threads per inch, thread form and series, and thread class. For example, the thread note ½-13

UNC-2A specifies ½″ nominal diameter, 13 threads per inch, coarse thread form, Class 2 fit, and external thread.

The nominal size of the diameter may be stated in fractional or decimal dimensions. The letters LH follow the thread note for left-hand threads. If not specified, the thread is a right-hand thread. The length of threaded fasteners is often included at the end of the thread note. See Figure 6-3.

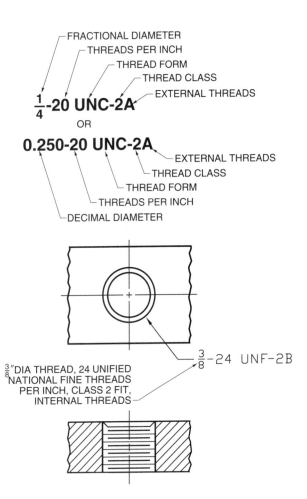

Figure 6-3. Threads are designated by thread notes.

Thread Representation

Thread representation is the method of drawing used to show a threaded part. Screw threads are represented on drawings by three methods: simplified representation, schematic representation, and detailed representation. One method of thread representation is commonly used throughout one drawing, although more than one method may be used on the same drawing for clarity. See Figure 6-4.

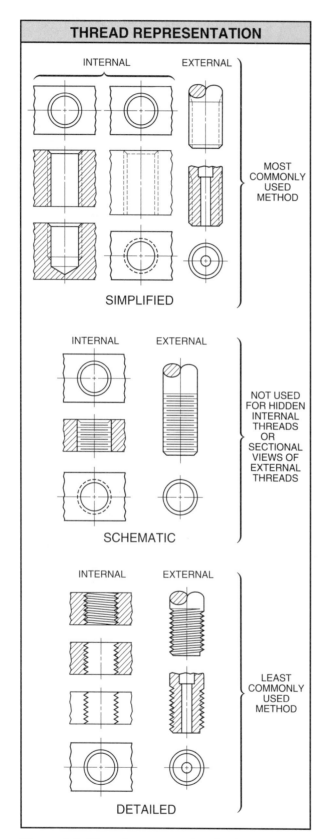

THREAD REPRESENTATION

INTERNAL   EXTERNAL

MOST COMMONLY USED METHOD

SIMPLIFIED

INTERNAL   EXTERNAL

NOT USED FOR HIDDEN INTERNAL THREADS OR SECTIONAL VIEWS OF EXTERNAL THREADS

SCHEMATIC

INTERNAL   EXTERNAL

LEAST COMMONLY USED METHOD

DETAILED

Figure 6-4. Thread representation is the method of drawing used to show a threaded part.

Simplified Representation. *Simplified representation* is a method of thread representation in which hidden lines are drawn parallel to the axis at the approximate depth of the thread. Simplified representation is the most commonly used method of thread representation. Various combinations of internal, external, and sectional views of male and female threads are shown with this method.

Schematic Representation. *Schematic representation* is a method of thread representation in which solid lines perpendicular to the axis represent roots and crests. This method is not used for hidden internal threads or sectional views of external threads.

Detailed Representation. *Detailed representation* is a method of thread representation in which the thread profiles are connected by helices. A *helix* is the curve formed by a line angular to the axis of a cylinder and in a plane wrapped around the cylinder. Detailed representation is the least commonly used method of thread representation because it is time-consuming to draw.

Bolts, Screws, and Nuts

Bolts, screws, and nuts are purchased parts. Consequently, they are generally shown on prints as thread representations with notes giving specific information. Bolts and screws are available in a wide range of sizes (diameters and lengths), hardnesses, head styles, etc. See Figure 6-5.

The length of a bolt or screw is the distance from the bearing surface of the head to the tip, measured parallel to the axis. The thread length for standard bolts is generally twice the diameter plus $1/4''$ for bolts up to 6″. For bolts over 6″, the thread length is generally twice the diameter plus $1/2''$.

Nuts are either square or hexagonal. Their distance across flats corresponds with standard English or metric dimensions to facilitate driving with wrenches, and sockets and ratchets.

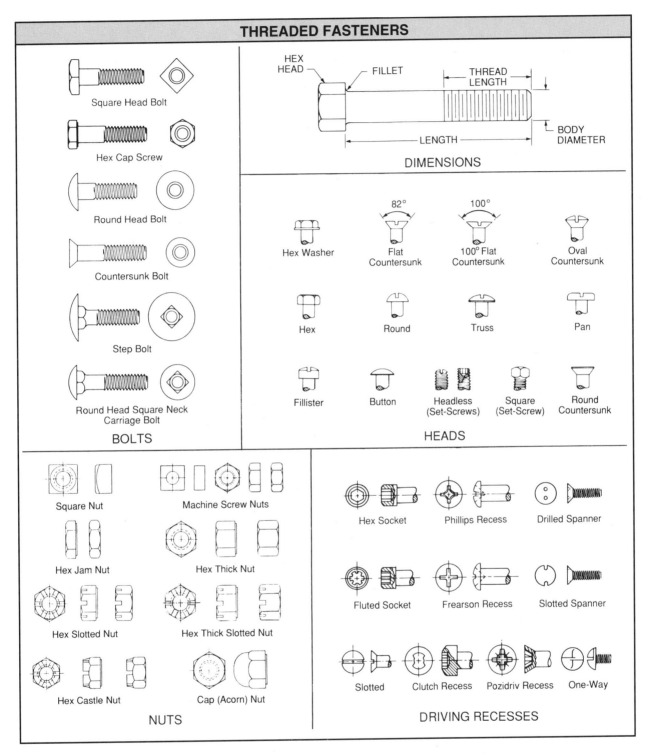

Figure 6-5. Bolts, screws, and nuts are threaded fasteners.

NON-THREADED FASTENERS

Non-threaded fasteners are devices that join or fasten parts together without threads. The most common non-threaded fastener is the rivet. A *rivet* is a cylindrical metal pin with a preformed head. The rivet shank is inserted through holes and pressed or beaten into a second head to hold the parts together.

The *shank* is the cylindrical body of a rivet. The riveting process can also be automated. The shape of the preformed head and the length and diameter of the shank distinguish one rivet from another.

Two parts are joined together by the grip of a rivet which fits through predrilled holes slightly larger than the shank of the rivet. The length of the shank must exceed the thickness of the two parts to be joined by enough material to allow the shank to be upset or shaped into the final form. The *grip* is the effective holding length of a rivet. The size of the rivet required is determined by the thickness of the parts being joined. See Figure 6-6.

Rivets are relatively inexpensive and are generally manufactured from ductile metals such as steel, aluminum, copper, brass, and bronze. A *ductile* metal is a metal that can be formed easily. Riveting can also be used to join materials that cannot be welded, such as dissimilar metals, plastics, or materials which could be damaged by heat.

A riveted joint is permanent. However, rivets can loosen under stress and become ineffective. Rivets are also subject to corrosion by liquids and generally cannot hold pressure because of the possibility of leaks.

Rivets are classified into three groups: large, small, and blind. *Large rivets* are rivets with a shank of ½″ or greater in diameter. The second head of large rivets can only be formed by applying force to the rivet after it has been heated red hot. *Small rivets* are rivets with a shank of ⁷⁄₁₆″ or less in diameter. *Blind rivets* are rivets with a hollow shank that join two parts with access from one side only.

Rivets are shown on prints with conventional representation. Shop rivets are shown as clear circles with slash marks indicating countersinking, flattening, near side, far side, and both sides. *Shop rivets* are rivets placed in the shop. Field rivets are shown as darkened circles with slash marks indicating countersinking, flattening, near side, far side, and both sides. *Field rivets* are rivets placed in the field.

Rivet placement is controlled by the thickness of the material being riveted, pitch, and margin. *Rivet pitch* is the distance from the center of one rivet to the center of the next rivet in the same row. *Back (transverse) pitch* is the distance from the center of one row of rivets to the center of the adjacent row of rivets. *Diagonal pitch* is the distance between the centers of rivets nearest each other in adjacent rows. *Margin* is the distance from the edge of the plate to the center line of the nearest row of rivet. See Figure 6-7.

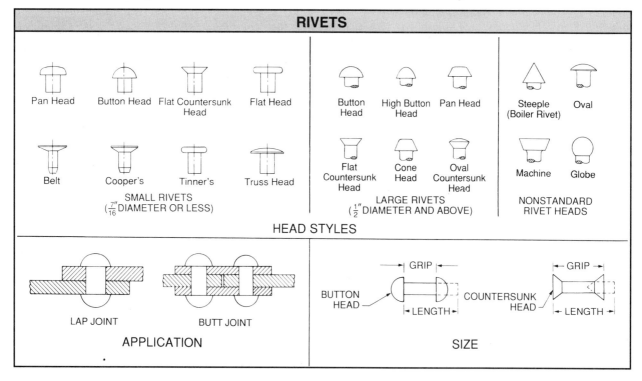

Figure 6-6. Rivets are non-threaded fasteners used to join or fasten parts.

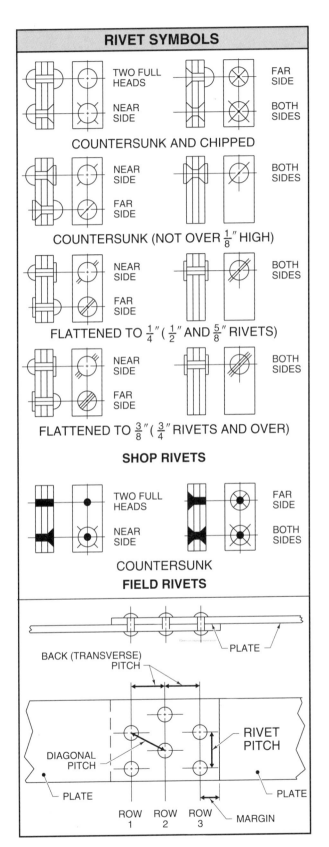

Figure 6-7. Conventional rivet symbols are used to show rivets on prints.

STRUCTURAL STEEL

Structural steel is steel used in the erection of structures. It is produced in a variety of shapes to serve as load-bearing units in building construction. Manufacturers of structural steel must roll the members with close tolerance and check the chemical properties constantly to ensure that the finished product will have the required strength. Standard specifications for structural steel are established by the American Society for Testing and Materials.

Available lengths of structural steel vary with suppliers. Standard structural steel shapes include beams (W, M, S, HP, and channel C and MC), angle (L), tee (WT, MT, and ST), and other structural steel shapes (tubing, pipe, bar, and plate). See Figure 6-8.

Beams

Beams are I-shaped structural steel. They are used for providing support over long expanses. Beams may be used as lintels, joists, piles, or girders, depending upon their size and the job requirements.

Standard I beams are designated S. Wide flange beams are designated W. Miscellaneous-shaped beams are designated M. Bearing piles are designated HP. Beams are described by their nominal dimensions over the flanges and the weight per foot. An example of a description for a W beam is W 14 × 34. The W indicates the type of beam. The size over the flanges is 14″. The weight per running foot is 34 pounds.

Channel. *Channel* is C-shaped structural steel used in conjunction with other structural shapes as support members or combined to serve as an I beam. American standard channel is designated C. Miscellaneous channel is designated MC. Channels can be welded back-to-back to form an I beam. Channel is dimensioned similar to beams with the flange width measured from the heel to the toe. For example, C 2½″ × ⅝ × 3/16 indicates a piece of channel 2½″ wide, ⅝″ deep, and 3/16″ thick.

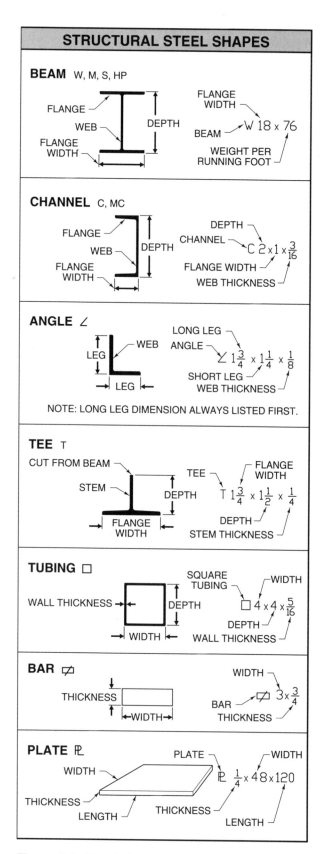

Figure 6-8. Standard structural steel shapes are drawn and noted on prints.

Angle

Angle is L-shaped structural steel consisting of two equal or unequal widths. Angle is widely used in support members where loads can be transferred to the 90° member. Angle is also used when rigidity of the members requires a 90° member. It can be notched, bent, and welded to provide a strong, reinforced joint without extensive labor. Angle is measured from the heel to the end of the leg. Unequal leg angle has different leg-size specifications, with the longer leg dimension always specified first, followed by the thickness. For example, L 4 × 3½ × ⅝ indicates unequal leg angle with the first leg 4″ wide and the other leg 3½″ wide. Each leg is ½″ thick.

Tee

Tee is T-shaped structural steel made from I beams cut to specifications by the mill or by suppliers. Tee is used where more strength is required than angle could provide. Tee is sized by flange width and nominal depth. For example, T 2 × 2 × ⁵⁄₁₆ indicates a piece of tee 2″ wide, 2″ deep, and ⁵⁄₁₆″ thick.

Other Structural Steel Shapes

Other structural steel shapes include tubing, pipe, bar, and plate. A wide variety of sizes is commercially available. Additionally, mills may produce custom sizes as required.

Tubing. *Tubing* is round-, square-, or rectangular-shaped structural steel. It is used for columns. Round tubing is sized based on the diameter and wall thickness. Square- and rectangular-shaped tubing are sized based upon the cross-sectional dimensions and wall thickness. The inside dimension (or diameter) is determined by subtracting the wall thicknesses from the cross-sectional dimension. An example of a description for rectangular tubing is 6 × 2 × ¼″. The piece of tubing is 6″ wide, 2″ deep, and has a wall thickness of ¼″.

Pipe. *Pipe* is round-shaped structural steel. It is measured using nominal inside diameter and nominal pipe size (NPS). See Figure 6-9. The NPS is the same as the inside diameter for pipe up to 12″. Pipe-wall thickness is specified using one of two standards. ANSI classifies pipe-wall thicknesses using schedule numbers (Schedule 40, 60, 80, etc.). For example, a 2″ nominal inside diameter pipe has a standard wall thickness of 0.154″ for Schedule 40, 0.218″ for Schedule 60, and 0.436 for Schedule 80.

ASTM and ASME classify pipe wall thickness as standard (STD), extra-strong (XS), and double extra-strong (XXS), using nominal inside diameter. The nominal inside diameter is determined using standard weight pipe. Extra-strong pipe and double extra-strong pipe have a reduced inside diameter as the wall thickness is increased. Outside wall thickness remains constant in the three weight classifications. For example, 3″ nominal inside diameter pipe has an outside diameter of 3.500″. In the standard weight, the pipe has an inside diameter

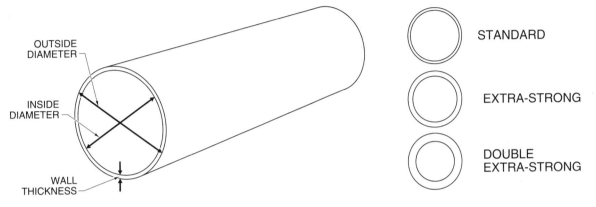

NOMINAL INSIDE DIAMETER SIZE (IN.)	OUTSIDE DIAMETER (BW GUAGE)	INSIDE DIAMETER (BW GUAGE)			NOMINAL WALL THICKNESS		
		STANDARD	EXTRA-STRONG	DOUBLE EXTRA-STRONG	SCHEDULE 40	SCHEDULE 60	SCHEDULE 80
1/8	0.405	0.269	0.215		0.068	0.095	
1/4	0.540	0.364	0.302		0.088	0.119	
3/8	0.675	0.493	0.423		0.091	0.126	
1/2	0.840	0.622	0.546	0.252	0.109	0.147	0.294
3/4	1.050	0.824	0.742	0.434	0.113	0.154	0.308
1	1.315	1.049	0.957	0.599	0.133	0.179	0.358
1 1/4	1.660	1.380	1.278	0.896	0.140	0.191	0.382
1 1/2	1.900	1.610	1.500	1.100	0.145	0.200	0.400
2	2.375	2.067	1.939	1.503	0.154	0.218	0.436
2 1/2	2.875	2.469	2.323	1.771	0.203	0.276	0.552
3	3.500	3.068	2.900	2.300	0.216	0.300	0.600
3 1/2	4.000	3.548	3.364	2.728	0.226	0.318	
4	4.500	4.026	3.826	3.152	0.237	0.337	0.674
5	5.563	5.047	4.813	4.063	0.258	0.375	0.750
6	6.625	6.065	5.761	4.897	0.280	0.432	0.864
8	8.625	7.981	7.625	6.875	0.322	0.500	0.875
10	10.750	10.020	9.750	8.750	0.365	0.500	
12	12.750	12.000	11.750	10.750	0.406	0.500	

Figure 6-9. Pipe is measured using the nominal inside diameter and nominal pipe size.

of 3.068″. In the extra-strong weight, the pipe has an inside diameter of 2.900″. In the double extra-strong weight, the pipe has an inside diameter of 2.300″.

Bar. *Bar* is round-, square-, or rectangular-shaped structural steel. Bar is used to reinforce sections of a fabricated weld part. Bar shapes are dimensioned based on their characteristics. An example of a description for square bar is $1\frac{1}{4}$ sq. The piece of square bar is $1\frac{1}{4}″$ wide and $1\frac{1}{4}″$ deep.

Plate. *Plate* is $\frac{3}{16}″$ or more thick structural steel used to cover large expanses of a structure. Plate can be joined with or without reinforcement mem-

bers. Plate is plain or patterned. The nominal thickness of plate with a raised pattern does not include the height of the raised pattern.

Plate is defined according to the rolling procedure used in its manufacture. *Sheared plate* is plate that is rolled between horizontal and vertical rollers and trimmed on all edges. *Universal plate* is plate that is rolled between horizontal and vertical rollers and trimmed only on the ends.

Sheet is $\frac{1}{8}″$ or less structural steel used to cover large expanses of a structure. Plate and sheet are commonly used in shipbuilding, pressure vessel construction, and building construction. Plate and sheet are specified by thickness, width, and length. For example, PL $\frac{3}{8} \times 48 \times 144$ indicates plate $\frac{3}{8}″$ thick, 48″ wide, and 144″ long.

FASTENERS AND STRUCTURAL STEEL

Review Questions

Name_____ Date _____

Completion

_____ **1.** Threaded fasteners are based upon an inclined _____ wrapped around a cylinder.

_____ **2.** Screw thread _____ are groups of diameter-pitch combinations.

_____ **3.** _____ is the distance between corresponding points on adjacent thread forms.

_____ **4.** Thread _____ is the method of drawing used to show a threaded part.

_____ **5.** The length of a bolt or screw is the distance from the bearing surface of the head to the tip, measured parallel to the _____.

_____ **6.** Rivet _____ is the effective holding length of a rivet.

_____ **7.** _____ metals are metals that can be formed easily.

_____ **8.** _____ rivets have a hollow shank that join two parts with access from one side only.

_____ **9.** Back pitch is also known as _____ pitch.

_____ **10.** _____ is the distance from the edge of the plate to the center line of the nearest row of rivets.

_____ **11.** _____ pitch is a standard screw thread series with a different number of threads per inch based on the diameter.

_____ **12.** The thread length of a standard bolt with a ⅜″ diameter and a 3″ length is _____″.

_____ **13.** The size of the rivet required is determined by the _____ of the parts being joined.

_____ **14.** UN threads are based on a(n) _____ thread angle.

_____ **15.** The pitch of a thread form having 20 threads per inch is _____″.

_____ **16.** A(n) _____ designates an external thread.

_____ **17.** _____ steel is steel used in the erection of structures.

_____ **18.** The letters in a beam designation indicate the _____.

_____ **19.** The _____ leg dimension of unequal leg angle is always given first.

_____ **20.** Rivets generally cannot hold _____.

Matching — Rivets

_____ 1. Shop rivet; countersunk and chipped; far side

_____ 2. Field rivet; countersunk; far side

_____ 3. Shop rivet; flattened to $\frac{3}{8}''$ for $\frac{3}{4}''$ rivets and over; near side

_____ 4. Shop rivet; countersunk; not over $\frac{1}{8}''$ high; near side

_____ 5. Field rivet; countersunk; both sides

_____ 6. Shop rivet; countersunk and chipped; both sides

_____ 7. Shop rivet; flattened to $\frac{1}{4}''$ for $\frac{1}{2}''$ and $\frac{5}{8}''$ rivets; far side

_____ 8. Shop rivet; countersunk and chipped; near side

_____ 9. Shop rivet; countersunk; not over $\frac{1}{8}''$ high; far side

_____ 10. Field rivet; countersunk; near side

True-False

T   F   1. Similar or dissimilar materials can be joined with threaded fasteners.

T   F   2. The Whitworth Thread had a standard thread angle of approximately 65°.

T   F   3. The basic profile of UN and UNR threads is the same as ISO metric threads, except for the diameter and number of threads per inch.

T   F   4. Screw thread series are distinguished from one another by the number of threads per inch for a series of specific diameters.

T   F   5. Tolerance is the difference between the design size and the basic size of a thread.

T   F   6. The tolerance for a Class 2A thread is less than the tolerance for a Class 1A thread.

T   F   7. The nominal diameter of a thread may be stated in fractional or decimal dimensions in a thread note.

T   F   8. The letters RH follow the thread note for right-hand threads.

T   F   9. The length of a threaded fastener may be specified at the end of the thread note.

T   F   10. The length of the shank of a rivet must be the same dimension as the two pieces to be joined.

T   F   11. Riveted joints are permanent.

T   F   12. A large rivet has a shank of $\frac{1}{2}''$ or greater in diameter.

T   F   **13.** Shop rivets are shown on prints as clear circles.

T   F   **14.** Standard I beams are designated with the letter S.

T   F   **15.** Schedule 40 pipe has a thicker wall than a piece of the same nominal inside diameter pipe in Schedule 80.

T   F   **16.** Plate is ¼″ or more thick structural steel used to cover large expanses.

T   F   **17.** Tee is cut from I beams.

T   F   **18.** The size of round tubing is based on the diameter and wall thickness.

T   F   **19.** Channel can be welded back-to-back to form an I beam.

T   F   **20.** Available lengths of structural steel vary with suppliers.

Matching — Thread Representation

_____ **1.** Simplified; external; section

_____ **2.** Schematic; external

_____ **3.** Schematic; internal; section

_____ **4.** Detailed; external

_____ **5.** Simplified; internal; section

_____ **6.** Simplified; internal

_____ **7.** Detailed; external; section

_____ **8.** Detailed; internal; section

_____ **9.** Detailed; internal

_____ **10.** Simplified; external

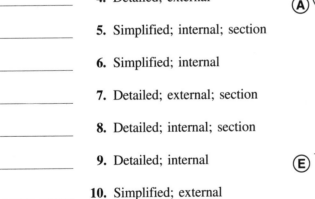
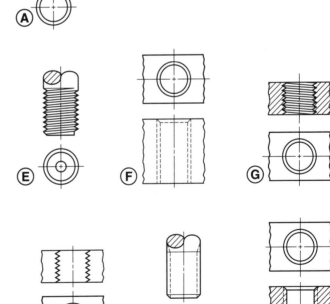

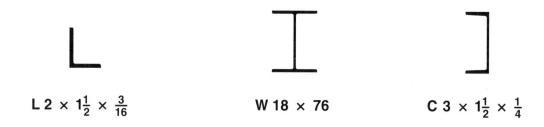

$L\ 2\ \times\ 1\frac{1}{2}\ \times\ \frac{3}{16}$ $W\ 18\ \times\ 76$ $C\ 3\ \times\ 1\frac{1}{2}\ \times\ \frac{1}{4}$

$T\ 2\frac{3}{4}\ \times\ 1\frac{3}{4}\ \times\ \frac{3}{8}$ $L\ 1\frac{1}{4}\ \times\ 1\frac{1}{4}\ \times\ \frac{1}{8}$

STRUCTURAL STEEL NOTES

Refer to Structural Steel Notes on page 144.

T   F   **1.** No note is given for an I beam.

_____   **2.** The depth of the channel is _____″.

_____   **3.** The weight per running foot of the beam is _____lb.

_____   **4.** The length of the longest leg on the largest angle is _____″.
      A. $\frac{3}{16}$
      B. $1\frac{1}{4}$
      C. $1\frac{1}{2}$
      D. 2

T   F   **5.** The smallest angle has a web thickness of $\frac{3}{16}″$.

_____   **6.** The width of the beam flange is _____″.

_____   **7.** The stem thickness of the tee is _____″.

T   F   **8.** Both angles have equal-length legs.

_____   **9.** The depth of the tee is _____″.

_____   **10.** The web thickness of the channel is _____″.

_____   **11.** The flange width of the channel is _____″.

_____   **12.** The flange width of the tee is _____″.

_____   **13.** The largest angle has a web thickness of _____″.

T   F   **14.** The largest angle weighs 2 lb per running foot.

_____   **15.** The leg length of the smallest angle is _____″.

Trade Competency Test

Name_____ Date _____

Refer to the Corner Weldment print on page 147.

_____ 1. _____ pieces are required to fabricate one Corner Weldment.

_____ 2. The detail part number for the Corner Weldment is _____.

_____ 3. The tolerance for all details is _____ unless otherwise specified.

T   F   4. The 1.06 DIA hole in Part No. 3 is centered on the piece.

_____ 5. Threaded holes specified ⅜-16 NC-2B are shown in Part No. _____ and Part No.
   _____.
   A. 1; 2
   B. 2; 3
   C. 3; 4
   D. 4; 5

T   F   6. The ½-20 NF-2B holes in Part No. 1 are equally spaced along the length of the piece.

_____ 7. Part No. 5 contains _____ sq in.

T   F   8. The ½-20 NF-2B holes are through holes.

T   F   9. Part No. 2 and Part No. 5 are the same size.

_____ 10. Disregarding saw cuts, one each of Part Nos. 2, 3, 4, and 5 could be cut from a _____″
   piece with the least waste.
   A. 24
   B. 28
   C. 30
   D. 36

T   F   11. All threaded holes of the Corner Weldment are National Fine threads.

_____ 12. The Corner Weldment contains _____ holes, _____ of which are threaded.
   A. five; three
   B. five; five
   C. seven; three
   D. seven; five

_____ 13. Part No. _____ is the smallest part of the assembly.

_____ 14. Part No. _____ requires the closest tolerance.

_____ 15. The pictorial drawing of the Corner Weldment is a(n) _____.

Refer to the One-Sway Frame-All print on page 148.

_____   1. The flange width of I Beam A is _____″.

_____   2. I Beam A is joined to Bar G with fillet welds having a(n) _____″ leg size.

T   F   3. The leg size of Angle Iron B is 3″.

_____   4. The thickness of Bar C is _____″.

_____   5. Angle Iron B is joined to Bar C with fillet welds having a(n) _____″ leg size.

_____   6. Bar C extends _____″ below Angle Iron D.

T   F   7. Angle Iron D leg sizes are $3\frac{1}{2}″ \times 3\frac{1}{4}″$.

_____   8. Bar E is 2″ wide and _____″ long.

_____   9. Bar E is joined to Angle Iron D with _____ welds.

_____   10. The leg thickness of Angle Iron F is _____″.

T   F   11. The length of Angle Iron F is not given.

_____   12. Angle Iron D is joined to Bar E with fillet welds _____″ long.

_____   13. Bar E is _____″ thick.

_____   14. Bar G is _____″ wide and 8″ long.

T   F   15. Bar G is centered 6′-6″ from the end of I Beam A.

Refer to the Support print on page 148.

_____   1. Hole A is centered _____″ from the hole on its immediate left.

T   F   2. The depth of Hole A is $\frac{1}{4}″$.

T   F   3. Hole B is drilled through and counterbored from the opposite side.

_____   4. The counterbore diameter of Hole B is _____″.

_____   5. The thread size specified for Hole A is _____.

_____   6. A total of _____ other holes have the same thread specification as Hole A.

_____   7. Hole B is drilled at a(n) _____″ diameter.

T   F   8. Hole B is drilled $\frac{3}{4}″$ off-center of the Support.

_____   9. The depth of the counterbore for Hole C is _____″.

_____   **10.** Hole C is counterbored from the _____ side.

_____   **11.** Hole C is drilled at a(n) _____″ diameter.

T   F   **12.** Hole D is drilled and tapped.

_____   **13.** The counterbore diameter of Hole D is _____″.

T   F   **14.** Hole E has the same specifications as Hole D.

_____   **15.** The center-to-center vertical distance from Hole D to Hole E is _____″.

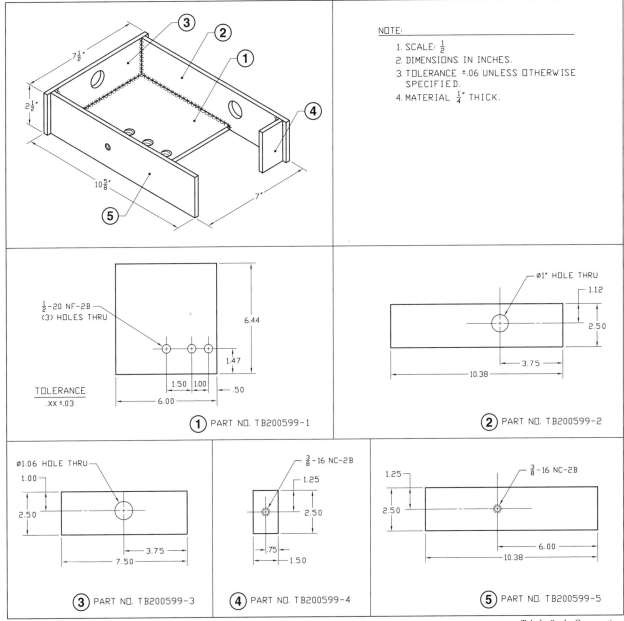

NOTE: _____
1. SCALE: $\frac{1}{2}$
2. DIMENSIONS IN INCHES.
3. TOLERANCE ±.06 UNLESS OTHERWISE SPECIFIED.
4. MATERIAL $\frac{1}{4}$″ THICK.

$\frac{1}{2}$-20 NF-2B
(3) HOLES THRU

6.44
1.47
1.50 1.00 .50
6.00

TOLERANCE
.XX ±.03

① PART NO. TB200599-1

∅1″ HOLE THRU
1.12
2.50
3.75
10.38

② PART NO. TB200599-2

∅1.06 HOLE THRU
1.00
2.50
3.75
7.50

③ PART NO. TB200599-3

$\frac{3}{8}$-16 NC-2B
1.25
2.50
.75
1.50

④ PART NO. TB200599-4

$\frac{3}{8}$-16 NC-2B
1.25
2.50
6.00
10.38

⑤ PART NO. TB200599-5

$7\frac{1}{2}$″
$2\frac{1}{2}$″
$10\frac{5}{8}$″
7″

Toledo Scale Corporation

CORNER WELDMENT

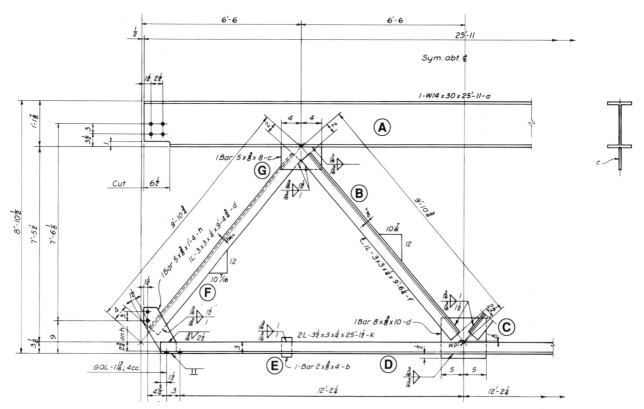

American Institute of Steel Construction, Inc.

ONE-SWAY FRAME-ALL

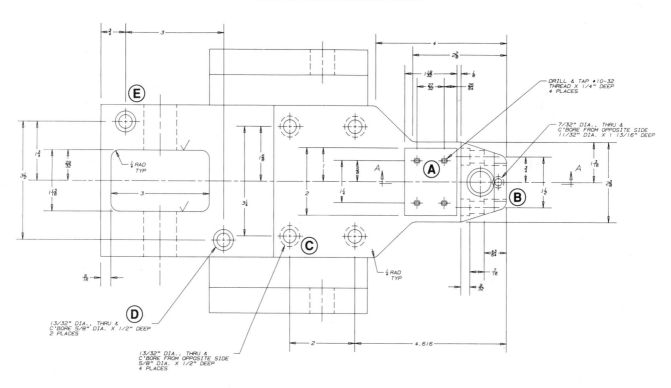

Worthington Industries

SUPPORT

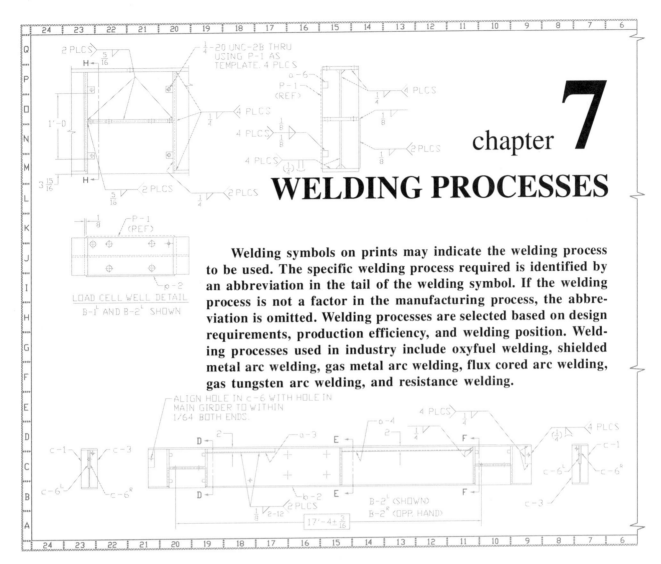

chapter 7
WELDING PROCESSES

Welding symbols on prints may indicate the welding process to be used. The specific welding process required is identified by an abbreviation in the tail of the welding symbol. If the welding process is not a factor in the manufacturing process, the abbreviation is omitted. Welding processes are selected based on design requirements, production efficiency, and welding position. Welding processes used in industry include oxyfuel welding, shielded metal arc welding, gas metal arc welding, flux cored arc welding, gas tungsten arc welding, and resistance welding.

WELDING PROCESSES

Welding processes commonly used in industry include oxyfuel welding (OFW), shielded metal arc welding (SMAW), gas metal arc welding (GMAW), flux cored arc welding (FCAW), gas tungsten arc welding (GTAW), and resistance welding (RW). Additionally, there are other, more specialized welding processes not as commonly used which are equally important. The required welding process for a particular weld is identified with an abbreviation in the tail of the welding symbol. See Figure 7-1.

Most welding processes use heat to join parts together. The equipment used to generate the heat required varies, depending on the welding process. The specific welding process used also determines the procedure followed to complete the welding task.

The applications of a welding process are based on the requirements of the weld, accessibility of the weld area, economic considerations, and available welding equipment. The welding process used may be determined by the welding position: flat (F), horizontal (H), vertical (V), or overhead (O).

The properties of the base metal must also be considered. For example, if excessive heat is applied when welding stainless steel, the metal will lose its rust-resistive qualities.

Oxyfuel Welding (OFW)

Oxyfuel welding (OFW) is a welding process that uses oxygen combined with a fuel to sustain a flame that generates the heat necessary for welding. Fuels commonly used with oxygen in the oxyfuel welding process include acetylene, MAPP (methylacetylene-propadiene) gas, natural gas, and propane. Oxyfuel welding, because of the gases used, is also known as gas welding.

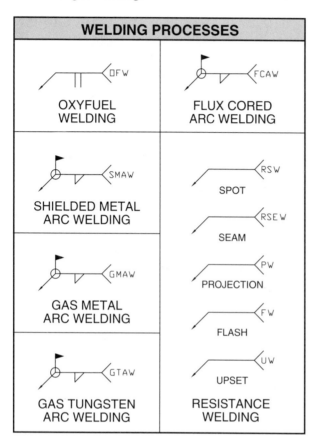

Figure 7-1. The welding process required is specified in the tail of the welding symbol.

Acetylene is the most common fuel used with oxygen. *Oxyacetylene welding (OAW)* is oxyfuel welding with acetylene. Oxyacetylene welding was first used in the late 1800s. It was discovered that acetylene, when combined with oxygen, produced a flame of approximately 6300°F. Other fuels can be used in place of oxygen for welding. The type of fuel used depends on availability, cost, and heat requirements. See Figure 7-2.

OFW Equipment. Oxyfuel welding equipment requires the same basic equipment regardless of the fuel used with oxygen. Tanks contain pressurized gases and provide storage until the gases are used in the welding process. Oxygen is pressurized to

2000 psi. Acetylene, which becomes unstable when pressurized in a free state above 15 psi, is stored in tanks filled with acetone. Acetylene tanks are commonly pressurized to 250 psi.

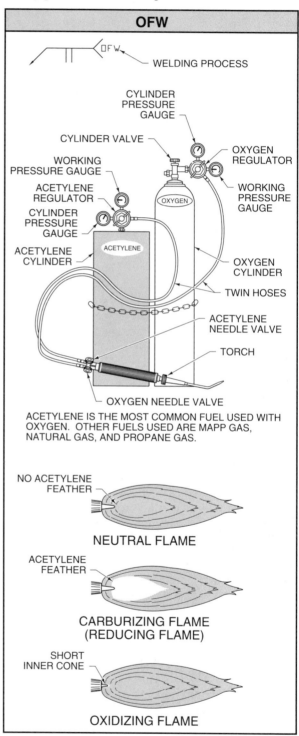

Figure 7-2. Oxyfuel welding (OFW) is a welding process that uses oxygen combined with a fuel to sustain a flame that generates the heat necessary for welding.

Warning: Serious explosions can occur if acetylene is pressurized in a free state above 15 psi.

Oxygen and acetylene tanks are available in various sizes. Regulators mounted on the tanks control the flow of gases through the hoses to the torch. Oxygen and acetylene are mixed together at the torch. Torch tips in various sizes direct the mixed gases out of the torch.

OFW Procedure. The oxyfuel welding torch provides the heat necessary to melt the parts to be welded. Before igniting the torch, the oxygen and acetylene regulator adjusting screws are backed out to relieve all pressure on the diaphragms of the regulators. The welder stands to one side and opens the acetylene and oxygen cylinder valves slowly. The acetylene cylinder valve is opened approximately one-fourth of a turn, and the oxygen cylinder valve is opened all the way. The regulator adjusting screws are turned to obtain the required working pressures.

The acetylene needle valve on the torch is opened approximately one-fourth of a turn in preparation for igniting the torch. A striker is used to ignite the flame, with the torch tip pointing down. The oxygen needle valve on the torch is slowly opened to adjust to a neutral flame. A *neutral flame* is an oxyfuel flame with a balanced mixture of oxygen and fuel. It is used for most welding operations. A *carburizing (reducing) flame* is an oxyfuel flame with an excess of fuel. An *oxidizing flame* is an oxyfuel flame with an excess of oxygen.

Heat from the oxyacetylene flame is directed to the weld parts. A depression of molten metal (puddle) is formed. The welding torch is manipulated as necessary to distribute the heat from the flame. Filler metal is added for weld reinforcement by dipping a welding rod into the puddle. The welding torch and the welding rod are held at 45° to the weld parts. The welding torch and welding rod are positioned, based on the weld joint and weld position, to provide maximum access and penetration.

OFW Applications. Oxyfuel welding is most commonly used in industry for maintenance and repair work. Oxyfuel welding can be performed in all positions. It is best suited for welding ¼″ or less thicknesses of metal.

The main advantage of oxyfuel welding is the ability to weld without electricity. This permits welding mobility and flexibility. However, set-up time, welding speed, and the amount of heat generated by the process limits the applications of the oxyfuel welding process.

Oxygen is also commonly used with other gases for oxyfuel cutting operations. Metal to be cut is preheated with an oxyfuel cutting torch. Oxygen is then fed into the molten metal to displace oxidized metal to quickly cut the metal.

Shielded Metal Arc Welding (SMAW)

Shielded metal arc welding (SMAW) is an arc welding process in which the arc is shielded by the decomposition of the electrode covering. The SMAW process is commonly used in industry because of its versatility, low cost, and minimal amount of equipment required. The SMAW process uses electricity to generate the heat necessary for melting the weld parts. Relatively high currents (55-200 A) and low voltage (18-35 V) are used.

An *ampere* or *amp (A)* is a unit of measure for electricity that expresses the quantity or number of electrons flowing through a conductor per unit of time. The flow of electrons measured in amps per unit of time is similar to the flow of water through a pipe measured in gallons per minute. A *volt (V)* is a unit of measure for electricity that expresses the electrical pressure differential between two points in a conductor. A *conductor* is any material through which electricity flows easily. The electrical pressure differential causes electrons to move. *Current* is the movement of electrons through a conductor. Voltage required to make electrons move in a conductor is similar to the water pressure required to make water flow in a pipe.

Current uses the weld parts as conductors in an electrical circuit. Heat generated from the resistance caused from electricity traveling through air melts the base metal. An electrode directs the flow of current and melts during the welding process to provide the filler metal required for weld strength. See Figure 7-3.

SMAW evolved from the carbon arc welding (CAW) process used in the mid-1800s. CAW uses heat generated from an electric arc jumping from one carbon electrode to another. If filler metal is required, a filler rod is dipped into the puddle formed in the base metal. The CAW process is seldom used today.

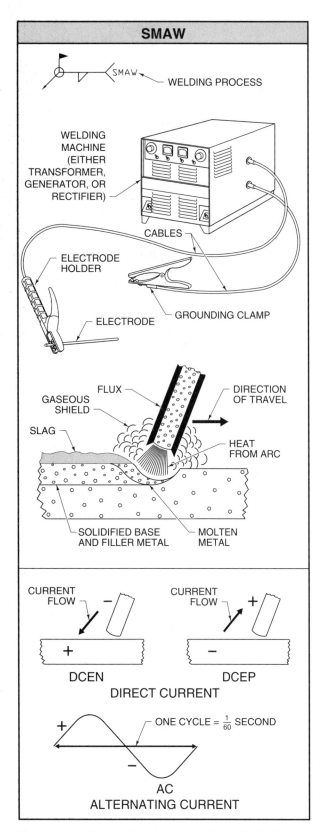

Figure 7-3. Shielded metal arc welding (SMAW) is an arc welding process in which the arc is shielded by the decomposition of the electrode covering.

SMAW was developed in the early 1900s. It was discovered that steel, when used to conduct electricity, melts from the heat of the arc while the arc is maintained. This allowed the welder to melt and penetrate the weld parts and add filler rod in the same operation. Experimentation with different electrode coating (flux) compositions lead to the different types of electrodes used today.

SMAW Equipment. SMAW equipment includes a welding machine, cables, electrode holder, and grounding device. The welding machine provides the electricity to generate the arc to be used when welding. Current used can be either direct current (DC) or alternating current (AC).

Direct current flows in one direction only, from the negative side to the positive side of the welding machine. The direction the current flows is determined by the connection of the cables in the circuit formed by the electrode and the weld parts (ground). *Direct current electrode negative (DCEN)* is the flow of current from the electrode (–) to the work (+). DCEN is also known as DC–, or direct current straight polarity (DCSP). DCSP is a nonstandard term for DCEN. *Direct current electrode positive (DCEP)* is the flow of current from the work (–) to the electrode (+). DCEP is also known as DC+ or direct current reverse polarity (DCRP). DCRP is a nonstandard term for DCEP.

Alternating current does not have a polarity as direct current does. In alternating current, the current alternates from DC– to DC+ to form a cycle. In the United States, AC completes 60 cycles per second (1 cycle = $\frac{1}{60}$ second).

Welding machines may be transformers, generators, or rectifiers. A *transformer* is a welding machine that produces AC only. A *generator* is a welding machine that produces DC only. A *rectifier* is a welding machine that produces either AC or DC.

Welding machines commonly require 220 or 440 V. All welding machine installations must comply to Article 630 of the National Electrical Code® (NEC®). Welding machines are rated by the National Electrical Manufacturers Association (NEMA) for output, input, and duty cycle.

Output is the maximum amperage and voltage of a welding machine. *Input* is the electrical requirements to operate a welding machine. *Duty cycle* is the length of time (expressed as a percentage) that a welding machine can operate at its rated output within a ten minute period. The duty cycle for manual welding machines is normally 60%. Automatic, continuous welding operations require 100% duty cycle.

Welding cables conduct electricity from the welding machine to the electrode holder and from the grounding device to the welding machine. The diameter of the welding cables required is determined by the ampacity of the welding operation. Welding cable conductors are based on AWG (American Welding Gauge) sizes. The length of the welding cables is determined by the distance from the welding machine to the parts being welded. See Figure 7-4.

WELDING CABLE (AWG)				
WELD TYPE	**AMPS**	**50'**	**100'**	**150'**
Manual or semiautomatic (up to 60% duty cycle)	75	6	6	4
	100	4	4	3
	150	3	3	2
	200	2	2	1
	250	2	2	1/0
	300	1	1	2/0
	350	1/0	1/0	3/0
	400	1/0	2/0	3/0
	450	2/0	3/0	4/0
	500	3/0	3/0	4/0
Semi or automatic (60% to 100% duty cycle)	400	4/0	4/0	—
	800	2-4/0	2-4/0	—
	1200	3-4/0	3-4/0	—
	1600	4-4/0	4-4/0	—

Figure 7-4. The size of the welding cable is based on the amperage and length.

Welding cables are attached to the welding machine on one end, and the electrode holder or grounding device on the other end. The *grounding device (ground)* is the connection between the welding cable and the weld parts in the welding circuit. The type of grounding device used depends on the size and shape of the parts being welded. Smaller weld parts are welded on welding stations with a grounding device permanently attached. Larger weld parts, which are welded and then transported to different locations, use a temporary grounding device such as a clamp.

Permanent and temporary grounding devices must be adequately secured to permit the proper flow of current. The *electrode holder* is a hand-held device that holds the electrode securely at the required angle for maximum access to the weld area.

The two parts of the SMAW electrode are the wire and flux. The *wire* is the part of the electrode that melts and forms the filler metal applied to the weld. The *flux* is the coating on the electrode. It serves a different purpose before, during, and after the welding operation.

Before welding, the flux aids in starting the arc. During welding, the flux is consumed and provides a gaseous shield surrounding the weld area. The gaseous shield prevents elements in the air, such as nitrogen or oxygen, from contaminating the molten metal in the weld. After welding, the flux turns into a crusty coating (slag) which covers the weld to allow slow cooling. This allows a controlled transition of the weld metal from the molten to the solid state, preserving the mechanical properties of the base metal and the weld area. After the weld area is sufficiently cooled, the slag is removed.

Electrodes are available in various sizes and types. They are identified using the AWS identification code. See Figure 7-5. The letter E indicates the electrode is to be used with an electric welding machine. The first two digits identify the tensile strength of the electrode wire. For example, 60 indicates 60,000 pounds of tensile strength. The third digit indicates the recommended positions for use with the electrode. For example, 1 indicates the electrode can be used in all welding positions. The fourth digit indicates the composition of the flux on the electrode. The flux composition determines the current and polarity best suited for the electrode.

SMAW electrodes are available in different sizes based on the diameter of the electrode wire. The flux covering on electrodes varies in thickness. For example, a 1/8″ E-7018 electrode has a thicker flux covering than a 1/8″ E-6010 electrode. However, the size of the wire (1/8″) is the same in both electrodes. The amperage required for welding increases with an increase in electrode diameter. The size and properties of the base metal determine the size of electrode to use. Larger diameter electrodes allow faster weld deposition, but may overheat the base metal due to higher amperages required. Smaller diameter electrodes deposit less weld metal more slowly and are useful on thinner structures which require low heat input.

AWS E60 SERIES ELECTRODES			
CLASSIFICATION	COVERING	POSITIONS	CURRENT
E6010	High cellulose sodium	F, V, OH, H	DCEP
E6011	High cellulose potassium	F, V, OH, H	AC or DCEP
E6012	High cellulose sodium	F, V, OH, H	AC or DCEN
E6013	High cellulose potassium	F, V, OH, H	AC or DC either polarity
E6020	High iron oxide	H-fillets	AC or DCEN
E6022	High iron oxide	F	AC or DC either polarity
E6027	High iron oxide, iron powder	H-fillets, F	AC or DCEN

WIRE

FLUX COVERING

TENSILE STRENGTH

E - 6 0 1 0

ELECTRIC WELDING

WELDING POSITION

FLUX COMPOSITION

Figure 7-5. Electrodes are identified using the AWS identification code.

SMAW Procedure. Current from the welding machine is conducted through the cables to the electrode holder. The current is adjusted to generate the heat required. For example, $3/8''$ steel may require 150 A and $1/4''$ steel may require 90 A. Electrode manufacturers recommend a general amperage range to be used with a specific electrode type and size on different metal thicknesses. However, a very accurate method of determining correct amperage is through trial and error by the welder. Voltage varies in the SMAW process, but does not have to be adjusted by the welder.

An arc is struck to begin the SMAW process. See Figure 7-6. The electrode touches the metal being welded to complete the electrical circuit. The electrode is withdrawn while still maintaining the arc between the electrode and the work. Electricity flows from the welding machine, through

the electrode cable, through the electrode holder, through the electrode, through air, to the weld parts, through the ground clamp, and ground cable, back to the welding machine.

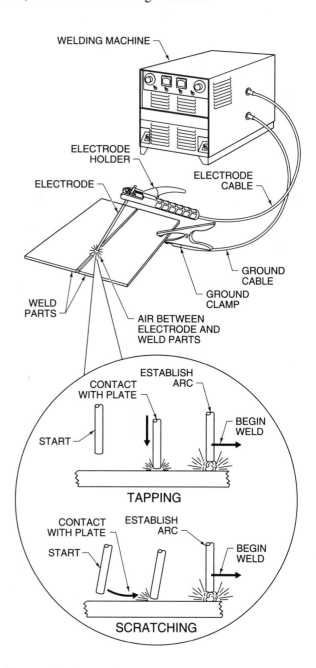

Figure 7-6. An arc is started by tapping or scratching the electrode on the weld parts.

The *arc length* is the distance from the electrode to the molten pool of the base metal. Correct arc length must be maintained throughout the welding operation, as the electrode melts and is fed into the weld. The *travel speed* is the speed at which the electrode is moved across the weld area. The travel speed must be correct and consistent to assure proper penetration and weld formation. In addition, the electrode must be held at the correct travel and work angles. The *travel angle* is the angle less than 90° of the electrode in relation to a perpendicular line from the weld and the direction of the weld. The *work angle* is the angle less than 90° of the electrode in relation to the workpiece. See Figure 7-7.

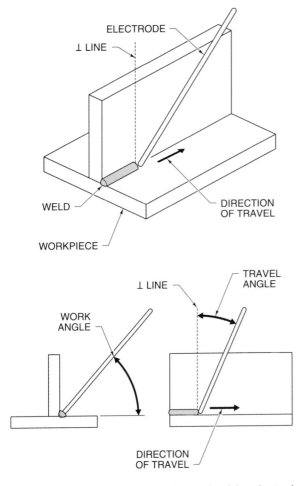

Figure 7-7. The travel angle is the angle of the electrode in relation to the direction of the weld. The work angle is the angle of the electrode in relation to the work.

SMAW Applications. SMAW is commonly used in industry for various weld requirements and positions. The electrode, because it is manually controlled, is extremely flexible. Initial equipment cost is minimal in comparison with other welding processes. The SMAW process is used to weld pipe because of the different positions required.

There are some limitations to the SMAW process. During the SMAW process, electrodes must be replaced as they are consumed in the weld. In addition, the slag that forms on the weld must be removed before additional weld passes or protective finish is applied.

Gas Metal Arc Welding (GMAW)

Gas metal arc welding (GMAW) is a welding process with a shielded gas arc between a continuous wire electrode and the weld metal. The wire electrode maintains the arc and is fed continuously by a wire feeder as it is consumed in the weld. The GMAW process does not require a flux covering on the electrode to provide a gaseous shield to protect the weld area. The weld area is protected by an inert gas shield. An *inert gas* is a gas that will not readily combine with other elements.

The process of gas shielded arc welding was experimented with in the early 1920s to aid weld purity and production efficiency. During the early 1950s, it was discovered that carbon dioxide could be used as an inexpensive shielding gas. This discovery, and the development of more versatile electrodes, increased the popularity of GMAW. See Figure 7-8.

GMAW Equipment. The equipment used in the GMAW process includes the welding machine, electrode, wire feeder, welding cable, welding gun, shielding gas regulator, and flowmeter. The specific welding task determines the type and configuration of equipment required.

The welding machine used for GMAW is similar to the welding machine used for SMAW. The SMAW welding machine has a single adjustment for amperage. The welding machine provides constant current over a range of voltages. Thus, the term constant power source is applied to the SMAW welding machine. On the GMAW welding machine, voltage is adjusted by the operator. The adjustment for amperage is determined by the speed at which the electrode is fed to the weld. The wire feed speed is located on the wire feeder. The GMAW welding machine produces constant voltage for potential with different amperages, depending on welding conditions. Thus, GMAW welding machines are known as constant potential welding machines.

The electrode used in the GMAW process is determined by the properties of the base metal and the requirements of the weld. The size of the electrode used depends on the deposition and heat requirements of the weld.

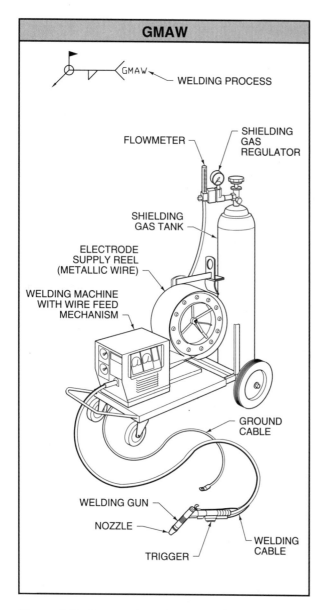

Figure 7-8. Gas metal arc welding (GMAW) is a welding process with a shielded gas arc between a continuous wire electrode and the weld metal.

The wire feeder feeds the electrode through the welding cable to the welding gun at the specified rate. The wire feeder can be a push, pull, or push-pull type, depending on the location of the drive

rollers. In the push type, the electrode is threaded through the drive rollers and pushed through the welding cable to the welding gun. In the pull type, the electrode is fed through the welding cable and pulled by drive rollers located on the welding gun. The push-pull type has drive rollers located before and after the welding cable.

The type of wire feeder used is determined by the characteristics of the electrode used. Soft aluminum electrodes must be pulled through the welding cable. Larger diameter electrodes often require the push-pull type feeder for consistent flow of wire. In all types of wire feeders, the drive rollers increase or decrease in speed as adjusted by the welder. The rate of wire speed is expressed in inches per minute (ipm). An inch button allows the welder to advance or retract the electrode at a slow speed when changing spools or if an electrode feeding problem occurs.

The welding cable conducts the flow of electricity, electrode, and shielding gas. The current flows through a copper cable within the welding cable. The welding cable also routes the electrode through a flexible metal cable. Shielding gas is transported through a separate hose within the welding cable. The welding cable must not be allowed to become kinked or damaged, as it may result in the restricted flow of electrode or shielding gas.

The welding gun directs the electrode to the weld area. Parts of the welding gun include the handle, contact tip, gas cup, and the trigger. The handle allows easy positioning of the gun by the operator. The contact tip conducts electricity from the welding cable to the electrode as it leaves the welding gun. The gas cup directs the flow of shielding gas to the weld. The gas cup size and shape may vary.

The trigger on the welding gun starts and stops the welding process. When the trigger is pulled, the current, shielding gas flow, and wire feed are activated. The gas flow regulator reduces shielding gas pressure from the tank to the working pressure required by the flowmeter.

The flowmeter regulates the flow of shielding gas as required by the welding operation. The working pressure is converted into gas flow which is expressed in cubic feet per hour (cfh). The amount of shielding gas required is determined by the type of gun, weld joint, base metal, and conditions of the weld area. For example, welding performed in windy conditions requires more shielding gas flowing to the weld. See Figure 7-9.

SHIELDING GAS AND WELDING POSITION						
GAS	SPRAY ARC STEEL	SHORT-CIRCUITING STEEL	SPRAY ARC STAINLESS STEEL	SHORT-CIRCUITING STAINLESS STEEL	SPRAY ARC ALUMINUM	SHORT-CIRCUITING ALUMINUM
Ar					All Positions	All Positions
Ar + 2% O_2	F, H		F, H, O			
Ar + 5% O_2	F, H		F, H, O			
Ar + 7% CO_2	F, H					
Ar + 25% CO_2		All Positions		All Positions		
Ar + 50% CO_2		All Positions		All Positions		
CO_2	F, H	All Positions				
He					All Positions	
Ar and He					All Positions	

Figure 7-9. Different shielding gases are required based on the type of base metal, welding position, and amperage.

Shielding gases commonly used in GMAW include argon, carbon dioxide, helium, and oxygen. Carbon dioxide and oxygen are not true inert gases, but can be used in mixtures or with electrodes specifically designed for use with these gases. The shielding gas used in GMAW is determined by the base metal, the welding position, and the amperage.

GMAW Procedure. The GMAW process requires three adjustments by the welder: shielding gas flow, voltage, and wire feed speed. Before welding, the welder should consult the manufacturer of the electrode used for the recommended settings. Final adjustments are made by the welder.

Shielding gas flow is adjusted by setting the flowmeter to the required cfh. The welding machine and wire feeder are adjusted to specifications. The trigger on the welding gun is depressed, which actuates shielding gas flow. Shielding gas flow is indicated by the reading on the flowmeter. Adjustment of the voltage and amperage is achieved with the aid of a person observing the welding (working) voltage and amperage while the weld is performed. Voltage is adjusted using a dial on the welding machine. Amperage is adjusted by increasing or decreasing the wire feed speed on the wire feeder. During the welding process, the

wire feeder feeds more electrode to the weld increasing the amperage if the stickout (electrode extension) is decreased. The wire feeder decreases the wire feed speed, decreasing the amperage if the electrode extension is increased.

Direct current electrode positive is most commonly used in the GMAW process. The flow of current and the voltage setting determines the type of filler metal transfer. The types of transfer which can occur in GMAW include spray, globular, and short-circuit transfer. See Figure 7-10.

Spray transfer is metal transfer in which molten metal from a consumable electrode is sprayed across the arc in small drops. It is used on thicker metals where faster weld deposition rates are required. Spray transfer is accomplished by using currents and voltages higher than those used in short-circuit transfer and can use any electrode diameter. Spray transfer is used primarily in flat and horizontal positions.

Globular transfer is metal transfer in which molten metal from a consumable electrode is spread across the arc in large drops. It is used on thin metals where shallow penetration is required. Globular transfer is accomplished by using currents and voltages lower than those used in short-circuit transfer, which results in a slow deposition rate with minimal heat input. Globular transfer can

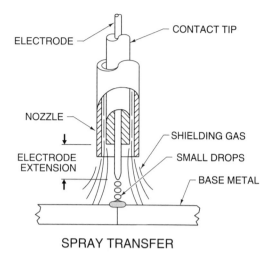

SPRAY TRANSFER

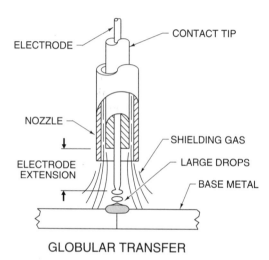

GLOBULAR TRANSFER

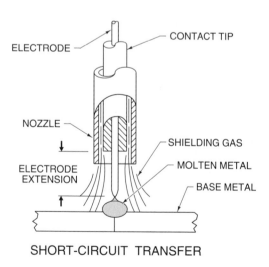

SHORT-CIRCUIT TRANSFER

Figure 7-10. The type of weld metal transfer is determined by the amperage and voltage used in the GTAW process.

be used in all positions; however, small electrode diameters must be used.

Short-circuit transfer is metal transfer in which molten metal from a consumable electrode is deposited during repeated short circuits. It is used on medium to larger thicknesses where moderate penetration and narrow beads are required. In short-circuit transfer, the electrode touches the base metal and short-circuits. Short-circuit transfer can be used in all positions with a variety of electrode diameters.

In the GMAW process, the speed of travel, work angle, and the travel angle (push or pull) vary, depending on the weld type, weld joint, metal thickness, and composition. Generally, a push angle provides less penetration than a drag (pull) angle.

GMAW Applications. GMAW has grown in popularity due to its versatility and capability for production welding. Because there is no interruption in the welding process due to electrode change, GMAW is well-suited for automated production applications. In addition, with no flux, there is less chance of slag inclusions or other weld defects.

The type of weld metal transfer permits a wide range of applications for position and base metal thickness. The GMAW process is used frequently in pipe welding because of the penetration and weld purity that can be achieved. Welds made using the GMAW process can be finished without extensive cleaning.

GMAW can be used on ferrous and nonferrous metals. *Ferrous metal* is any metal that contains iron. *Nonferrous metal* is any metal that does not contain iron.

Flux Cored Arc Welding (FCAW)

Flux cored arc welding (FCAW) is a welding process that uses an arc shielded by gas from within the electrode. It is an arc welding process very similar to GMAW in principle of operation and equipment used. The basic difference between GMAW and FCAW is that FCAW uses a tubular wire with flux in its core as an electrode. The FCAW electrode is often referred to as an "inside-out coated electrode." In the FCAW process, the weld metal is transferred as in the GMAW short-circuit transfer process. See Figure 7-11.

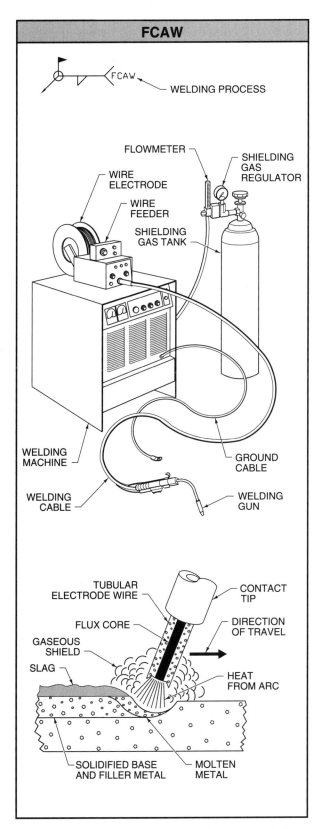

FCAW

WELDING PROCESS

FLOWMETER

SHIELDING GAS REGULATOR

WIRE ELECTRODE

WIRE FEEDER

SHIELDING GAS TANK

WELDING MACHINE

WELDING CABLE

GROUND CABLE

WELDING GUN

TUBULAR ELECTRODE WIRE

CONTACT TIP

FLUX CORE

DIRECTION OF TRAVEL

GASEOUS SHIELD

SLAG

HEAT FROM ARC

SOLIDIFIED BASE AND FILLER METAL

MOLTEN METAL

Figure 7-11. Flux cored arc welding (FCAW) is a welding process that uses an arc shielded by gas from within the electrode.

Greater weld metal deposition and deeper penetration are possible with the FCAW process than with the GMAW process. FCAW and GMAW processes became popular in the early 1950s. The SMAW process was the most widely used and accepted process until that time.

A breakthrough in the FCAW process occurred in the early 1960s with the development of a tubular wire electrode that did not require shielding gas. Shielding gas was generated by the flux contained in the core of the electrode consumed during the welding process. This reduced the cost of the process by eliminating the need for shielding gas and the required equipment. Shielding gas can be used in the FCAW process for increased penetration and filler metal deposition.

FCAW Equipment. FCAW equipment required includes the welding machine, electrode, wire feeder, welding cable, welding gun, shielding gas regulator, and flowmeter. The equipment is basically the same equipment required in the GMAW process.

Welding machines used in the FCAW process must have a capability for the higher amperages and voltages required, as compared to the GMAW process. A push-type wire feeder is most commonly used due to the rigidity of the flux cored electrode. The wire feeder drive rollers used for electrodes greater than $\frac{1}{16}''$ in diameter are knurled to prevent slippage when resistance from long or bent welding cables occurs.

The FCAW welding gun is larger than the GMAW gun. Generally, welding at greater than 600 A requires a water-cooled gun to prevent overheating. A water circulator maintains the flow of water to the gun. In addition, guns used without a shielding gas are more likely to overheat without the cooling effect of the shielding gas. Some FCAW welding guns have fume extractors to remove smoke and/or toxic fumes caused by the welding process.

Smoke extractors increase the visibility of the weld, in addition to reducing the air pollution of the welding environment. However, smoke extractors add weight and bulk to the welding gun. Smoke extractors do not remove any shielding gas from the weld area. Shielding gas, if used, is controlled by the same equipment as used in the GMAW process.

FCAW Procedure. The FCAW process follows essentially the same procedure as the GMAW process. FCAW electrodes are designed for use with either DCEP or DCEN. Shielding gas may or may not be required depending on the electrode used. A drag angle is usually recommended to eliminate possible slag inclusions caused by entrapment of flux in the weld.

FCAW Applications. FCAW combines the production efficiency of the GMAW process and the penetration and deposition rates of the SMAW process. In addition, FCAW is useful where shielding gas is unavailable.

The most common application of the FCAW process is structural fabrication. High deposition rates achieved in a single pass make FCAW very popular in the railroad, shipbuilding, and automotive industries. The FCAW process can be used in all positions with the proper electrode and the required shielding gas.

Gas Tungsten Arc Welding (GTAW)

Gas tungsten arc welding (GTAW) is a welding process in which shielding gas protects the arc between a tungsten electrode and the weld area. It is an arc welding process that uses a nonconsumable tungsten electrode and a shielding gas for welding. The electrode directs the arc. The shielding gas protects the weld area against contamination from nitrogen and oxygen.

The GTAW process can be used to weld with or without filler metal. Filler metal may be added for reinforcement if necessary. See Figure 7-12. The GTAW process is sometimes referred to as TIG (tungsten inert gas) welding.

The GTAW process was developed in the late 1930s primarily for welding aluminum and magnesium in the aircraft industry. Initially, the GTAW process used DCEP, until DCEN was introduced. A breakthrough in GTAW occurred during World War II when alternating current, used with high frequency, produced high-quality welds on aluminum. Helium was used as a shielding gas, but was later replaced by less expensive argon.

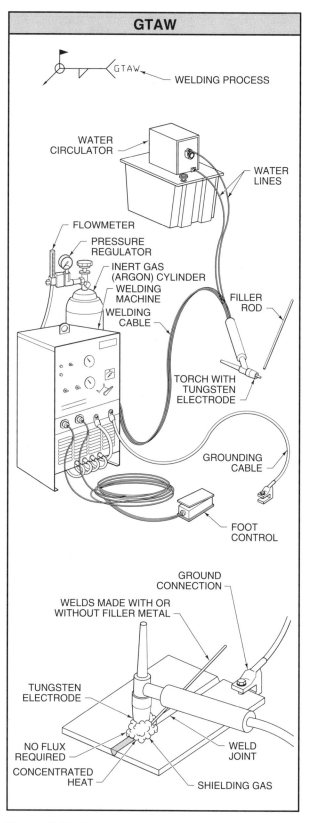

Figure 7-12. Gas tungsten arc welding (GTAW) is a welding process in which shielding gas protects the arc between the tungsten electrode and the weld area.

GTAW Equipment. GTAW equipment required includes a welding machine, welding cable, torch, tungsten electrode, water circulator, shielding gas regulator, flowmeter, and foot control. The specific welding task determines the type and configuration of equipment required.

The welding cable transports welding current and shielding gas to the torch. On water-cooled torches, the welding cable also transports cooling water to and from the torch. The GTAW torch is designed to hold the electrode, direct the shielding gas to the weld, and allow easy positioning to the weld by the welder.

The tungsten electrode is held in place by the collet of the torch. The size of the tungsten electrode used determines the size of the collet necessary. The gas cup directs the flow of shielding gas to the weld. Different shaped gas cups may be used for different joints and weld types to ensure adequate shielding gas coverage.

The GTAW welding machine may provide DC, AC, or both. In addition, AC high frequency (ACHF) may be used to start and maintain the arc in the welding process. Most ferrous metals are welded using DCEN. In DCEN, electrons flow from the electrode to the work, and gas ions flow away from the work to the electrode. DCEN provides deep penetration in the weld.

DCEP is rarely used in the GTAW process. In DCEP, electrons flow from the work to the electrode, and gas ions flow from the electrode to the work. DCEP provides an excellent cleansing action in the weld area with the flow of gas ions.

Most nonferrous metals are welded using ACHF. Alternating current provides both DCEN and DCEP at 60 cycles per second. However, at this rate, the arc has a tendency to be inconsistent because of the slow transition from DCEN to DCEP. This produces an arc that sputters and is difficult to control. The use of ACHF permits the arc to be maintained without interruption between the DCEN and DCEP side of the AC cycle. ACHF provides a combination of the penetrating qualities of DCEN and the cleansing action of DCEP. See Figure 7-13.

The GTAW torch can be either air-cooled or water-cooled, depending on the amperage used. Air-cooled torches are generally used for welding up to 200 A. Water-cooled torches are used when welding above 200 A.

The water circulator provides cooling water to the torch and removes heated water to maintain a safe torch operating temperature. The water circulator consists of a tank, pump, regulator, and feed and return lines. The flow rate of cooling water required depends on the welding operation. Torch manufacturers provide recommended settings for cooling-water flow.

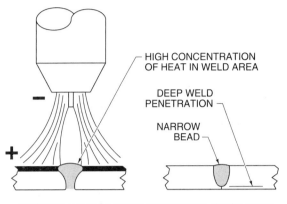

DIRECT CURRENT ELECTRODE NEGATIVE
(DCEN)

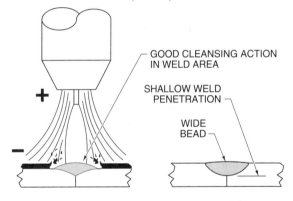

DIRECT CURRENT ELECTRODE POSITIVE
(DCEP)

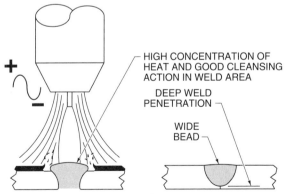

ALTERNATING CURRENT HIGH FREQUENCY
(ACHF)

Figure 7-13. Alternating current high frequency (ACHF) produces the penetration of direct current electrode negative (DCEN) and the cleansing action of direct current electrode positive (DCEP).

GTAW electrodes are made of tungsten. Tungsten, which has the highest melting point of all metals, does not melt when correct procedures are followed. Incorrect current, excessive amperage, and/or electrode contamination can result in melting or deformation of tungsten electrodes.

Pure tungsten electrodes are the least expensive of the tungsten electrodes used and are identified with a green marking. Pure tungsten electrodes are most commonly used for welding aluminum. They are designed for use with alternating current. Tungsten electrodes can be alloyed with one or two percent thorium. Thoriated (one or two percent alloy) tungsten electrodes can conduct higher amperages and provide a more stable arc than a pure tungsten electrode. Thoriated electrodes are designed for use with direct current. Tungsten electrodes can also be striped (sandwiched) with pure tungsten and thorium for achieving benefits of both the pure and thoriated tungsten electrodes.

One percent thoriated tungsten electrodes are identified with a yellow marking. Two percent thoriated tungsten electrodes are identified with a red marking. Striped tungsten electrodes are identified with a blue marking. Tungsten electrodes with 1/2% zirconia added have a brown marking. Alloyed tungsten electrodes are used for some aluminum welding and other types of metal. The type and size of the tungsten electrode required is determined by the current, amperage, and base metal.

Argon and helium are the shielding gases most commonly used in the GTAW process. Argon is less expensive. It is heavier than air. This facilitates efficient coverage of the weld area. It is easier to control in drafty conditions. Helium produces deeper weld penetration, but is lighter than air. This requires more gas to be used for adequate shielding.

Different percentages of argon and helium are used to obtain required penetration at the lowest cost. Shielding gas is reduced to working pressure from the source by a gas flow regulator. The flowmeter accurately controls the amount of gas flowing to the weld area. The type and amount of shielding gas used is determined by current, amperage, type of weld, base metal, and welding conditions.

Shielding gas is required before, during, and after the welding operation. Before welding, shielding gas is directed to the weld area to displace the air in the weld area. During welding, shielding gas flow continues. After welding, a postflow timer controls the time shielding gas flows after the arc

is stopped. This protects the weld until the weld is no longer subject to contamination.

The foot control or torch-mounted control allows the welder to start and vary current flow without interfering with the weld operation. For example, the welder can increase the amperage at the beginning of the weld to obtain necessary penetration on the cold base metal.

GTAW Procedure. The GTAW procedure varies slightly, depending on the base metal, weld joint, weld type, and position. Adjustments required for all GTAW operations include selecting current, adjusting amperage, selecting the tungsten electrode, adjusting cooling water flow, selecting the shielding gas, adjusting shielding gas flow rate, and adjusting electrode extension. This information is available from the manufacturer. Fine adjustments can be made by the welder.

The tungsten electrode tip is prepared differently, depending on the composition of the electrode. Pure tungsten electrodes require a spherical end formed by striking the arc on a copper plate until a ball-shaped end is formed. The electrode should never touch the plate when striking an arc and welding. Thoriated tungsten electrodes must be ground to a point, with the length of the ground surface approximately $2\frac{1}{2}$ times the diameter of the electrode. The prepared end is opposite of the end with the color marking to allow for future identification of the electrode. See Figure 7-14.

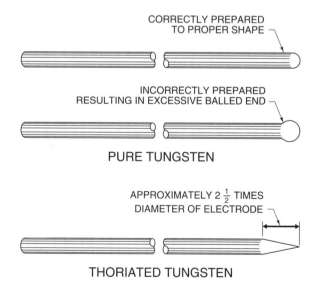

CORRECTLY PREPARED TO PROPER SHAPE

INCORRECTLY PREPARED RESULTING IN EXCESSIVE BALLED END

PURE TUNGSTEN

APPROXIMATELY $2\frac{1}{2}$ TIMES DIAMETER OF ELECTRODE

THORIATED TUNGSTEN

Figure 7-14. Tungsten electrode tip preparation is determined by the composition of the electrode.

The GTAW technique is similar to the OFW technique. The torch is manipulated to distribute the heat evenly in the weld area. Filler metal, if required, is added to the puddle with an in-and-out motion. A push or drag angle is used depending on the size and type of metal welded. Care is taken to avoid contamination of the electrode by touching the weld metal.

GTAW Applications. GTAW is used where accurate control of weld penetration and weld purity are critical. The weld metal deposited is free of slag and/or spatter, allowing metal finishing without cleaning or extensive preparation. Carbon steels can be welded using the GTAW process. However, the most common application is on aluminum, magnesium, stainless steel, and other metals which cannot be welded satisfactorily using other welding processes.

GTAW is commonly used for joining metals in the aerospace and aircraft industries. The low-heat input of GTAW permits welding on very thin metals with minimal distortion and/or alteration of base metal properties. In addition, because of the weld penetration and purity, GTAW is widely used in welding pressure vessels and piping systems in nuclear power plants.

More skill is required for GTAW than is required for GMAW or SMAW. Filler metal must be added manually, reducing filler metal deposition rates and increasing welding time, resulting in higher labor costs for GTAW than other arc welding processes.

Resistance Welding (RW)

Resistance welding (RW) is a group of welding processes in which welding occurs from the heat obtained by resistance to the flow of current through the workpieces. It is widely used in industry for joining metals less than ¼″ thick.

Resistance welding joins weld parts by passing a high electric current through the weld parts. The resistance offered by the weld parts to the current generates the heat necessary to melt the metal. Pressure is applied joining the molten metal of the weld parts at the point of resistance. The molten metal cools, leaving a weld nugget. See Figure 7-15.

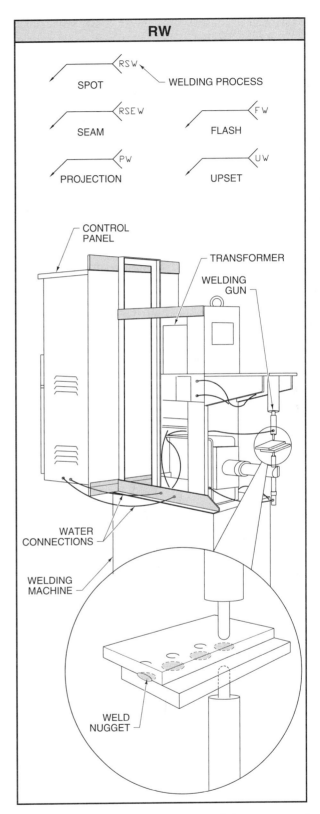

Figure 7-15. Resistance welding (RW) is a group of welding processes in which welding occurs from the heat obtained by resistance to the flow of current through the workpieces.

RW was developed from the 1880s discovery that a current passed through a fine wire produced heat. RW is classified into different types based on the configuration of equipment required. The most common types of resistance welding are: resistance spot welding (RSW), resistance seam welding (RSEW), projection welding (PW), flash welding (FW), and upset welding (UW). The RW type is noted in the tail of the welding symbol.

RW Equipment. RW equipment includes a power source, holding mechanism, electrodes, and weld time controller. The power source provides the current necessary to generate heat at the point of resistance (weld area). The holding mechanism secures the weld parts during the welding process. Electrodes conduct electricity from the power source to the weld area and back to the power source to complete the electrical circuit. The weld time controller regulates the time current is passed through the weld parts.

RW Procedure. The specific RW procedure is determined by the RW type performed. In RSW, the electrodes serve as a holding mechanism in addition to conducting electricity to the weld parts. The welder positions the weld parts and activates the flow of the current. The welded parts are then removed from the electrodes.

In RSEW, a continuous weld is made using rotary electrodes. If required, a series of successive welds can be made with intermittent passage of current through the rotary electrodes. The welder is responsible for feeding the weld parts into the welding machine for welding in the proper location.

In PW, the weld area is predetermined by forming the weld parts before welding. Projections on the weld parts act as points of resistance. They are melted into the weld area to form the weld nugget. PW is usually automated, with the welder observing the operation.

In FW, the weld parts are secured in a holding device and are drawn toward each other as electric current is passed through the parts. An arc is established between the parts melting the metal in the weld area. The weld parts are then pressed together for fusion.

UW is similar to FW. In UW, the weld parts are in contact and are pressed against each other while current is passed through. Resistance at the weld joint generates the heat necessary for fusion.

RW Applications. RW is widely used in industry to join ferrous and nonferrous sheet metal rapidly. Surface preparation other than cleaning is minimal. The weld area is easily controlled by electrode size and the amperage and duration of the current applied. Welds can be made in any position, depending on the RW equipment. Weld joints most commonly used include the lap joint and butt joint. Multiple welds can be made using several electrodes positioned at the required locations, with the welding current applied at the same time.

The major advantage of RW is the speed at which welds can be made. Operator training is minimal. Hand-held welding machines offer more flexibility than stationary floor welding machines. Larger structures must be welded using portable welding machines. Smaller weld parts are more quickly welded using stationary welding machines, which are more mechanized than portable welding machines.

Other Welding Processes

Special considerations, such as weld purity, heat input, and base metal properties, may necessitate less common welding processes than GTAW, GMAW, etc. These include submerged arc welding (SAW), stud arc welding (SW), electron beam welding (EBW), laser beam welding (LBW), ultrasonic welding (USW), and friction welding (FRW).

In SAW, the welding process occurs under a shield of granular flux which covers the arc and filler metal. This allows deep penetration and heavy deposition rates using DCEP. The SAW process is usually automated, with welds possible only in the flat and horizontal positions.

SW is a type of arc welding that welds studs or various fasteners to weld parts. The stud is used as the electrode. Current is sent through the stud causing an arc. Pressure is applied which results in the stud being fused to the weld part in one operation.

In EBW, fusion of the weld parts is achieved by focusing a high-power, dense beam of electrons on the area to be joined. The kinetic energy of the electrons is changed to thermal energy at the weld area, causing the weld metal to melt and fuse together. The heat input is low, minimizing distortion.

In LBW, a concentrated beam of light is focused on the weld area. Heat is generated by the beam of light, providing a great degree of welding accuracy.

In USW, welding is performed by applying vibratory energy to bring the weld parts into close contact and form a bond. The USW process is most commonly used on small, thin weld parts which can be moved as necessary to achieve fusion.

In FRW, welding is performed by rotating one of the weld parts and bringing a stationary second weld part in contact with the rotating weld part. Pressure is applied, and heat from the friction causes the weld parts to fuse together. The FRW process is commonly used to join dissimilar metals.

WELD APPLICATION SPECIFICATION

The weld application is specified with letters in the tail of the welding symbol. Weld applications are classified into manual (MA), semiautomatic (SA), machine (ME), and automatic (AU) processes. The abbreviations are included in the tail of the welding symbol with the welding process designation. See Figure 7-16.

Manual welding is welding with equipment that is completely controlled by the welder. The electrode holder, torch, or gun is moved manually during the welding operation. Semiautomatic welding uses manual control in addition to equipment that controls one or more of the welding conditions. For example, in semiautomatic welding using the GMAW process, the gun is advanced manually, and the electrode is fed to the weld by a wire feeder.

Machine welding or mechanized welding uses equipment to hold and manipulate the electrode holder, torch, or gun that is under constant observation and control by the operator. Adjustments are made as required in response to changes in welding conditions. Automatic welding uses equipment to completely control the welding operation. This minimizes the need for observation or adjustment of controls. A variation of automatic welding is adaptive control welding. Adaptive control welding uses equipment to control the welding process and adjust or adapt automatically in response to changing welding conditions.

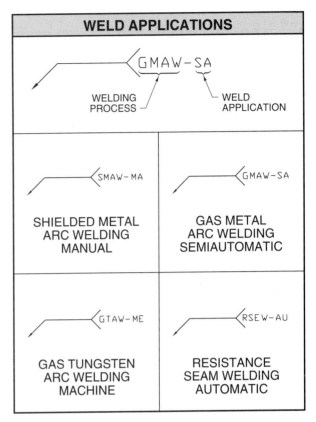

Figure 7-16. The weld application is specified with the welding process in the tail of the welding symbol.

BRAZING (B) AND SOLDERING (S)

Weld parts that must be joined without melting the base metal commonly use the brazing (B) or soldering (S) process. The parts are joined by heating the joint area and adding filler metal.

Brazing is a group of welding processes in which metal is joined by heating the filler metal at temperatures greater than 840°F, but less than the melting point of the base metal. *Soldering* is a group of welding processes in which metal is joined by heating the filler metal at temperatures less than 840°F and less than the melting point of the base metal.

In both brazing and soldering, the filler metal acts as an adhesive. Filler metal used in brazing is a brass or bronze alloy. Soldering uses a lead-tin alloy as filler metal.

Joints to be brazed or soldered are designed for strength by maximizing the surface area of the joint receiving filler metal. Joints commonly used in brazing and soldering are the lap joint and butt joint. See Figure 7-17.

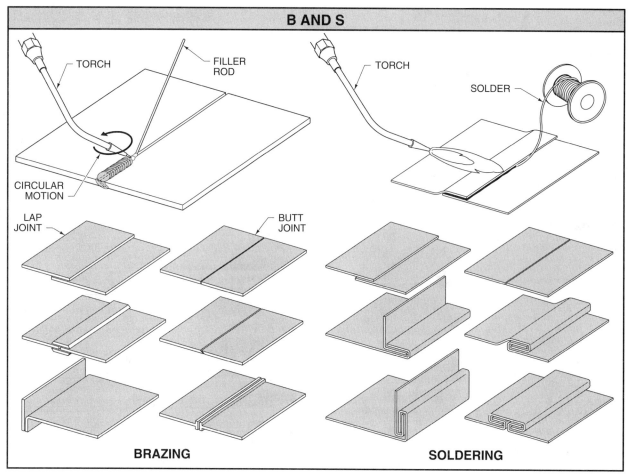

Figure 7-17. Brazing and soldering are groups of welding processes in which metal is joined by heating the filler metal at temperatures greater than 840°F for brazing and less than 840°F for soldering, but at less than the melting point of the base metal.

Review Questions

Name_____ Date _____

Completion

_____ 1. A(n) _____ is a welding machine that produces either AC or DC.

_____ 2. A(n) _____ flame is an oxyfuel flame with a balanced mixture of oxygen and fuel.

_____ 3. A(n) _____ is any material through which electricity flows easily.

_____ 4. _____ is the flow of current from the work (–) to the electrode (+).

_____ 5. A(n) _____ is a welding machine that produces DC only.

_____ 6. The _____ angle is the angle less than 90° of the electrode in relation to a perpendicular line from the weld and the direction of the weld.

_____ 7. The prepared end of a tungsten electrode should be opposite of the end with the _____.

_____ 8. RW is commonly used for joining metals less than _____″ thick.

_____ 9. A(n) _____ flame is an oxyfuel flame with an excess of oxygen.

_____ 10. A(n) _____ regulates the flow of shielding gas during the GMAW process.

_____ 11. A(n) _____ is a welding machine that produces AC only.

_____ 12. In the United States, AC completes _____ cycles per second.

_____ 13. Oxyacetylene tanks are commonly pressurized to _____ psi.

_____ 14. The amperage required for welding increases with an increase in electrode _____.

_____ 15. The _____ is the distance from the electrode to the molten pool of the base metal.

_____ 16. _____ is the maximum amperage and voltage of a welding machine.

_____ 17. _____ is the flow of current from the electrode (–) to the work (+).

_____ 18. GMAW welding machines are known as constant _____ welding machines.

_____ 19. _____ transfer is metal transfer in which molten metal from a consumable electrode is spread across the arc in large drops.

_____ 20. GTAW air-cooled torches are generally used for welding up to _____ A.

True-False

T   F   **1.** Oxyfuel welding may use natural gas, propane, or other gases as a fuel.

T   F   **2.** The oxyacetylene process can produce a flame of approximately 6300°F.

T   F   **3.** The oxyacetylene welding torch and welding rod are held at an approximate 30° angle to the weld parts.

T   F   **4.** Relatively low currents and high voltage are used in SMAW.

T   F   **5.** SMAW current can be either AC or DC.

T   F   **6.** The duty cycle for manual welding machines is normally 100%.

T   F   **7.** The wire is the part of the SMAW electrode that melts and forms the filler metal.

T   F   **8.** The GMAW process requires a flux covering on the electrode to provide a gaseous shield to protect the weld area.

T   F   **9.** More skill is required for GTAW than is required for GMAW or SMAW.

T   F   **10.** The first two letters of an electrode's AWS identification code identify the tensile strength of the electrode wire.

T   F   **11.** The GTAW process can be used to weld with or without filler metal.

T   F   **12.** Slag in a weld provides for slow cooling of the weld.

T   F   **13.** When lighting an oxyacetylene torch, the oxygen needle valve is opened first.

T   F   **14.** Current uses the weld parts as conductors in an electrical circuit.

T   F   **15.** The rate of wire speed for a GMAW wire feeder is expressed in feet per minute (fpm).

T   F   **16.** One percent thoriated tungsten electrodes are identified with a red marking.

T   F   **17.** The welding process, if specified, is indicated by a symbol in the body of the welding symbol.

T   F   **18.** Brazing is a group of welding processes in which metal is joined by heating the filler metal at temperatures greater than 1840°F but less than the melting point of the base metal.

T   F   **19.** The main advantage of OFW is the ability to weld without electricity.

T   F   **20.** The ground is the connection between the welding cable and the weld parts in the welding circuit.

Trade Competency Test

Name_____ Date _____

Refer to the Punch print on page 171.

_____ 1. The scale for the drawing is 1″ = _____″.

_____ 2. Collar E is joined to the punch by _____.

_____ 3. Callout D specifies carbide grade _____.

_____ 4. The maximum dimension for G is _____″.

_____ 5. Collar E fits on the shaft having a(n) _____″ diameter.

_____ 6. The maximum width of I is _____″.

_____ 7. The steel used to manufacture the punch is AISI type _____.

_____ 8. The type of weld specified at C is a(n) _____ weld.

_____ 9. The maximum angle dimension at H is _____.

_____ 10. The maximum wall thickness of collar E is _____″.

_____ 11. Collar E is _____″ wide.

T   F   12. The tolerance for fractional dimensions is ±$\frac{1}{16}$″.

T   F   13. The project is a punch for a slug press.

T   F   14. The minimum diameter of I is 2.875″.

_____ 15. Dimension A specifies a(n) _____ of .015″.

_____ 16. The right edge of collar E is specified to be _____.

_____ 17. The drawing was completed on _____.

_____ 18. The minimum dimension for F is _____″.

T   F   19. The drawing number is C-707-27863.

T   F   20. The width of the undercut area to the right of I is .375″.

Refer to the Floor Plate Bracket print on page 172.

_____ **1.** The type of weld specified at A is a(n) _____ weld.

_____ **2.** Weld C specifies that the _____ welding process be used.

_____ **3.** Weld B is applied _____.

_____ **4.** A fillet weld with unequal legs is specified for weld _____.

_____ **5.** Weld A specifies that the _____ welding process be used.

_____ **6.** The drawing was drawn at _____ size.

_____ **7.** Weld D is applied by _____.

T   F **8.** Weld B specifies a fillet weld with a ½″ leg.

T   F **9.** Weld C is applied semiautomatically.

T   F **10.** Weld A is applied by machine.

Refer to the Support Assembly print on page 172.

_____ **1.** The type of weld specified at A is a(n) _____ weld.

T   F **2.** The support assembly is symmetrical about the vertical center line.

T   F **3.** The support assembly is symmetrical about the horizontal center line.

_____ **4.** The diameter of the RSW welds is _____.

_____ **5.** A total of _____ SMAW welds are required.

_____ **6.** _____ welds are completed using the SMAW process.

_____ **7.** The SMAW welds are made from the _____ side.

_____ **8.** A total of _____ RSW welds are required.

T   F **9.** There is no side significance to either RSEW weld.

_____ **10.** The diameter of the RSEW welds is _____″.

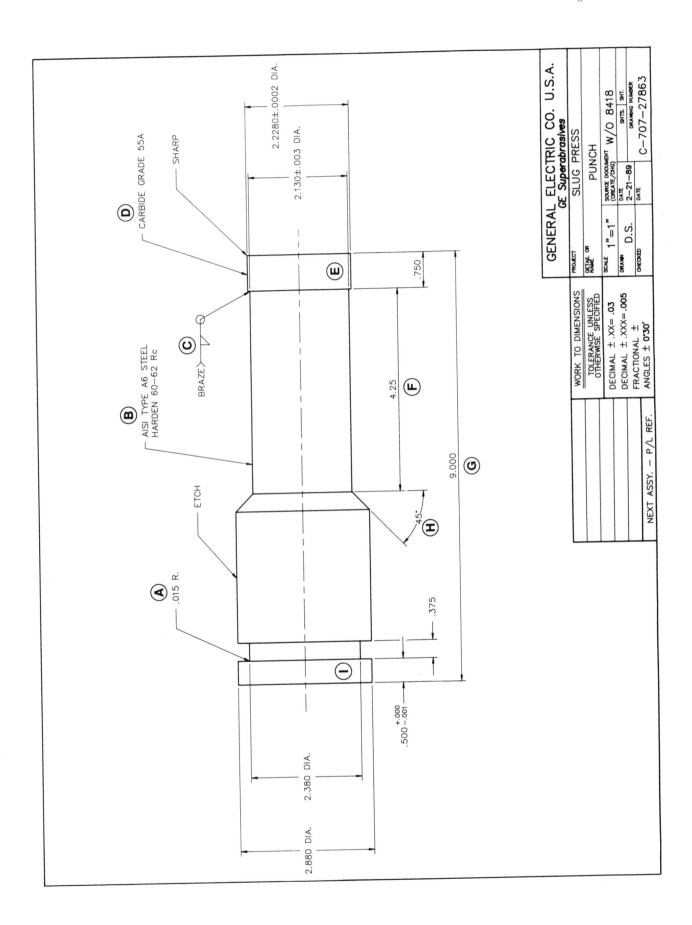

CARBIDE GRADE 55A

SHARP

Ⓓ

Ⓔ

BRAZE ▷

Ⓒ

AISI TYPE A6 STEEL
HARDEN 60-62 Rc

Ⓑ

ETCH

Ⓐ

.015 R.

.750

4.25

Ⓕ

9.000

Ⓖ

45°

Ⓗ

2.2280±.0002 DIA.

2.130±.003 DIA.

Ⓘ

.375

.500 +.000 -.001

2.380 DIA.

2.880 DIA.

GENERAL ELECTRIC CO. U.S.A.		
GE Superabrasives		
PROJECT	SLUG PRESS	
DETAIL OR NAME	PUNCH	
SCALE **1"=1"**	SOURCE DOCUMENT (CREATE/CHG)	W/O 8418
DRAWN D.S.	DATE 2-21-89	SHTS. SHT.
CHECKED	DATE	DRAWING NUMBER C-707-27863

WORK TO DIMENSIONS	
TOLERANCE UNLESS OTHERWISE SPECIFIED	
DECIMAL ± .XX= .03	
DECIMAL ± .XXX= .005	
FRACTIONAL ±	
ANGLES ± 0°30'	

NEXT ASSY. — P/L REF.

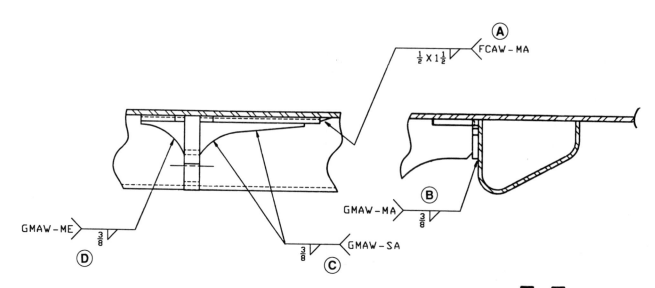

Ⓐ FCAW–MA

½ X 1½

Ⓑ GMAW–MA $\frac{3}{8}$

GMAW–ME $\frac{3}{8}$ Ⓓ

$\frac{3}{8}$ Ⓒ GMAW–SA

SECTION **E - E**
QUARTER SIZE

VME AMERICAS INC.

FLOOR PLATE BRACKET

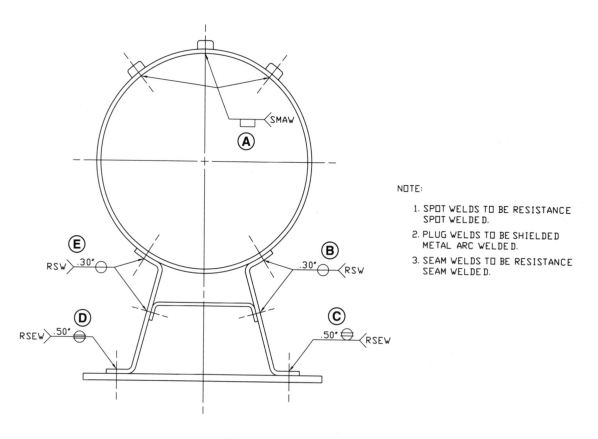

SMAW
Ⓐ

Ⓔ
RSW .30′

Ⓓ
RSEW .50′

Ⓑ
.30′ RSW

Ⓒ
.50′ RSEW

NOTE:

1. SPOT WELDS TO BE RESISTANCE SPOT WELDED.

2. PLUG WELDS TO BE SHIELDED METAL ARC WELDED.

3. SEAM WELDS TO BE RESISTANCE SEAM WELDED.

SUPPORT ASSEMBLY

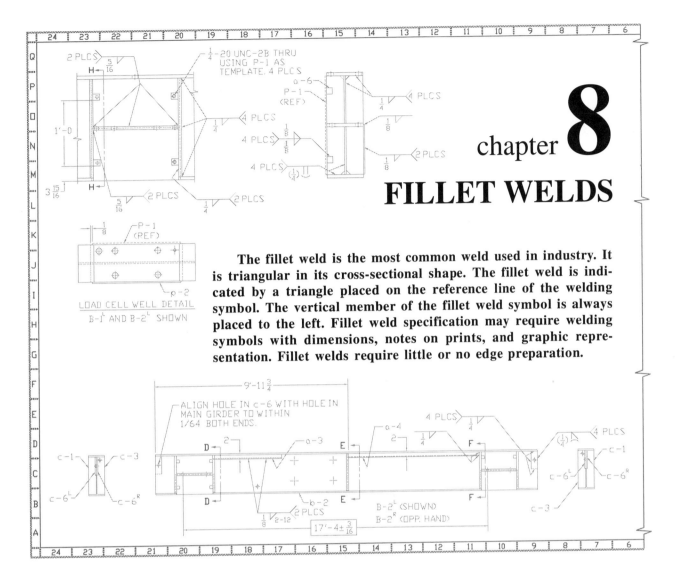

chapter 8

FILLET WELDS

The fillet weld is the most common weld used in industry. It is triangular in its cross-sectional shape. The fillet weld is indicated by a triangle placed on the reference line of the welding symbol. The vertical member of the fillet weld symbol is always placed to the left. Fillet weld specification may require welding symbols with dimensions, notes on prints, and graphic representation. Fillet welds require little or no edge preparation.

SPECIFICATION

A *fillet weld* is a weld type in the cross-sectional shape of a triangle. It joins two surfaces at approximately a right angle, to form a lap joint, T-joint, or corner joint. Parts of the fillet weld include the weld root, weld face, weld toe, fillet weld leg, and fillet weld throat. See Figure 8-1.

The *weld root* is the area where the filler metal intersects the base metal opposite the weld face. It is shown in cross-sectional shape and is the deepest point of the fillet weld triangle. The *weld face* is the exposed surface of a weld, bounded by the weld toes of the side on which welding was done. The weld face of a fillet weld may be concave or convex. A *concave* weld face is curved inward. A *convex* weld face is curved outward.

The *weld toe* is the intersection of the base metal and the weld face. It is the point at which the weld face meets the base metal. The *fillet weld leg* is the distance from the joint root to the weld toe. The *joint root* is the part of a joint to be welded where the members are the closest to each other.

The fillet weld throat may refer to the actual throat, effective throat, or theoretical throat. The *actual throat* is the shortest distance from the face of a fillet weld to the weld root after welding. The *effective throat* is the shortest distance from the face of a fillet weld to the weld root, minus any convexity after welding. The *theoretical throat* is the distance from the face of a fillet weld to the weld root before welding.

Fusion in the fillet weld occurs where the base metal and the filler metal are melted together. The *weld interface* is the area where the filler metal

173

and the base metal mix together. The *fusion face* is the surface of the base metal that is melted during welding. *Depth of fusion* is the distance from the fusion face to the weld interface.

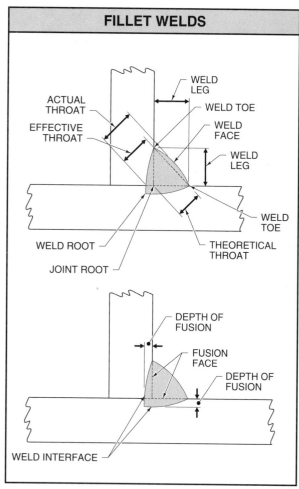

Figure 8-1. Parts of a fillet weld include the weld root, weld face, weld toe, fillet weld leg, and fillet weld throat.

Size

The fillet weld size is determined by the leg sizes of the fillet weld. The *fillet weld leg size* is the dimension from the root of the weld to the toes of the weld after welding. Fillet weld sizes are indicated by notes on prints, by dimensions included on the left side of the weld symbol, and by graphic representation. See Figure 8-2.

Notes on prints may be general or specific. *General notes* are notes that apply a given specification to all items on the prints. For example, a note such as BRUSH ALL FILLET WELDS TO REMOVE SLAG is a general note that applies to all fillet

welds on a set of prints. *Specific notes* are notes that apply a given specification to specific items. For example, a note such as ¼″ LEG ON MEMBER A is a specific note which applies to Member A only.

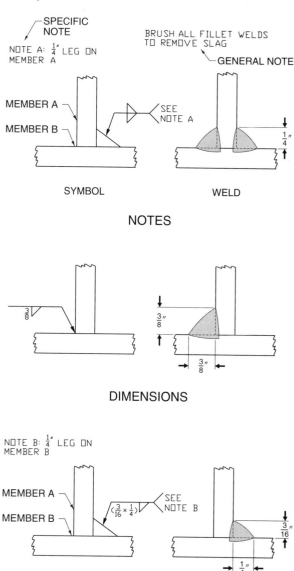

Figure 8-2. Fillet weld sizes are indicated by notes, dimensions, and graphic representation.

The dimension giving the size of the leg is shown on the left side of the fillet weld symbol. See Figure 8-3. For example, a dimension such as ⁵⁄₁₆″ to the left of the fillet weld symbol indicates that both legs of the fillet weld are ⁵⁄₁₆″. A single dimension indicates leg sizes to be equal.

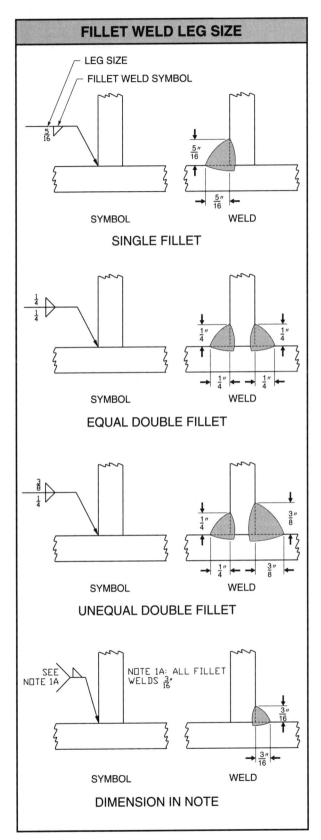

Figure 8-3. The dimension giving the size of the weld is shown on the left side of the fillet weld symbol.

On double-fillet weld joints, the leg size for each side of the weld joint is indicated next to each fillet weld symbol. Fillet weld size is required for both sides of the double-fillet weld, even if they are the same size on both sides. The welder may apply several passes to meet size requirements. The number of passes will vary with the welding process used.

Most fillet welds have the same or equal leg sizes for a given weld. However, if unequal legs are required, the dimensions are specified in the welding symbol. Unequal leg dimensions are shown in parentheses to the left of the weld symbol. The orientation of unequal leg dimensions is not included in the welding symbol and must be shown on the print using a note or graphic representation of the weld joint. See Figure 8-4.

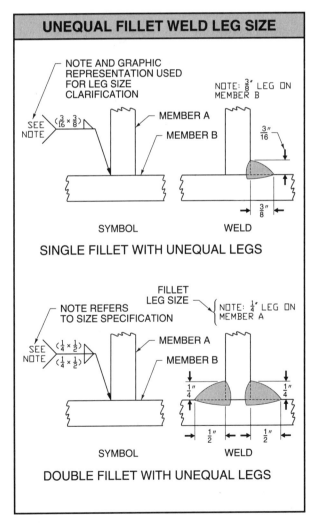

Figure 8-4. Unequal leg dimensions are shown in parenthesis to the left of the weld symbol.

Length

Fillet weld length is indicated using the shape of the weld parts and the information included on the welding symbol. If there are no abrupt changes in the direction of the weld applied, the weld is continuous and extends the full length of the part. Specific weld lengths are indicated by a number to the right of the weld symbol on the welding symbol. See Figure 8-5.

The number to the right of the weld symbol shows the length of weld for the given side of the joint. Welds on both sides of a joint of equal length require a dimension on both sides of the reference line. The absence of a dimension indicates the weld is to be applied the full length of the member. If the weld is less than the entire length of the member, dimensions on the weld part can be used to indicate the start and end of the weld.

Graphic representation can also be used with dimensions and welding symbols to indicate the length and location of fillet welds. The welds can be represented on the weld part using a solid area or hatched lines with dimensions providing exact length and location information. Welding symbols detail the joint and weld type specifications. The welding symbol indicates the location of the weld on the part within the dimension lines provided.

The *weld-all-around symbol* is a supplementary symbol indicated by a circle at the intersection of the arrow and reference line, which specifies that the weld extends completely around the joint. Changes in direction of the weld require multiple arrows on the welding symbol to indicate the location of the weld. Multiple arrows on the welding symbol are not required if the weld-all-around symbol can be used.

Intermittent Fillet Welds

Intermittent fillet welds are short sections of fillet welds applied at specified intervals on the weld parts. Intermittent fillet welds are indicated on the welding symbol by length and pitch. See Figure 8-6. The length specified is the length of each weld section applied. The pitch specified is the center of each weld section. For example, an intermittent fillet weld with the dimension 3 - 6 specifies that 3″ weld sections are to be centered on 6″ intervals.

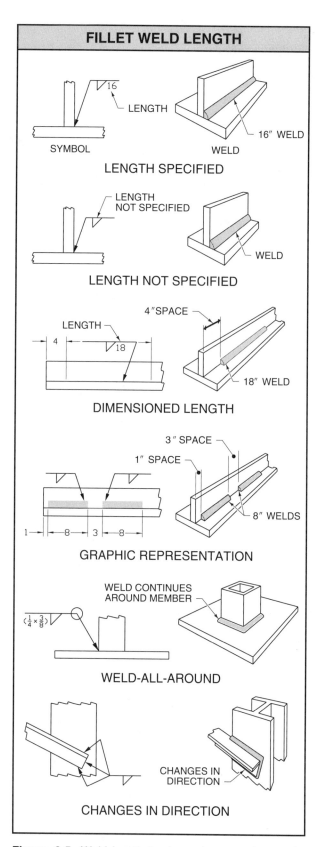

FILLET WELD LENGTH

LENGTH SPECIFIED

LENGTH NOT SPECIFIED

DIMENSIONED LENGTH

GRAPHIC REPRESENTATION

WELD-ALL-AROUND

CHANGES IN DIRECTION

Figure 8-5. Weld length is shown by a number to the right of the weld symbol.

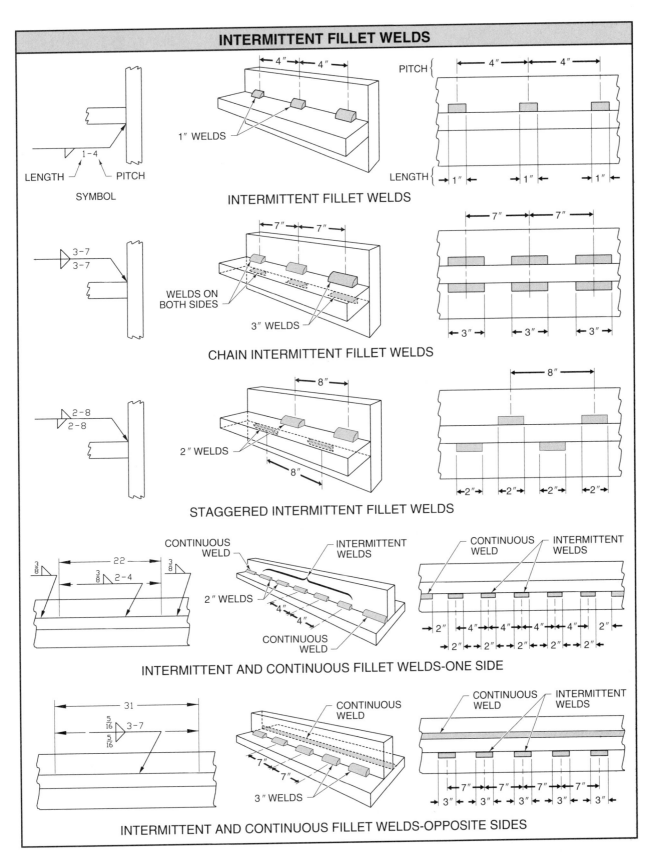

Figure 8-6. Length and pitch dimensions of intermittent fillet welds are shown by numbers to the right of the weld symbol.

The length and pitch of intermittent welds is indicated to the right of the weld symbol. Length and pitch dimensions can also be shown on the weld part, or can be included in a note on the print.

Chain intermittent fillet welds are intermittent fillet welds that have the same specified length and pitch and are applied to both sides of the weld joint. Chain intermittent fillet welds are spaced symmetrically on both sides unless otherwise noted on the print.

Staggered intermittent fillet welds are intermittent fillet welds that have a staggered pitch and are applied to both sides of the weld joint. Staggered intermittent fillet welds are indicated on the welding symbol by a double-fillet weld symbol placed out of alignment on the reference line of the welding symbol. The length and pitch for staggered intermittent fillet welds are indicated in the same way as intermittent fillet welds.

Intermittent fillet welds, unless otherwise specified, are applied beginning at the edge of the dimensioned weld part. If intermittent fillet welds are to be applied with continuous fillet welds, dimensions are used to indicate the location of the required welds. Intermittent and continuous fillet welds applied on the same side of the weld part require separate welding symbols.

CONTOUR AND FINISH

Weld contour is the cross-sectional shape of the completed weld face. *Weld finish* is the method used to achieve the surface finish. Both the fillet weld contour and finish are specified on the welding symbol. The contour symbol is placed next to the angled line of the fillet weld symbol to indicate a flat, convex, or concave contour. See Figure 8-7.

Fillet welds requiring mechanical finishing after welding to obtain the desired contour have a letter next to the weld contour symbol to show the finishing method. The finishing methods that may be specified are C - chipping, H - hammering, G - grinding, M - machining, R - rolling, or U - unspecified. See Figure 8-8.

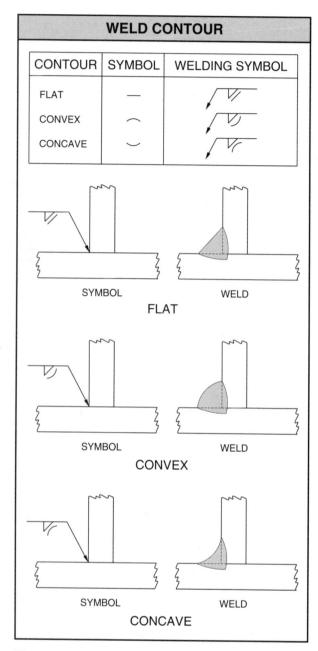

Figure 8-7. Weld contour is the cross-sectional shape of the completed weld face.

COMBINED WELD SYMBOLS

Combined weld symbols are weld symbols used when the weld joint, weld type, and welding operation require more information than can be specified with one weld symbol. The fillet weld is most commonly used in combination with groove welds to specify weld size, joint member configuration, root opening, and edge preparation. The fillet weld

FINISHING METHODS

LETTER	MECHANICAL METHOD	SYMBOL		
		FLAT	CONVEX	CONCAVE
C	CHIPPING			
H	HAMMERING			
G	GRINDING			
M	MACHINING			
R	ROLLING			
U	UNSPECIFIED			

FLAT CONTOUR OBTAINED BY GRINDING WELD

SYMBOL   WELD

Figure 8-8. Weld finish is the method used to achieve the surface finish.

COMBINED WELD SYMBOLS

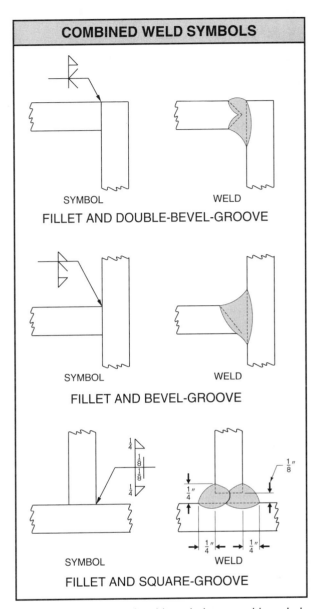

SYMBOL   WELD

FILLET AND DOUBLE-BEVEL-GROOVE

SYMBOL   WELD

FILLET AND BEVEL-GROOVE

SYMBOL   WELD

FILLET AND SQUARE-GROOVE

Figure 8-9. Combined weld symbols are weld symbols used when the weld joint, weld type, and welding operation require more information than can be specified with one weld symbol.

symbol alone could not specify all of this required information. For example, a T-joint with a ⅛″ root opening is specified on the print as a fillet and square-groove weld. Leg size is indicated as ¼″ to the left of the weld symbol. If no weld penetration dimension is specified, the weld penetration required is based on the accepted tolerances of the job. See Figure 8-9.

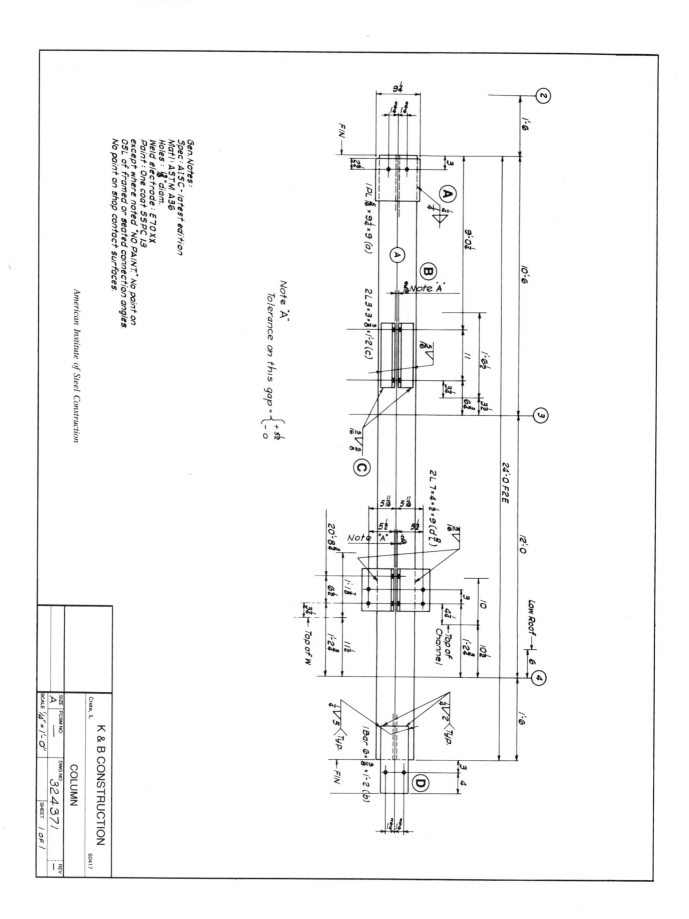

Review Questions

Name_____ Date _____

Matching — Fillet Weld Symbols

_____ 1. Weld-all-around

_____ 2. $\frac{1}{4}''$ leg on Member A; $\frac{5}{16}''$ leg on Member B

_____ 3. Concave; grinding

_____ 4. Staggered; intermittent

_____ 5. $\frac{3}{8}''$ leg length near side; $\frac{1}{4}''$ leg length other side

_____ 6. Flat contour

_____ 7. Convex; grinding

_____ 8. Chain intermittent

_____ 9. $\frac{1}{4}''$ leg length near side; $\frac{3}{8}''$ leg length other side

_____ 10. $\frac{5}{16}''$ leg on Member A; $\frac{1}{4}''$ leg on Member B

Completion

_____ 1. A(n) _____ weld is a weld type in the cross-sectional shape of a triangle.

_____ 2. _____ notes are notes that apply a given specification to all items on a set of prints.

_____ 3. Specific weld lengths are indicated by a number to the _____ of the weld symbol on the welding symbol.

_____ 4. The weld-all-around symbol is a supplementary symbol indicated by a(n) _____ at the intersection of the arrow and reference line, which specifies that the weld extends completely around the joint.

_____ 5. Intermittent fillet welds are indicated on the welding symbol by length and _____.

_____ 6. Weld _____ is the cross-sectional shape of the completed weld face.

_____ 7. The _____ member of the fillet weld symbol is always placed to the left.

_____ 8. Weld _____ is the method used to achieve the surface finish.

_____ 9. For combined weld symbols, the fillet weld is most commonly used in conjunction with _____ welds.

_____ 10. The fillet weld is indicated by a(n) _____ placed on the reference line of the welding symbol.

Matching — Fillet Welds

_____ 1. Weld toe

_____ 2. Theoretical throat

_____ 3. Depth of fusion

_____ 4. Weld interface

_____ 5. Weld leg

_____ 6. Effective throat

_____ 7. Fusion face

_____ 8. Joint root

_____ 9. Weld face

_____ 10. Actual throat

True-False

T   F   1. Fillet welds require little or no edge preparation.

T   F   2. Fillet weld sizes may be indicated by notes on prints.

T   F   3. The orientation of unequal leg dimensions of a fillet weld is included in the welding symbol.

T   F   4. Pitch for intermittent fillet welds is specified at the end of each weld section.

T   F   5. Fillet weld length and pitch dimensions may be indicated by notes on prints.

T   F   6. A single dimension left of the weld symbol indicates that fillet weld leg sizes are equal.

T   F   7. No dimension to the right of a fillet weld symbol indicates that the welder may determine the length of the weld.

T   F   8. Chain intermittent fillet welds are spaced symmetrically on opposite sides unless otherwise noted on the print.

T   F   9. The method of finishing the surface of a fillet weld must be specified.

T   F   10. The fillet weld is the most common weld used in industry.

√‾ ‾‾

Trade Competency Test

Name _____ Date _____

Refer to the Column print on page 180.

_____ 1. The total length of the column is _____.

_____ 2. Weld A specifies a(n) _____ weld, both sides.

_____ 3. The maximum gap at dimension B is _____″.

_____ 4. The minimum gap at dimension B is _____″.

T   F   5. Weld C specifies a fillet weld, other side.

_____ 6. Weld C specifies a(n) _____″ fillet weld leg size.

_____ 7. Weld C specifies a(n) _____″ fillet weld length.

_____ 8. The length of the angle at weld C is _____.

T   F   9. The angle iron at weld C has a web thickness of $\frac{3}{16}$″.

_____ 10. Bar D is joined at the left edge with $\frac{1}{4}$″ welds that are _____″ long.

_____ 11. Bar D is joined at the top and bottom with $\frac{1}{4}$″ welds that are _____″ long.

_____ 12. The diameter of the holes in bar D is _____″.

T   F   13. The holes in bar D are located $3\frac{1}{2}$″ from center.

T   F   14. Bar D has a web thickness of $\frac{3}{8}$″.

_____ 15. The length of bar D is _____″.

_____ 16. The leg size of the angle joined by weld C is _____″.

_____ 17. A(n) _____ electrode is used for all welds.

_____ 18. Bar D extends _____″ beyond the I beam.

_____ 19. The plate joined by weld A is _____″ thick.

_____ 20. The material used for the column is ASTM _____.

Refer to the Motor Housing print on page 187.

_____ 1. The length of the motor housing is _____″.

_____ 2. The outside diameter of the motor housing is _____″.

T   F 3. Weld A specifies a ⅛″ groove weld.

_____ 4. Weld A specifies that the airbox rods are to be _____ welded at both ends.

_____ 5. Weld B specifies a(n) _____ weld.

_____ 6. Weld B specifies a(n) _____ contour.

_____ 7. Weld C specifies fillet welds with a(n) _____″ leg size.

_____ 8. Weld C specifies 1″ fillet welds spaced on _____″ centers.

_____ 9. The mounting bracket joined by weld C is centered _____″ from the left side.

T   F 10. Weld D specifies a fillet weld with a ½″ leg size.

T   F 11. Weld D specifies welds that are ⅛″ in length.

T   F 12. Weld D specifies fillet welds located around the outside of the can.

_____ 13. Weld E specifies a(n) _____ contour.

_____ 14. Weld F specifies a(n) _____ intermittent fillet weld.

_____ 15. Weld F fillet welds on both sides are _____″ in length.

T   F 16. The can seam is located under the airbox.

T   F 17. The mounting bracket joined by weld C is mounted after the stator is inserted in the can.

T   F 18. The mounting bracket joined by weld C is to be stamped with white paint.

T   F 19. There are two rods located in the airbox.

T   F 20. The inside diameter of the can is 11¾″.

_____ 21. The center of the can is _____″ above the bottom of the base.

_____ 22. Diameter A is to be _____ to diameter B within .015″.

T   F 23. Four provisions are made for motor leads.

T   F 24. Leads T1 and T2 exit on the lower-left side of the can.

T   F 25. The motor housing is symmetrical in the front view.

Refer to the Motor Hinge Base print on page 188.

_____ 1. The horizontal shaft is _____″ at center above the bottom of the base plate.

_____ 2. The base plate is _____″ thick.

_____ 3. The threaded upright is welded to the base plate with a(n) _____″ fillet weld.

_____ 4. The threaded upright is _____″ in diameter.

_____ 5. The maximum overall length of the horizontal shaft is _____″.

_____ 6. The minimum diameter of the drilled hole in the horizontal shaft is _____″.

T   F 7. The bottom of the base plate is finished.

T   F 8. The motor hinge base is stress relieved after welding.

T   F 9. The drawing was completed by E.C.S.

_____ 10. The 1/4-20 UNC hole is described in the _____ view.

_____ 11. The center of the horizontal shaft is _____″ from the front edge of the base plate.

T   F 12. The top view contains hidden lines showing the diameter of the counterbored holes.

T   F 13. The drawing was drawn on a size C sheet.

T   F 14. The welder should allow 1/4″ for finish where marked V.

T   F 15. Both ends of the horizontal shaft are finished.

T   F 16. Weld A is a fillet weld.

_____ 17. Weld A is finished to a(n) _____ contour.

_____ 18. The center of the horizontal shaft is _____″ to the right of the back edge of the base plate.

_____ 19. The overall height of the motor hinge base is _____″.

_____ 20. The centers of the four counterbored holes are _____″ from the sides.

_____ 21. The vertical welds joining the center brace to the threaded upright are _____″ fillet welds on both sides.

_____ 22. Weld C is finished to a(n) _____ contour.

T   F 23. The threaded upright has a 1/2-13 UNC thread form.

T   F 24. The drawing is done to a scale of 1/4″ = 1′-0″.

_____ 25. The centers of the two counterbored holes are _____″ from the front edge of the base plate.

_____ 26. The center brace is _____″ wide.

T   F   27. The $\frac{5}{16}$″ fillet welds joining the center brace to the base plate are welded to the threaded upright.

_____ 28. The depth of the four counterbored holes is _____″.

_____ 29. The center to center dimension from the threaded upright to the horizontal shaft is _____″.

_____ 30. Weld B is a $\frac{3}{16}$″ _____ weld.

_____ 31. Weld D specifies a $\frac{3}{16}$″ fillet weld on the _____ side.

_____ 32. The four counterbored holes are located on $2\frac{1}{4}$″ and _____″ centers.

T   F   33. The motor hinge base is symmetrical in the front and top views.

T   F   34. The length of the horizontal shaft may vary by as much as .009″.

_____ 35. The diameter of the four counterbored holes is _____″.

_____ 36. The threaded upright is _____ in shape.

_____ 37. The center brace can be fabricated from a piece of $\frac{1}{2}$″ × 3″ × _____″ stock.

_____ 38. The back plate supporting the horizontal shaft is positioned at a(n) _____° angle.

T   F   39. All threaded holes have coarse threads.

_____ 40. The horizontal shaft extends _____″ on each end beyond the back plate.

T   F   41. The bottom of the plate is finished.

T   F   42. The drawing number of the part is 3591.

_____ 43. The counterbored holes are shown with object lines only in the _____ view.

T   F   44. The $\frac{1}{4}$-20 UNC threaded hole is shown with hidden lines in all three views.

_____ 45. The height of the threaded upright before welding is _____″.

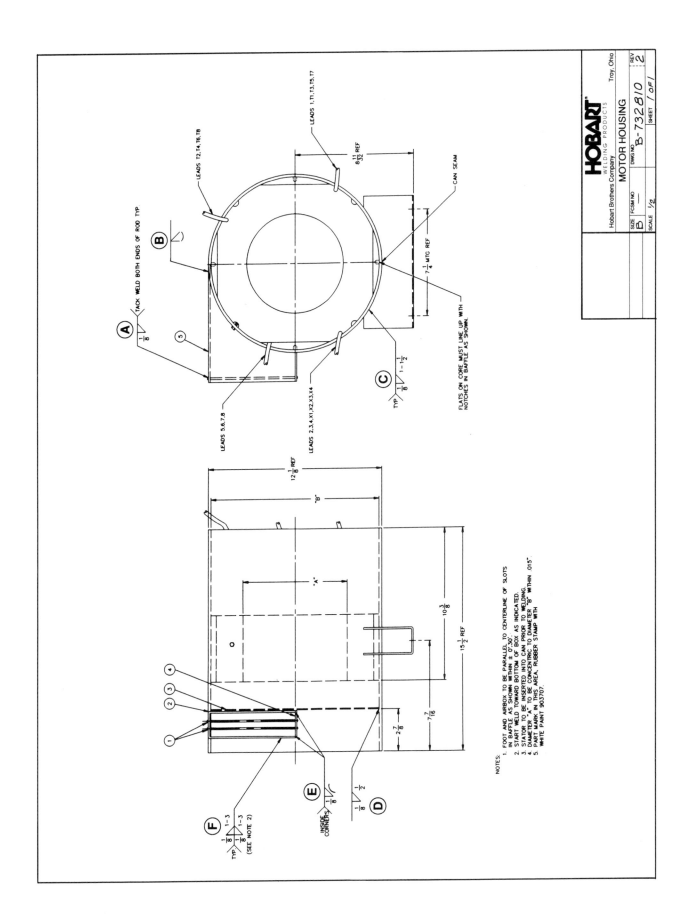

NOTES:
1. FOOT AND AIRBOX TO BE PARALLEL TO CENTERLINE OF SLOTS IN BAFFLE AS SHOWN WITHIN ± 0°.30'.
2. START WELD TOWARD BOTTOM OF BOX AS INDICATED.
3. STATOR TO BE INSERTED INTO CAN PRIOR TO WELDING.
4. DIAMETER "A" TO BE CONCENTRIC TO DIAMETER "B" WITHIN .015".
5. PART MARK IN THIS AREA. RUBBER STAMP WITH WHITE PAINT 903707.

LEADS 12,14,16,18

LEADS 1,11,13,15,17

TACK WELD BOTH ENDS OF ROD TYP

CAN SEAM

FLATS ON CORE MUST LINE UP WITH
NOTCHES IN BAFFLE AS SHOWN.

LEADS 5,6,7,8

LEADS 2,3,4,X1,X2,X3,X4

$8\frac{11}{32}$ REF

$7\frac{1}{4}$ MTG REF

$12\frac{1}{8}$ REF

"B"

"A"

$10\frac{3}{8}$

$15\frac{1}{2}$ REF

$7\frac{7}{16}$

$2\frac{7}{8}$

INSIDE CORNERS

(SEE NOTE 2)

Hobart Brothers Company   Troy, Ohio

HOBART
WELDING PRODUCTS

MOTOR HOUSING

SIZE B   FCSM NO —   DWG NO B-732810   REV 2

SCALE 1/2   SHEET 1 of 1

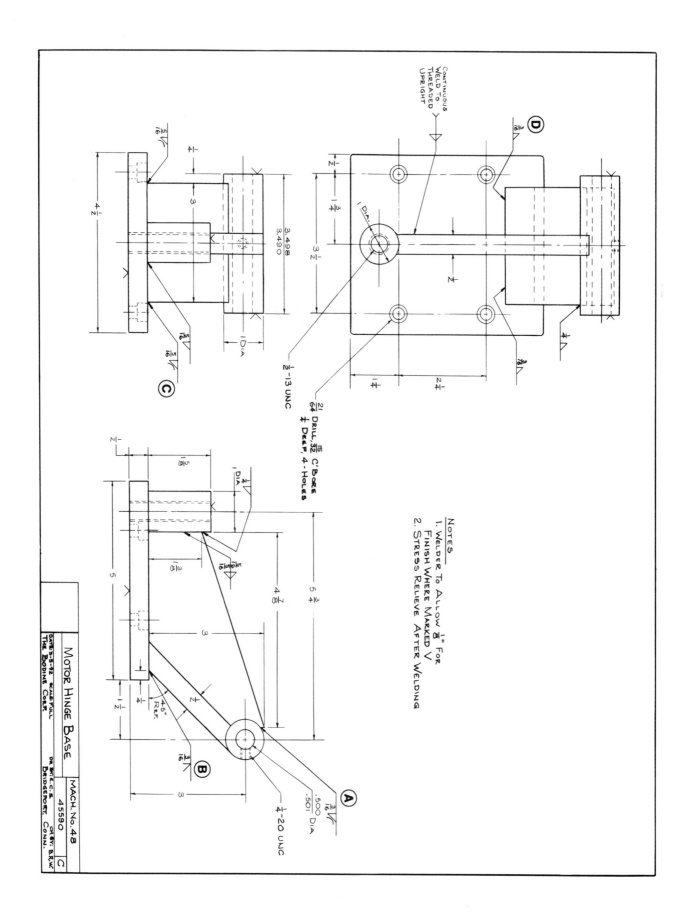

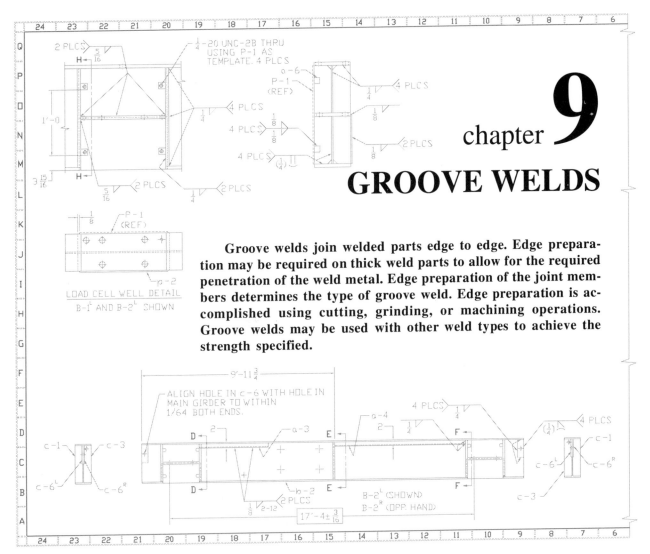

chapter 9

GROOVE WELDS

Groove welds join welded parts edge to edge. Edge preparation may be required on thick weld parts to allow for the required penetration of the weld metal. Edge preparation of the joint members determines the type of groove weld. Edge preparation is accomplished using cutting, grinding, or machining operations. Groove welds may be used with other weld types to achieve the strength specified.

SPECIFICATION

A *groove weld* is a weld type made in the groove of the pieces to be welded. Groove welds specified on prints include information regarding edge preparation, effective throat, and root opening. This information gives the location, shape, and penetration necessary to meet the design requirements of the groove weld. Parts of a groove weld include the weld face, weld toes, weld root, face reinforcement, and root reinforcement. See Figure 9-1.

The *weld face* is the exposed surface of a weld, bounded by the weld toes of the side on which welding was done. The *weld toe* is the intersection of the base metal and the weld face. The *weld root* is the area where the filler metal intersects the base metal opposite the weld face. *Face reinforcement* is filler metal which extends above the surface of

the joint member on the side of the joint on which the welding was done. *Root reinforcement* is filler metal which extends above the surface of the joint member on the opposite side of the joint on which welding was done.

The *root face* is the surface of the groove next to the root. The *groove face* is the surface of the joint member included in the groove of the weld. The *root edge* is a face that comes to a point and has no width.

Fusion in the groove weld occurs where the base metal and the filler metal are melted together. The *weld interface* is the area where the filler metal and the base metal mix together (interface). The *fusion face* is the surface of the base metal that is melted during welding. *Depth of fusion* is the distance from the fusion face to the weld interface.

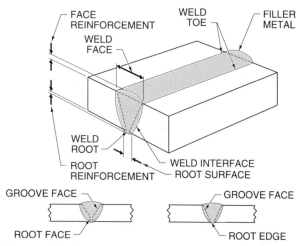

Figure 9-1. Parts of a groove weld include the weld face, weld toes, weld root, face reinforcement, and root reinforcement.

Edge Preparation

Edge preparation of the weld parts is accomplished by cutting, grinding, or machining. Groove welds are identified by joint member configuration and/or by edge preparation of the joint members as they appear in cross-sectional shape. See Figure 9-2.

Single-Square-Groove Weld. A *single-square-groove weld* is a groove weld having square-edged joint members with the weld made from one side. Single-square-groove welds are identified by two parallel, vertical lines on the reference line of the welding symbol.

Double-Square-Groove Weld. A *double-square-groove weld* is a groove weld having square-edged joint members with the weld made from both sides. Double-square-groove welds are identified by two parallel, vertical lines placed on both sides of the reference line of the welding symbol.

Single-V-Groove Weld. A *single-V-groove weld* is a groove weld having both joint members angled on the same side with the weld made from that side. Single-V-groove welds are identified by a V placed on the reference line of the welding symbol.

Double-V-Groove Weld. A *double-V-groove weld* is a groove weld having joint members angled on both sides with the weld made from both sides. Double-V-groove welds are identified by placing a

GROOVE WELDS					
TYPE	SYMBOL	WELD	TYPE	SYMBOL	WELD
SINGLE-SQUARE-GROOVE WELD			DOUBLE-SQUARE-GROOVE WELD		
SINGLE-V-GROOVE WELD			DOUBLE-V-GROOVE WELD		
SINGLE-BEVEL-GROOVE WELD			DOUBLE-BEVEL-GROOVE WELD		
SINGLE-U-GROOVE WELD			DOUBLE-U-GROOVE WELD		
SINGLE-J-GROOVE WELD			DOUBLE-J-GROOVE WELD		
SINGLE-FLARE-V-GROOVE WELD			DOUBLE-FLARE-V-GROOVE WELD		
SINGLE-FLARE-BEVEL-GROOVE WELD			DOUBLE-FLARE-BEVEL-GROOVE WELD		

Figure 9-2. Groove welds are identified by the joint member configuration and/or by the edge preparation of the joint members as they appear in cross-sectional shape.

V placed on both sides of the reference line of the welding symbol.

Single-Bevel-Groove Weld. A *single-bevel-groove weld* is a groove weld having one joint member beveled with the weld made from that side. Single-bevel-groove welds are identified by a V placed on the reference line of the welding symbol with the vertical leg facing left.

Double-Bevel-Groove Weld. A *double-bevel-groove weld* is a groove weld having joint members beveled on both sides with the weld made from both sides. Double-bevel-groove welds are identified by a V placed on both sides of the reference line of the welding symbol with the vertical leg facing left.

Single-U-Groove Weld. A *single-U-groove weld* is a groove weld having joint members grooved in a U shape on one side with the weld made from that side. Single-U-groove welds are identified by a U placed on the reference line of the welding symbol.

Double-U-Groove Weld. A *double-U-groove weld* is a groove weld having joint members grooved in a U shape on both sides with the weld made from both sides. Double-U-groove welds are identified by a U placed on both sides of the reference line of the welding symbol.

Single-J-Groove Weld. A *single-J-groove weld* is a groove weld having joint members grooved in a J shape on one side with the weld made from that side. Single-J-groove welds are identified by a J placed on the reference line of the welding symbol.

Double-J-Groove Weld. A *double-J-groove weld* is a groove weld having joint members grooved in a J shape on both sides with the weld made from both sides. Double-J-groove welds are identified by a J placed on both sides of the reference line of the welding symbol.

Single-Flare-V-Groove Weld. A *single-flare-V-groove weld* is a groove weld having radiused joint members with the weld made from one side. Sin-gle-flare-V-groove welds are identified by two arcs placed on the reference line of the welding symbol.

Double-Flare-V-Groove Weld. A *double-flare-V-groove weld* is a groove weld having radiused joint members with the weld made from both sides. Dou-ble-flare-V-groove welds are identified by two arcs placed on both sides of the reference line of the welding symbol.

Single-Flare-Bevel-Groove Weld. A *single-flare-bevel-groove weld* is a groove weld having one straight joint member and one radiused joint member with the weld made from one side. Single-flare-bevel-groove welds are identified by a vertical line and a disconnected arc placed on the reference line of the welding symbol.

Double-Flare-Bevel-Groove Weld. A *double-flare-bevel-groove weld* is a groove weld having two radiused joint members with the weld made from both sides. Double-flare-bevel-groove welds are identified by a vertical line and a disconnected arc placed on both sides of the reference line of the welding symbol.

Edge preparation dimensions. Edge preparation dimensions indicate the location, depth of preparation, and groove angle. Edge preparation dimensions are dictated by user's standards or are indicated on the welding symbol on the same side as the weld symbol. The dimensions shown pertain to the arrow side or other side. Dimensions for both sides of the joint must be included. See Figure 9-3.

The joint member requiring edge preparation is indicated by a break in the direction of the arrow of the welding symbol. The arrow points to the member requiring edge preparation. Depth of penetration is indicated with a number to the left of the weld symbol. Depth of penetration dimensions must be included with the corresponding weld symbol for both sides, even if the dimensions are the same.

Groove angle is indicated by placing the included angle outside the weld symbol on the welding symbol. Different groove angles required for arrow side and other side are indicated on both sides as required.

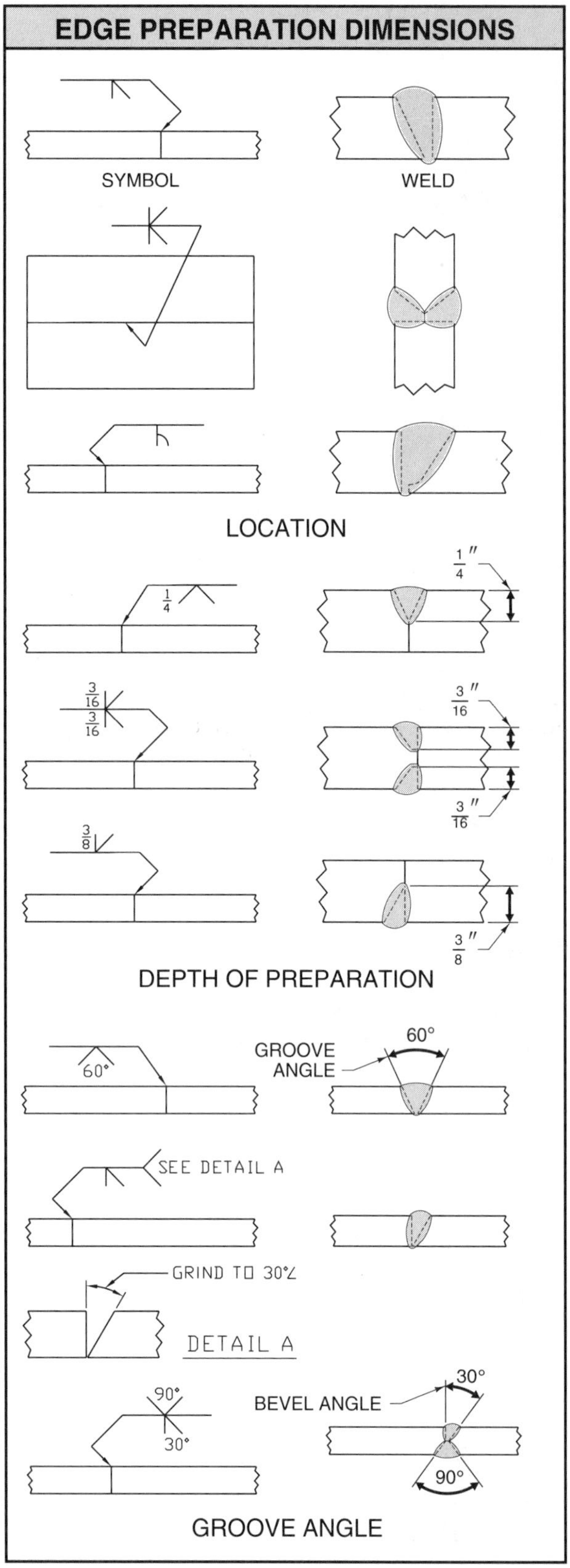

EDGE PREPARATION DIMENSIONS

SYMBOL

WELD

LOCATION

DEPTH OF PREPARATION

GROOVE ANGLE

Figure 9-3. Edge preparation dimensions indicate the location, depth of penetration, and groove angle.

When U- and J-groove welds are not specified by the user's standards, the weld can be shown in cross-sectional view, as a detail, or with notes on prints. Notes can apply a given specification for all or selected welds included in the note. A weld dimension is not included on the welding symbol if a note provides sufficient dimension information.

Penetration

Depth of penetration of the filler metal in the groove weld is indicated by the dimension given in parenthesis to the left of the weld symbol. See Figure 9-4. Penetration may be less or greater than the groove size, depending on the weld specifications. Depth of penetration of a flare-V-groove weld is the distance from the point of tangency to the top of the weld part. Depth of penetration smaller than the groove size indicates penetration less than the groove size. Depth of penetration larger than the groove size indicates penetration exceeding the groove size.

If the depth of penetration is specified with the edge preparation optional, the depth of penetration is shown in parenthesis without the use of a weld symbol. No depth of preparation or depth of penetration used with a weld symbol indicates complete penetration by the weld metal in the weld parts. When weld type and edge preparation are optional, the letters CJP are included in the tail of the welding symbol.

Root Opening

The *root opening* of a weld joint is the distance between joint members at the root of the weld before welding. See Figure 9-5. The size of the root opening has a direct effect on the amount of penetration in the joint members.

The root opening of the weld is specified as a dimension inside the weld symbol. A root opening of 0 (zero) indicates that joint members are butted together. They are in contact with one another and have no root opening. If no root opening dimension is indicated, the root opening is dictated by the user's standards.

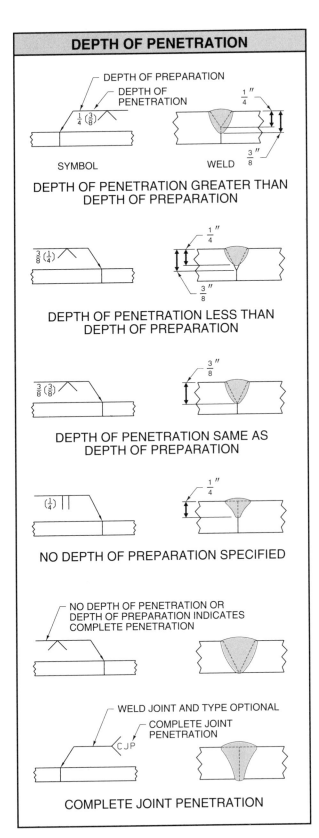

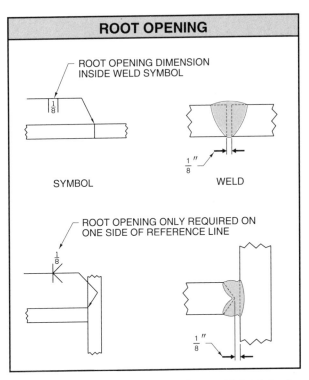

Figure 9-5. The root opening of a weld joint is the distance between joint members at the root of the weld before welding.

Flare-Groove Welds

Flare-groove welds are either flare-V-groove or flare-bevel-groove welds and are specified by the groove weld size and by the groove dimensions. *Groove weld size* is the distance from the weld face to the weld root. Groove dimensions are formed by the distance from the point of tangency or by the radius of one or two joint members. Flare-V-groove welds are specified by groove weld size and the groove dimensions formed by the curved groove faces of two joint members. Flare-bevel-groove welds are specified by groove weld size and groove dimensions formed by the curved groove face on one joint member. See Figure 9-6.

COMBINED WELD SYMBOLS

The groove weld is used in combination with fillet welds and other welds to specify weld size, root opening, weld penetration, and weld edge preparation. The groove weld symbol alone could not specify all of this required information. Multiple weld symbols are combined on one welding symbol

Figure 9-4. Depth of penetration of the filler metal in the groove weld is indicated by the dimension given in parenthesis to the left of the weld symbol.

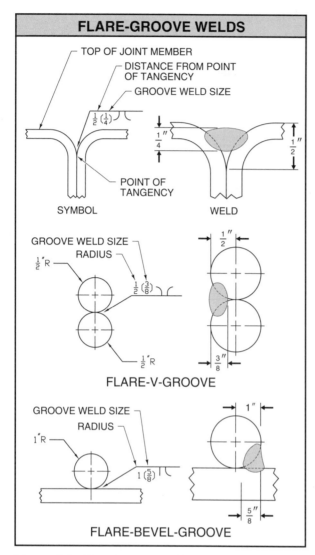

Figure 9-6. Flare-groove welds are specified by indicating groove size and groove dimensions formed by joint members.

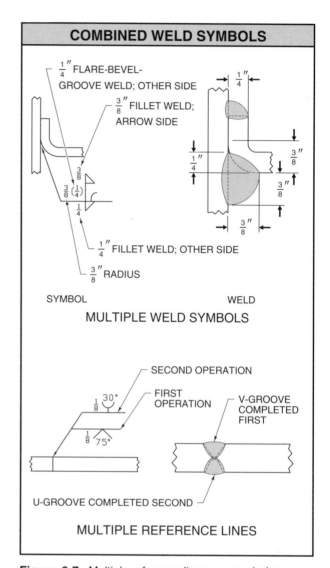

Figure 9-7. Multiple reference lines are used when more than one weld type is required.

reference line when multiple welds are required. Multiple reference lines are used when more than one weld type or operation is required in a specific order. See Figure 9-7.

CONTOUR AND FINISH

Groove weld contour and finish are specified on the welding symbol above or below the weld symbol. Groove welds approximately flush without fin-

ishing use the flush contour weld symbol. If the print specifies a flush, convex, or concave contour, a contour symbol is used.

Groove welds requiring mechanical finishing after welding to obtain the desired contour have a letter above or below the weld contour symbol. The finishing methods that may be specified are C - chipping, H - hammering, G - grinding, M - machining, R - rolling, or U - unspecified. Contour and finish are specified in the welding symbol with the quality or degree of finish specified by the manufacturer.

Review Questions

Name_____ Date _____

Completion

_____ 1. The _____ is the exposed surface of a weld bounded by the weld toes of the side on which the welding was done.

_____ 2. The _____ is the surface of the groove next to the root.

_____ 3. _____ reinforcement is filler metal which extends above the surface of the joint member on the side of the joint on which welding was done.

_____ 4. _____ reinforcement is filler metal which extends above the surface of the joint member on the opposite side of the joint on which welding was done.

_____ 5. The _____ edge is the part of a weld that is a face that comes to a point and has no width.

_____ 6. The weld _____ is the area where the filler metal intersects the base metal opposite the weld face.

_____ 7. The fusion _____ is the surface of the base metal that is melted during welding.

_____ 8. Depth of penetration of a flare-V-groove weld is the distance from the point of _____ to the top of the weld part.

_____ 9. Flare-groove welds are specified by indicating groove sizes and _____.

_____ 10. _____ in a weld occurs when the base metal and the filler metal are melted together.

True-False

T   F   1. The weld toe is the intersection of the base metal and the weld face.

T   F   2. Groove welds join parts edge to edge.

T   F   3. Depth of penetration is indicated with a number to the right of the weld symbol.

T   F   4. Groove angle is indicated by placing the included angle inside the weld symbol.

T   F   5. When weld type and edge preparation are optional, the letters CJP are included in the tail of the welding symbol.

T   F   6. The root opening of a weld is the distance between joint members at the root of the weld after welding.

T   F   **7.** Groove weld contour and finish are specified on the welding symbol above or below the weld symbol.

T   F   **8.** On round joint members welded to other round joint members, groove size is the radius of the round joint member.

T   F   **9.** On round joint members welded to flat members, groove size is the diameter of the round joint member.

T   F   **10.** Multiple reference lines are used when more than one weld type or operation is required in a specific order.

Matching — Groove Weld Symbols

_____ **1.**

_____ **2.**

_____ **3.**

_____ **4.**

_____ **5.**

_____ **6.**

_____ **7.**

_____ **8.**

_____ **9.**

_____ **10.**

Matching — Groove Weld Parts

_____ **1.** Weld interface

_____ **2.** Root reinforcement

_____ **3.** Weld face

_____ **4.** Filler metal

_____ **5.** Weld root

_____ **6.** Root surface

_____ **7.** Face reinforcement

_____ **8.** Weld toe

_____ **9.** Base metal

Trade Competency Test

Name_____ Date _____

Refer to the Bucket Assembly print on page 201.

_____ 1. Weld H specifies a(n) _____" fillet weld leg size.

_____ 2. Dimension D specifies a maximum angle of _____°.

_____ 3. Weld I specifies that a(n) _____ weld be completed first.

_____ 4. Weld J specifies a(n) _____ weld on both sides.

_____ 5. Weld B specifies that the weld shall be _____.

_____ 6. Weld E specifies a(n) _____ mm fillet weld leg size.

_____ 7. The maximum diameter of hole E is _____ mm.

_____ 8. Section A-A is a(n) _____-joint.

T   F 9. Weld A specifies a square-groove weld, bevel-groove weld, and fillet weld on both sides.

_____ 10. Weld G specifies a(n) _____ weld.

T   F 11. The bucket capacity is rated for 8 cu yd.

T   F 12. Weld F has a 9.7 mm leg size.

T   F 13. The radius of K is 533".

_____ 14. The contour of weld G is _____.

_____ 15. The vertical distance from center line of hole D to center line of hole E is _____ mm.

_____ 16. The maximum diameter of hole E is _____ mm.

_____ 17. Weld A specifies a(n) _____" bevel-groove depth of preparation.

_____ 18. Weld J specifies a(n) _____ fillet weld leg size.

_____ 19. Weld I specifies a(n) _____ weld after the bevel-groove weld.

_____ 20. Hole E is drilled through both _____ at two places.

_____ 21. Weld areas are to be _____ to 350° − 400° where designated.

_____ **22.** Weld C specifies a(n) _____″ fillet weld leg size.

_____ **23.** The threaded hole above hole D has _____ tpi.

T   F   **24.** Weld G specifications are per E1-042 MS208.

T   F   **25.** Dimension L is the vertical distance from the center line of hole D to the center line of hole E.

Refer to the Vault X-32 print (Views, Section Views) on page 202.

_____ **1.** Weld J specifies a(n) _____ weld.

_____ **2.** Weld B specifies a(n) _____″ fillet weld leg size.

_____ **3.** Section C-C provides a view from the _____ side.

_____ **4.** Weld A specifies fillet welds _____″ in length.

_____ **5.** Weld G specifies a(n) _____ contour.

_____ **6.** Weld C specifies a(n) _____ weld, arrow side.

T   F   **7.** Section B-B provides a view from the right side.

T   F   **8.** Weld B specifies welds $13\frac{1}{4}″$ in length.

T   F   **9.** Weld I specifies a $\frac{1}{8}″$ root opening.

_____ **10.** Weld D specifies a flush contour by _____.

_____ **11.** Weld F specifies fillet welds spaced on _____″ centers.

_____ **12.** The overall width of the vault is _____″.

_____ **13.** Weld E specifies a(n) _____ weld.

_____ **14.** Section _____ shows the overall depth of the vault.

_____ **15.** Weld D is _____″ long.

_____ **16.** Weld G is completed on the _____ side of the vault.

_____ **17.** The overall height of the vault is _____″.

_____ **18.** Weld K specifies the same weld as weld _____.

T   F   **19.** Weld H is completed on the front side of the vault.

T   F   **20.** Section A-A shows the vault to be symmetrical.

Refer to the Vault X-32 print (Details) on page 203.

T F **1.** Detail II specifies intermittent fillet welds to secure the interior door stops.

_____ **2.** The detail of the hinge tab weldment is drawn to _____ scale.

_____ **3.** The interior doors are _____" wide.

_____ **4.** Detail II is drawn to _____ scale.

_____ **5.** The hinge tab is joined with _____ welds.

T F **6.** Detail II shows the side and bottom plate joints.

T F **7.** The hinge tab contains a hole of unspecified diameter.

T F **8.** The height of the interior doors is 6⅛".

_____ **9.** The interior door stops are joined with fillet welds spaced on _____" centers.

_____ **10.** Detail I specifies a(n) _____" fillet weld leg size.

_____ **11.** The interior doors are _____" thick.

_____ **12.** The width of the top door is _____".

_____ **13.** Detail I specifies welds _____" in length.

_____ **14.** The top door is _____" tall.

_____ **15.** Detail II specifies a(n) _____" fillet weld leg size.

T F **16.** The top view of the interior doors is shown.

T F **17.** The width of the top and bottom doors is the same.

T F **18.** The thickness of the top door is 1⁷⁄₁₆".

_____ **19.** Detail I specifies fillet welds on _____" centers.

_____ **20.** The height of the bottom door is _____".

_____ **21.** Part number 21 shows a(n) _____ view.

_____ **22.** The interior door stops are set back _____".

T F **23.** Detail I is drawn to full scale.

T F **24.** Four interior door stops are required.

T F **25.** A ¼" groove weld joins the door stop to the vertical divider.

Refer to the Vault X-32 print (Bill of Materials, Welding Symbol Notes, and Title Block) on page 204.

_____ 1. A total of _____ different items are listed on the Bill of Materials.

_____ 2. The drawing was completed on _____.

_____ 3. A total of _____ shelf standards are required.

_____ 4. Part number 17 specifies _____ gauge metal.

_____ 5. The drawing scale is _____, unless noted.

T   F   6. All door stops are made from ½″ square stock.

T   F   7. A total of two horizontal spacers are required.

T   F   8. Note 8 specifies the leg size of all fillet welds.

_____ 9. The drawing was completed on a(n) _____ size sheet.

_____ 10. The top and _____ pieces of the vault are the same size.

_____ 11. The drawing was drawn by _____.

_____ 12. Note 3 specifies _____ welds.

_____ 13. Note 7 specifies welds at the same location in both _____.

_____ 14. The interior door stops are _____″ in length.

_____ 15. Note 2 refers to the location of _____″ welds.

T   F   16. The drawing number for the vault is D-001004-S.

T   F   17. The interior shims are ¾″ wide.

T   F   18. All tolerances for fraction dimensions are ±1/16″.

_____ 19. The hinge tabs are _____″ thick.

_____ 20. Note 5 refers to welds made on the _____ of the safe.

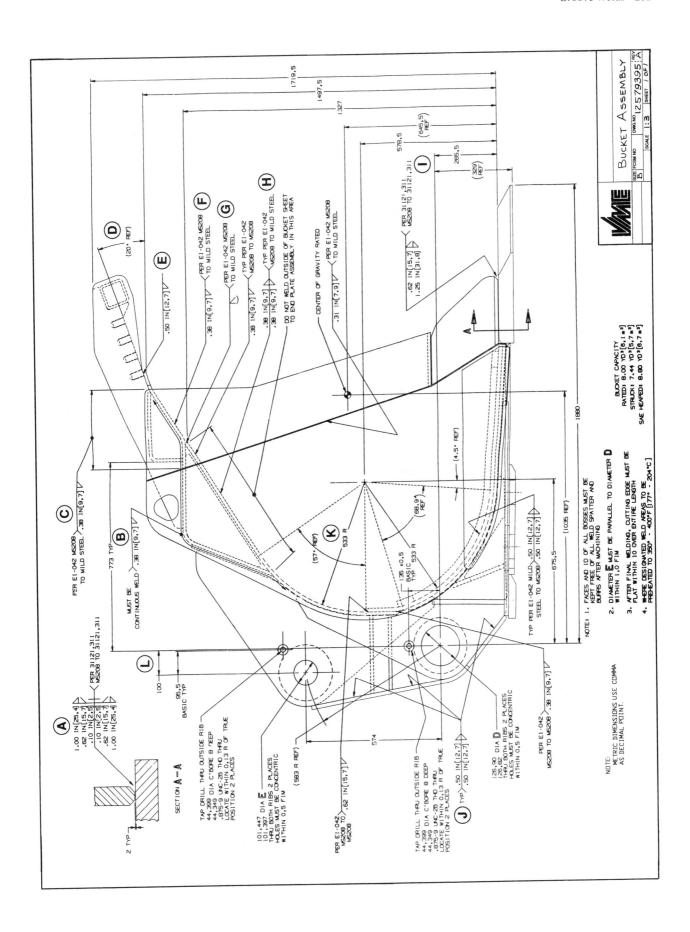

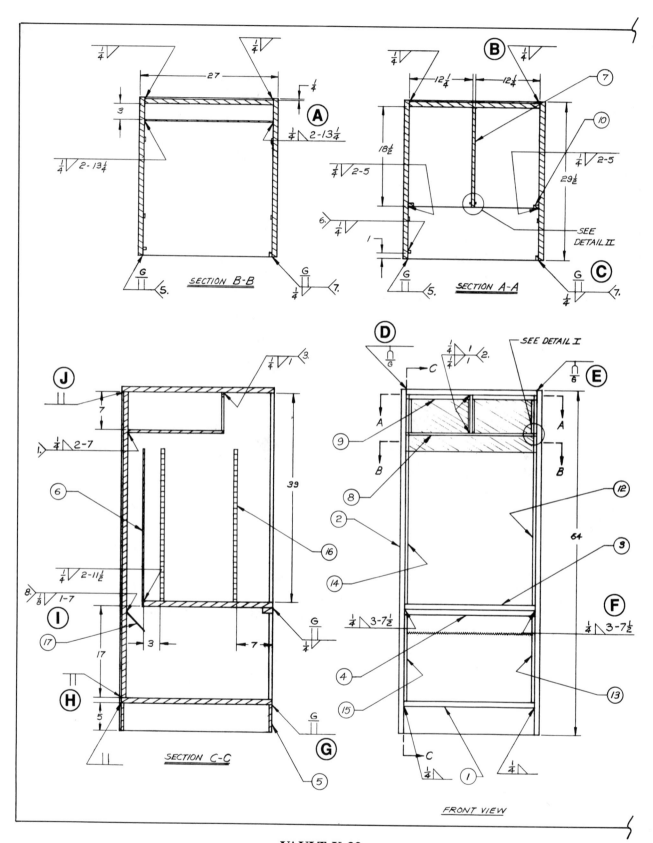

VAULT X-32
(Views, Section Views)

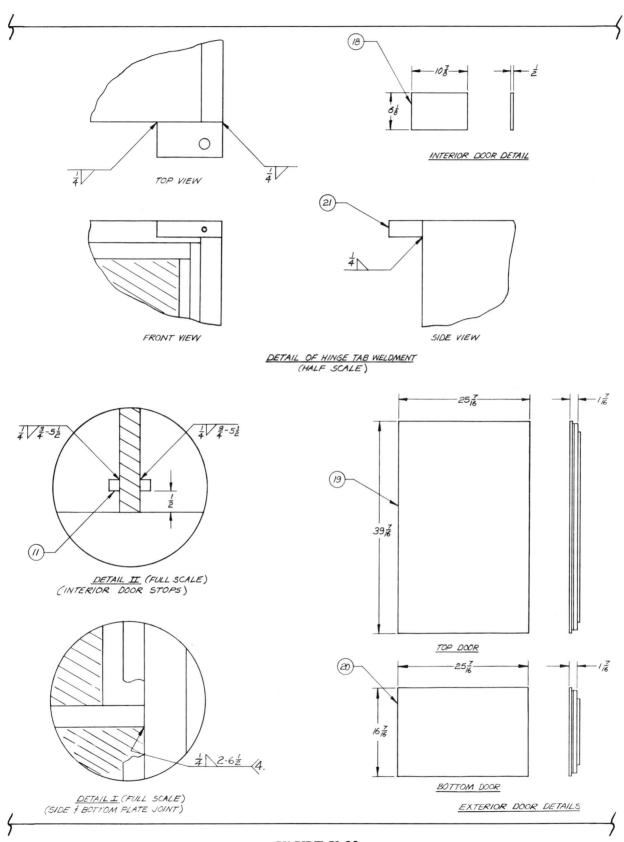

TOP VIEW

INTERIOR DOOR DETAIL

FRONT VIEW

SIDE VIEW

DETAIL OF HINGE TAB WELDMENT
(HALF SCALE)

DETAIL II (FULL SCALE)
(INTERIOR DOOR STOPS)

TOP DOOR

DETAIL I (FULL SCALE)
(SIDE & BOTTOM PLATE JOINT)

BOTTOM DOOR

EXTERIOR DOOR DETAILS

VAULT X-32
(Details)

ITEM	SIZE, DESCRIPTION, MATERIAL	QTY.
1	1 X 29 1/2 X 24 15/16 - TOP & BOTTOM, STL. PLT.	2
2	1 X 29 1/2 X 64 - SIDE, STL. PLT.	2
3	1 X 25 1/2 X 24 15/16 - HORIZONTAL DIV., STL. PLT.	1
4	1 X 2 X 24 15/16 - HORIZONTAL SPACER, STL. PLT.	1
5	1/2 X 4 7/8 X 24 7/8 - BOTT. FRONT & BACK PLT., STL. PLT.	2
6	1/4 X 28 1/2 X 24 7/8 - CHUTE DIV., STL. PLT.	1
7	1/2 X 7 X 18 1/2 - VERTICAL DIV., STL. PLT.	1
8	1/2 X 18 1/2 X 24 15/16 - INT. COMPT. BOTT. PLT., STL. PLT.	1
9	1/2 X 3/4 X 12 3/16 - INT. SHIM, H.R.S.	2
10	1/2 X 1 X 6 3/16 - INT. SIDE SHIM., C.R.S.	2
11	1/4 X 1/4 X 6 1/2 - INT. DOOR STOP, C.R.S.	2
12	1/2 X 3/4 X 38 15/16 - EXT. DOOR SHIM, C.R.S.	1
13	1/2 X 3/4 X 15 15/16 - EXT. DOOR SHIM, C.R.S.	1
14	1/2 X 1/2 X 38 15/16 - EXT. DOOR STOP, C.R.S.	1
15	1/2 X 1/2 X 15 15/16 - EXT. DOOR STOP, C.R.S.	1
16	SHELF STD. X 28 1/2	4
17	16 GA. X 24 7/8, ANTI FISH	1
18	1/2 X 10 7/8 X 6 1/8, - INT. DOOR, H.R.S. PLT.	2
19	1 7/16 X 25 7/16 X 39 7/16 - EXT. DOOR, LAMINT.	1
20	1 7/16 X 25 7/16 X 16 7/16 - EXT. DOOR, LAMINT.	1
21	3/4 X 3 1/8 X 1 5/8 - HINGE TAB, C.R.S.	4

WELDING SYMBOL NOTES:

1.) THE CENTER LINE OF EACH WELD IS 7" IN FROM EACH SIDE.

2.) (2), 1" WELDS, EACH ONE STARTING IN, 2" FROM THE FRONT & BACK.

3.) (3), 1" WELDS ON 5 1/2" CENTERS, BOTH SIDES OF THE VERTICAL DIVIDER.

4.) TYPICAL AT SAME LOCATION ON OTHER SIDE OF SAFE.

5.) TYPICAL OF ALL WELDED SURFACES ON THE FACE OF THE SAFE.

6.) (3) PLACES FREE OF WELD, (1) 2" SPACE CENTERED ON THE ℄ OF THE COMPARTMENT & THE OTHER (2) SPACES CENTERED ON 17 1/2" ℄'S ABOVE & BELOW THE ℄ OF THE COMPARTMENT.
-AT SAME LOCATION IN LOWER COMPARTMENT:
(3) PLACES FREE OF WELD, (1) 2" SPACE CENTERED AT THE ℄ OF THE COMPARTMENT & THE OTHER (2) SPACES CENTERED ON 5 1/2" ℄'S ABOVE & BELOW THE ℄ OF THE COMPARTMENT.

7.) TYPICAL AT SAME LOCATION IN BOTH COMPARTMENTS.

8.) SIDES OF BAFFLE RECEIVE 1/8" CONTINUOUS FILLET WELD.

REV.	DESCRIPTION	DATE	BY
	CHANGE		

DRAWN BY K.A. DATE 10/17/91

APPROVED BY DATE

QUOTE REF. -5-253-84(B)

SPECIAL NOTE:

UNLESS OTHERWISE SPECIFIED DIMENSIONS ARE IN INCHES. TOLERANCES ARE:
FRACTIONS ±. DECIMALS .XX ± .XXX ± ANGLES ±

DO NOT SCALE DWG.

ALLIED/GARY SAFE & VAULT CO., INC.
SPOKANE, WASHINGTON - CINCINNATI, OHIO

REF. DWG. NO. C-000470-S

VAULT X-32

SCALE 1/8, UNLESS NOTED | D | DRAWING NO. D-007004-S

VAULT X-32
(Bill of Materials, Welding Symbol Notes, and Title Block)

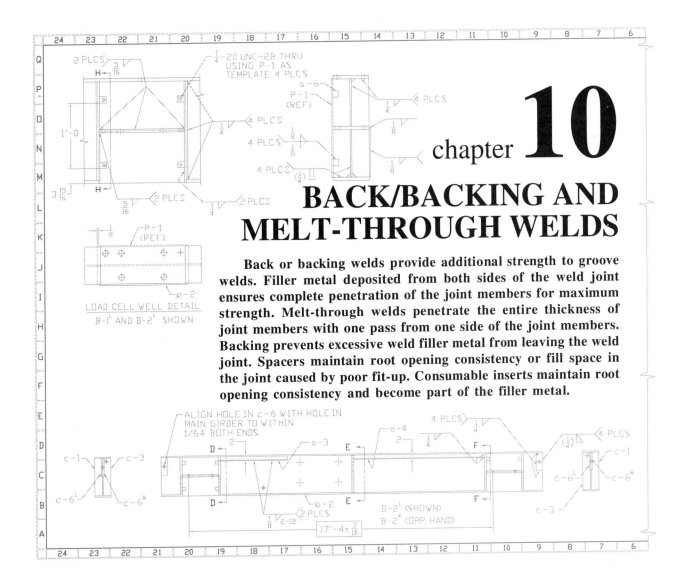

LOAD CELL WELL DETAIL
B-1^L AND B-2^L SHOWN

chapter 10

BACK/BACKING AND MELT-THROUGH WELDS

Back or backing welds provide additional strength to groove welds. Filler metal deposited from both sides of the weld joint ensures complete penetration of the joint members for maximum strength. Melt-through welds penetrate the entire thickness of joint members with one pass from one side of the joint members. Backing prevents excessive weld filler metal from leaving the weld joint. Spacers maintain root opening consistency or fill space in the joint caused by poor fit-up. Consumable inserts maintain root opening consistency and become part of the filler metal.

BACK OR BACKING WELDS

Back or backing welds are welds made on the opposite side of a groove weld. A *back weld* reinforces a groove weld and provides complete penetration through the thickness of the joint members. It is completed after the groove weld is completed. A *backing weld* supports and retains weld filler metal deposited on the opposite side of the groove joint. It is completed before the groove weld is completed. The backing weld is fused with subsequent groove welds for complete penetration through the joint members. See Figure 10-1.

Back welds and backing welds are indicated on prints using a weld symbol in the shape of a hollow half circle. The flat side of the weld symbol is positioned on the reference line directly opposite of the groove weld symbol. Because the weld symbol is the same for the back weld and the backing weld, a note designating the weld type is included in the tail of the welding symbol. For example, the note BACK WELD placed in the tail of the welding symbol indicates a back weld. The note BACKING WELD placed in the tail of the welding symbol indicates a backing weld. A multiple reference line or note on the print is also used to indicate back or backing welds by the sequence of welding operations required. See Figure 10-2.

The height of a back or backing weld may be specified with a dimension placed to the left of the weld symbol. For example, $^3/_{16}$ placed to the left of the back or backing weld symbol indicates a height of $^3/_{16}''$. If any other dimensions are required, they are shown as part of the drawing. See Figure 10-3.

Edge preparation required for the joint members is specified in a note placed in the tail of the welding symbol or in a note on the print. For example, BACK GOUGE placed in the tail of the welding symbol indicates that the groove weld is back gouged. See Figure 10-4.

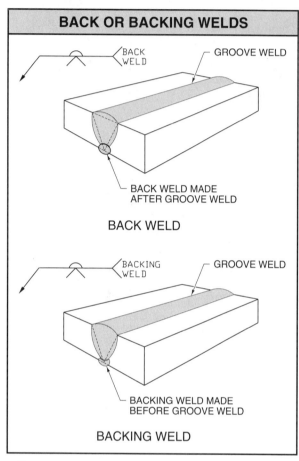

BACK OR BACKING WELDS

BACK WELD

BACKING WELD

Figure 10-1. Back welds are made after groove welds. Backing welds are made before groove welds.

CONTOUR AND FINISH

Weld contour and finish symbols are placed above or below the back or backing weld symbol. If the joint design requires a flush or convex weld contour, a contour symbol is used. Mechanical finishing methods used after welding to obtain the desired contour are designated by a letter placed above or below the weld contour symbol.

The mechanical finishing methods are C - chipping, H - hammering, G - grinding, M - machining, R - rolling, and U - unspecified. Contour and finish are specified in the welding symbol with the quality or degree of finish specified by the manufacturer.

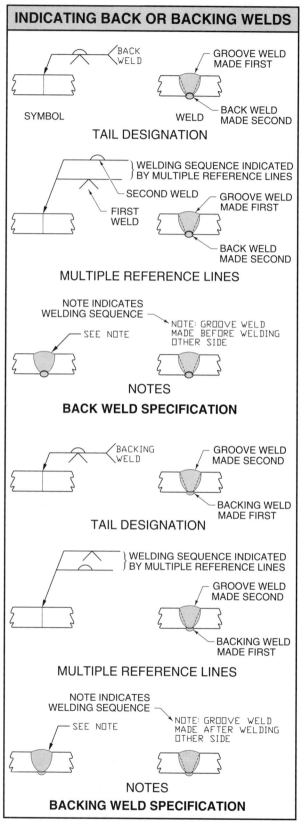

INDICATING BACK OR BACKING WELDS

TAIL DESIGNATION

MULTIPLE REFERENCE LINES

NOTES

BACK WELD SPECIFICATION

TAIL DESIGNATION

MULTIPLE REFERENCE LINES

NOTES

BACKING WELD SPECIFICATION

Figure 10-2. A back or backing weld is indicated by a weld symbol in the shape of a hollow half circle.

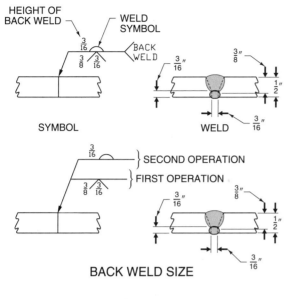

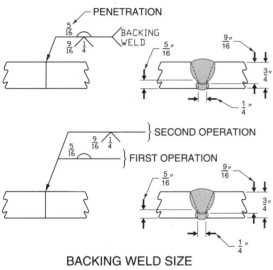

Figure 10-3. The height dimension of a back or backing weld is placed to the left of the weld symbol.

MELT-THROUGH WELDS

A *melt-through weld* is a weld specified to have complete penetration through the joint members to the opposite side of the joint members. The melt-through weld symbol is the same shape as the back or backing weld symbol and is solid instead of hollow. Like the back or backing weld symbol, the melt-through weld symbol is located opposite the weld symbol.

Melt-through welds are commonly specified with groove welds. The height of root reinforcement of the melt-through weld is specified with a dimen-

sion placed to the left of the melt-through weld symbol. Contour of the melt-through weld is specified using contour symbols. See Figure 10-5.

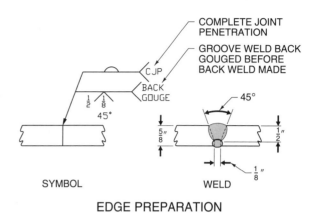

Figure 10-4. Edge preparation of joint members may be specified by a note in the tail of the welding symbol or in a note on the print.

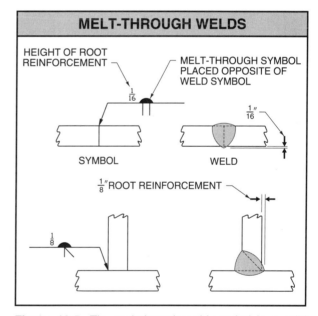

Figure 10-5. The melt-through weld symbol is a solid half circle.

BACKING OR SPACERS

Backing is a ring, strip, plate, or bar used on the opposite side of the weld to prevent excessive penetration during the welding process and/or to reinforce the weld joint. Backing is used in combination with a groove weld to prevent excessive weld filler metal from leaving the weld joint. Backing may or may not be fused with the weld and can be removed after welding as required.

Spacers are metal pieces placed in the root of the weld joint to maintain consistent root opening, provide consistent weld filler metal characteristics, and compensate for poor fit-up conditions. Spacers are consumed in the weld metal during the welding process.

Backing and spacers are indicated on the welding symbol with a rectangle. The location of the rectangle determines whether a spacer or backing is specified. A spacer symbol is centered on the reference line of the welding symbol.

A backing symbol is located either above or below the reference line of the weld symbol. The backing symbol is used with a groove weld symbol to prevent confusion with a plug or slot weld. See Figure 10-6.

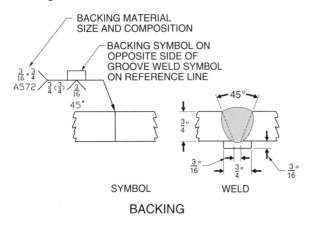

BACKING

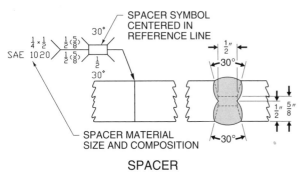

SPACER

Figure 10-6. Backing and spacers are indicated on the welding symbol with a rectangle.

Spacer specifications regarding size may also be specified in a note. A joint with spacer is specified by placing the spacer symbol inside the groove weld symbol. The spacer material consumed in the weld is indicated by a note keyed to the welding symbol.

Backing specifications are indicated in the tail of the welding symbol or with a note to specify materials and dimensions. The letter R in the backing symbol indicates the backing is to be removed after the weld is complete. See Figure 10-7.

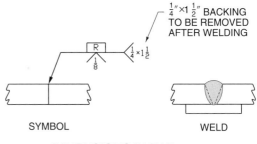

DIMENSIONS IN TAIL

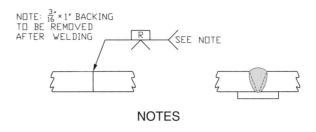

NOTES

Figure 10-7. Backing specifications are indicated by dimensions in the tail of the welding symbol or in a note.

CONSUMABLE INSERTS

A *consumable insert* is a spacer that provides the proper opening of the weld joint and becomes part of the filler metal during the welding process. A consumable insert is placed in the root of the weld to provide the proper fit-up and root opening of the joint members. During the welding process, the consumable insert melts to become part of the filler metal. This reduces the possibility of weld defects and improves the mechanical characteristics of the weld joint.

Consumable inserts are primarily used with the GTAW process in pipe welding. They are also used in structural and pressure vessel applications where welding is performed from one side of the joint.

Edge preparation of the joint members must be compatible with the consumable insert specified. Weld joint dimensions, such as root openings, are sized to include the consumable insert. Generally, a smaller root opening is used with a consumable insert because root reinforcement is already in place. This reduces the possibility of undesired melt-through and improves weld integrity.

The consumable insert symbol is indicated by a square on the opposite side of the groove weld symbol on the reference line. The location of the consumable insert symbol on the welding symbol is the same as the backing weld.

The AWS consumable insert class and other information is included in the tail of the welding symbol. For example, the classification numbers 1, 2, 3, 4, or 5 may be placed in the tail of the welding symbol to indicate the shape of the consumable insert. Consumable inserts are specified in the AWS publication "Specification for Consumable Inserts" (ANSI/AWS A5.30). See Figure 10-8.

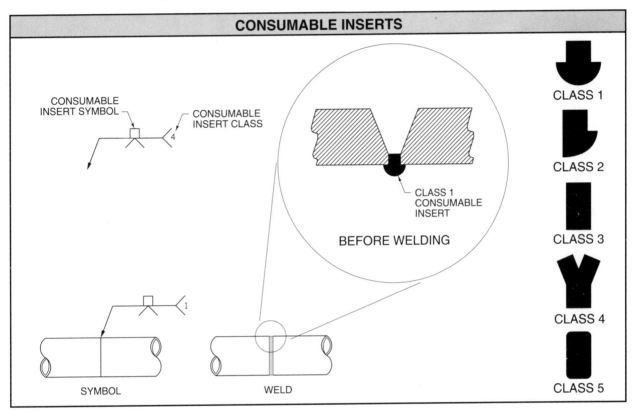

Figure 10-8. Consumable inserts are designed for specific edge preparations and are commonly used in pipe welding.

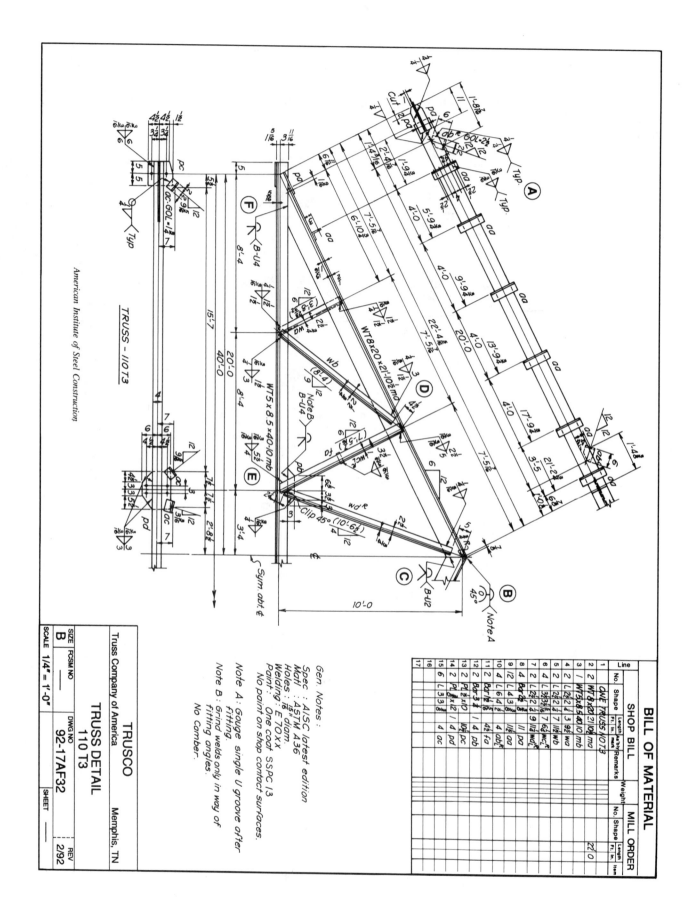

American Institute of Steel Construction

TRUSS — 110 T3

Gen. Notes :

Spec : AISC latest edition
Matl : ASTM A36
Holes : 15/16" diam.
Welding : E 70xx
Paint : One coat SSPC 13
No paint on shop contact surfaces.
Note A: Gouge single U groove after fitting.
Note B : Grind welds only in way of fitting angles.
No Camber.

BILL OF MATERIAL

	SHOP BILL				MILL ORDER		
Line	No.	Shape	Length Ft. In.	Ar'b'r/Remarks	Weight No. Shape	Length Ft. In.	Item
1	1			ONE TRUSS 110T3			
2	2			W18x20 21'-10½ ma		22'-0	
3	1			WT5x8.5 40.10 mb			
4	2	L 2½		wa			
5	2	L 2½ 2¼		wb			
6	4	L 2½ 2¼		wc			
7	2	L 2½ 2¼		wd			
8	4	Pd ½		aa			
9	2	L 4 3		pa			
10	4	L 6 4		ab			
11	2	Bar½		fa			
12	2	Pl ½x10		pb			
13	1	Bar-3		pc			
14	2	Pl ½x12		pd			
15	4	L 3 3		oc			
16							
17							

SIZE	FCSM NO	DWG NO	REV
B	—	**92-17AF32**	**2/92**

TRUSCO
Truss Company of America   Memphis, TN

TRUSS DETAIL
110 T3

SCALE	SHEET
1/4" = 1'-0"	—

Review Questions

Name _____ Date _____

Completion

_____ 1. Back or backing welds provide additional strength to _____ welds.

_____ 2. Back or backing welds are made on the _____ side of a groove weld.

_____ 3. A back weld is completed _____ the groove weld is completed.

_____ 4. A backing weld is completed _____ the groove weld is completed.

_____ 5. The height of a back or backing weld may be specified with a dimension placed to the _____ of the weld symbol.

_____ 6. The melt-through weld symbol is the same _____ as the back or backing weld symbol.

_____ 7. The height of root reinforcement of a melt-through weld is specified with a dimension placed to the _____ of the weld symbol.

_____ 8. _____ is used on the opposite side of a weld to prevent excessive penetration.

_____ 9. A consumable _____ is a spacer that provides the proper opening of the weld joint and becomes part of the filler metal during the welding process.

_____ 10. _____ are metal pieces placed in the root of the weld joint to maintain a consistent root opening.

True-False

T   F   1. A back weld reinforces a groove weld and provides complete penetration through the thickness of the joint members.

T   F   2. The same weld symbol is used for back welds and backing welds.

T   F   3. Mechanical finishing methods used after welding to obtain the desired contour are designated by a letter placed on the right side of the weld contour symbol.

T   F   4. Consumable inserts are primarily used with the GTAW process in pipe welding.

T   F   5. Spacer specifications regarding size are always given in the tail of the welding symbol.

T   F   6. The consumable insert symbol is indicated by a triangle on the opposite side of the groove weld symbol on the reference line.

T   F   7. Backing prevents excessive weld filler metal from leaving the weld joint.

T　　F　　**8.** Multiple reference lines may be used to indicate back or backing welds by the sequence of welding operations required.

T　　F　　**9.** The mechanical finishing method may be unspecified.

T　　F　　**10.** Backing may or may not be fused with the weld.

T　　F　　**11.** Melt-through welds penetrate the entire thickness of joint members with one pass.

T　　F　　**12.** Edge preparation required for a joint member may be specified in a note placed in the tail of the welding symbol.

T　　F　　**13.** Weld contour and finish symbols are always placed above the back or backing weld symbol.

T　　F　　**14.** Backing and spacers are indicated on the welding symbol with a rectangle.

T　　F　　**15.** Six classes of consumable inserts are commercially available.

Matching — Welding Symbols

_____　**1.** $3/16''$ root reinforcement

_____　**2.** $3/16''$ root opening of groove weld

_____　**3.** Back weld

_____　**4.** Spacer symbol

_____　**5.** First weld

_____　**6.** Second weld

_____　**7.** Melt-through weld

_____　**8.** $3/16''$ height of back or backing weld

_____　**9.** Backing symbol

_____　**10.** Reference line

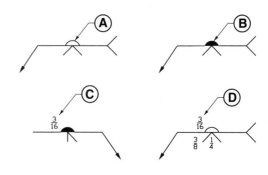

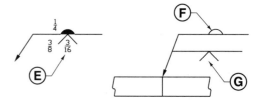

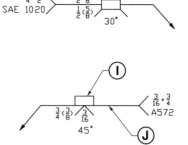

Trade Competency Test

Name_____ Date _____

Refer to the Truss Detail print on page 210.

_____ **1.** Parts aa are joined to part ma with _____ welds.

_____ **2.** Parts aa are _____″ in length.

T   F   **3.** The structural shape of parts aa is angle.

_____ **4.** Weld A specifies fillet welds with a(n) _____″ leg size, both sides.

_____ **5.** Weld B specifies a(n) _____″ root opening.

_____ **6.** Weld B specifies a(n) _____° groove angle.

_____ **7.** Weld B specifies a(n) _____ weld after the U-groove weld.

_____ **8.** Weld C specifies welds to be _____ only in way of fitting angles.

_____ **9.** Weld C specifies a(n) _____ weld opposite of the back weld.

T   F   **10.** Weld D specifies ¼″ welds to be 3″ in length and ³⁄₁₆″ welds to be 1½″ in length.

T   F   **11.** Weld E specifies ³⁄₁₆″ fillet welds.

_____ **12.** A(n) _____ electrode is specified for joining truss parts.

_____ **13.** Weld F specifies a back weld and a(n) _____ weld.

_____ **14.** Part pc is joined to part mb with fillet welds _____″ in length.

_____ **15.** Part ac is joined to part pc with ¼″ fillet welds welded-_____.

_____ **16.** Holes in part pc are _____″ in diameter.

_____ **17.** Part pc is _____″ thick.

_____ **18.** Part wc rl is angle with _____″ legs.

_____ **19.** Part pd is _____″ thick.

_____ **20.** Weld e joins part mb and part _____.

T   F   **21.** Shop bill line number 9 calls for 14 pieces of angle.

T   F   **22.** ASTM A36 material is used on all truss parts.

T   F   **23.** Weld E specifies welds $5\frac{1}{2}''$ in length, both sides.

T   F   **24.** Part wa and part ma are joined with fillet welds $\frac{3}{16}''$ in length.

T   F   **25.** Part pb and mb are joined with a bevel-groove weld.

_____ **26.** The height of the truss at the center line is _____ $''$.

_____ **27.** The truss rises _____ $''$ per foot.

T   F   **28.** A mill order is specified for all structural steel.

T   F   **29.** Six $3 \times 3 \times \frac{3}{8}$ angles are required.

_____ **30.** The truss is _____ about the center line.

T   F   **31.** The truss is fabricated by TRUSCO.

_____ **32.** The identifying number of the truss is _____.

_____ **33.** _____ lines are used to discontinue truss parts to the right of the center line.

_____ **34.** The size of the print sheet is $11'' \times$ _____ $''$.

_____ **35.** Each piece of $2\frac{1}{2} \times 2 \times \frac{1}{4}$ angle is _____ in length.

_____ **36.** The WT 8×20 beams have a finished length of _____.

_____ **37.** Parts aa are spaced on _____ centers.

T   F   **38.** One coat of paint is required.

_____ **39.** The WT 5×8.5 beam is _____ in length.

T   F   **40.** The largest web thickness of angle used is $\frac{1}{2}''$.

T   F   **41.** No paint shall be applied to shop contact surfaces.

_____ **42.** The scale of the drawing is _____.

_____ **43.** A total of _____ pieces of $3\frac{1}{2} \times 3\frac{1}{2} \times \frac{5}{16}$ angle are required.

T   F   **44.** The Truss Company of America is located in Nashville, TN.

_____ **45.** The Truss Detail print was last reused on _____.

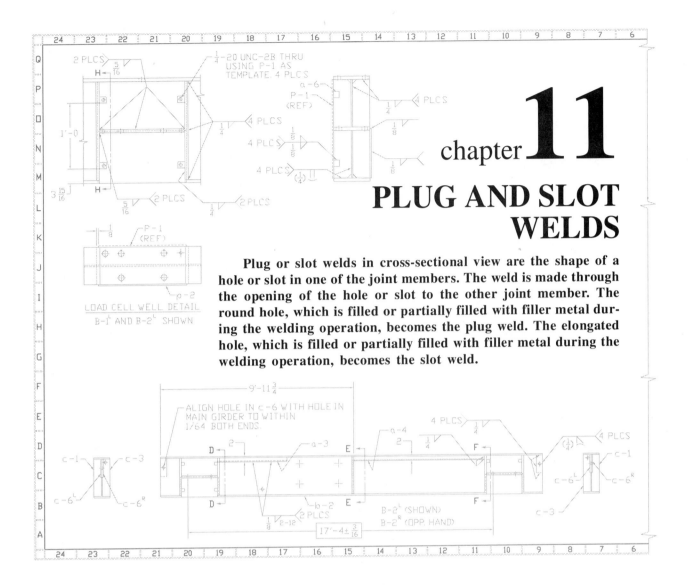

chapter 11

PLUG AND SLOT WELDS

Plug or slot welds in cross-sectional view are the shape of a hole or slot in one of the joint members. The weld is made through the opening of the hole or slot to the other joint member. The round hole, which is filled or partially filled with filler metal during the welding operation, becomes the plug weld. The elongated hole, which is filled or partially filled with filler metal during the welding operation, becomes the slot weld.

APPLICATIONS

A *plug weld* is a weld type in the cross-sectional shape of a hole in one of the joint members. A *slot weld* is a weld type in the cross-sectional shape of a slot (elongated hole) in one of the joint members. Plug and slot welds are determined by their hole shapes.

Plug and slot welds are commonly used when the weld must not be seen from one side. Additionally, plug and slot welds are used where access is limited to one side of the weld joint.

Plug and slot welds should not be confused with the fillet weld, as the base of the hole or slot is filled. In a fillet weld, filler metal is deposited in a triangular shape on the perimeter of the hole or slot. See Figure 11-1.

SPECIFICATION

Plug or slot welds are indicated on the welding symbol with the same weld symbol. The weld symbol for a plug or slot weld is in the shape of a rectangle. Joint member dimensions and notes included on the prints indicate whether a plug or slot weld is required.

The plug or slot weld symbol is placed on the arrow side or other side, depending on the joint member preparation and configuration. The plug or slot weld symbol is placed below the reference line if the hole or the slot is in the arrow side member of the joint. The plug or slot weld symbol is placed above the reference line if the hole or the slot is in the other side member of the joint. See Figure 11-2.

215

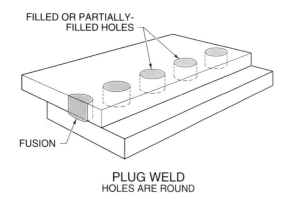

PLUG WELD
HOLES ARE ROUND

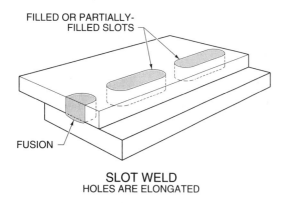

SLOT WELD
HOLES ARE ELONGATED

Figure 11-1. Plug or slot welds are determined by their hole shape.

For clarity, detail drawings and section-view drawings may be used to enhance or to provide additional information to the information provided in the welding symbol. For example, section-view drawings can clearly show the cross-sectional shape of the object being welded.

Plug and slot welds generally are used on joints with two joint members. Where more than one joint member is welded to another joint member, section views provide joint member preparation requirements. See Figure 11-3.

Plug Weld Dimensions

All plug weld dimensions are provided on the side of the reference line with the weld symbol. Dimensions that apply to plug welds include plug weld size, angle of countersink, depth of filling, and pitch. See Figure 11-4.

Size. *Plug weld size* is the diameter of the hole through the joint member at the faying surface of the weld joint. *Faying surface* is the part of the joint member which is in full contact prior to welding.

Plug weld size is specified with a number placed to the left of the plug weld symbol. For example, a $\frac{3}{8}$ dimension placed to the left of the plug weld symbol specifies a $\frac{3}{8}''$ diameter plug weld.

Angle of Countersink. If the hole of a plug weld is tapered, the angle of countersink dimension is specified on the welding symbol. The angle of countersink is the included angle of the taper.

The angle of countersink is specified on the welding symbol above or below the plug weld symbol. For example, a 60° dimension placed below the plug weld symbol specifies a 60° included angle of the plug weld. If the angle of countersink is not specified, users standards apply to the weld.

Depth of Filling. Depth of filling of the plug weld is complete unless specified with a dimension. Depth of filling that is less than complete has a depth of filling dimension specified inside the plug weld symbol. For example, a $\frac{1}{2}$ dimension placed inside the plug weld symbol specifies a $\frac{1}{2}''$ depth of filling.

Pitch. The pitch of plug welds in a straight line is specified by placing a dimension to the right of the plug weld symbol. For example, a 3 placed to the right of the plug weld symbol specifies that plug welds are 3″ center to center. Successive plug welds not in a straight line are specified using dimensions on the print.

Slot Weld Dimensions

Slot weld dimensions specified on the welding symbol include only the depth of filling if the depth of filling is less than complete. Depth of filling (if required) is indicated inside the slot weld symbol on the side of the reference line with the slot weld symbol. More information is required to specify slot welds than plug welds. The welding symbol, if used to specify all required slot weld information, would be confusing and hard to interpret.

JOINT MEMBER PREPARATION

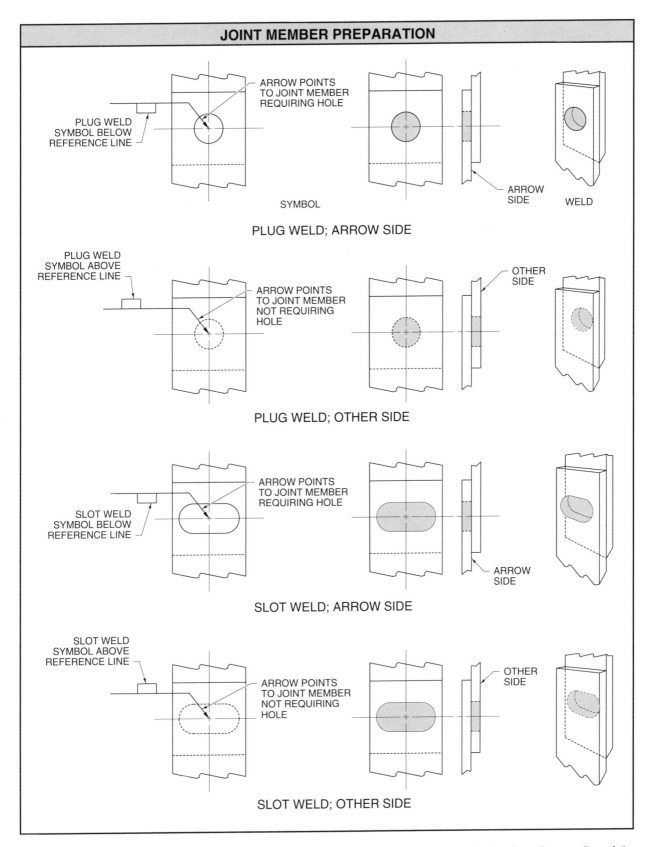

Figure 11-2. The weld symbol for a plug or slot weld is a rectangle placed above or below the reference line of the welding symbol.

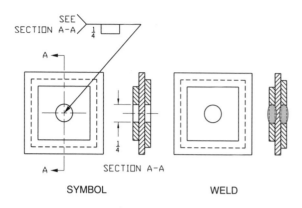

THREE JOINT MEMBERS

Figure 11-3. Section views provide joint member preparation for three or more joint members.

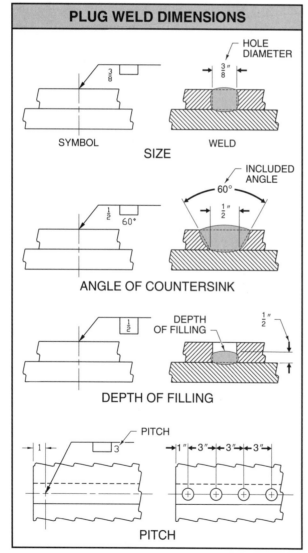

Figure 11-4. Plug weld dimensions are included with the plug weld symbol on the welding symbol.

Slot weld size, pitch, and included angle (angle of bevel) are specified using weld part dimensions, section views, and details on the prints. See Figure 11-5.

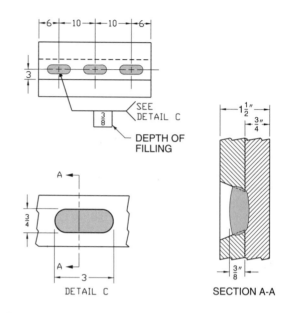

Figure 11-5. Slot weld dimensions are specified with the slot weld symbol and drafting techniques.

CONTOUR AND FINISH

Standard weld contour symbols are placed above or below the plug or slot weld symbol if the weld requires a flush or slightly convex surface. Welds requiring a flat but not flush surface are specified with a special note in the tail.

Mechanical finishing methods used after welding to obtain the desired contour (postweld finishing) have a letter above or below the weld contour symbol. These mechanical finishing methods are C - chipping, H - hammering, G - grinding, M - machining, R - rolling, and U - unspecified. Contour and finish are specified in the welding symbol with the quality or degree of finish specified by the manufacturer. See Figure 11-6.

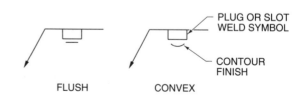

Figure 11-6. Weld contour and finish symbols are used to specify contour and finish of plug and slot welds.

Review Questions

Name_____ Date _____

Completion

_____ 1. A(n) _____ weld is a weld type in the cross-sectional shape of a hole in one of the joint members.

_____ 2. A(n) _____ weld is a weld type in the cross-sectional shape of an elongated hole in one of the joint members.

_____ 3. The weld symbol for a plug or slot weld is in the shape of a(n) _____.

_____ 4. The plug or slot weld symbol is placed _____ the reference line if the hole or slot is in the arrow side member of the joint.

_____ 5. The plug or slot weld symbol is placed _____ the reference line if the hole or slot is in the other side member of the joint.

_____ 6. Plug weld size is the diameter of the hole through the joint member at the _____ surface of the weld joint.

_____ 7. The pitch of plug welds in a straight line is specified by placing a dimension to the _____ of the plug weld symbol.

_____ 8. Mechanical finishing methods may be used after welding to obtain the desired _____.

_____ 9. All plug weld dimensions are provided on the side of the reference line with the weld _____.

_____ 10. A dimension placed to the right of the plug weld symbol specifies _____.

True-False

T   F   1. Plug and slot welds are determined by their hole shapes.

T   F   2. Plug and slot welds generally are used on joints with two joint members.

T   F   3. Joint member dimensions and notes included on the prints indicate whether a plug or slot weld is required.

T   F   4. The angle of countersink is always specified above the plug weld symbol.

T   F   5. The depth of filling dimension is placed inside the plug or slot weld symbol.

T   F   6. Plug welds are generally stronger than slot welds.

T F **7.** The size of a plug weld is specified with a number placed to the left of the plug weld symbol.

T F **8.** A dimension placed below the plug weld symbol specifies the width of the plug weld.

T F **9.** More information is required to specify slot welds than plug welds.

T F **10.** Plug and slot welds are commonly used when the weld must not be seen from the other side.

T F **11.** The faying surface is the part of the joint member which is in full contact prior to welding.

T F **12.** The angle of countersink should always be specified for a plug weld.

T F **13.** Depth of filling, if required, is indicated inside the slot weld symbol.

T F **14.** A $\frac{3}{8}$ dimension placed to the left of a plug weld symbol specifies $\frac{3}{8}''$ depth of filling.

T F **15.** The same weld symbol is used to indicate plug or slot welds.

Matching — Plug and Slot Weld Symbols

_____ **1.** Plug weld; 60° included angle

_____ **2.** Plug weld; other side

_____ **3.** Plug weld; $\frac{3}{8}''$ size

_____ **4.** Plug weld; $\frac{3}{8}''$ depth of filling

_____ **5.** Plug weld; 3″ pitch

_____ **6.** Plug weld; arrow side

_____ **7.** Slot weld; convex contour

_____ **8.** Slot weld; flush contour

_____ **9.** Slot weld; flat contour, machine finish

_____ **10.** Slot weld; flush contour; unspecified finish

Trade Competency Test

Name_____   Date _____

Refer to the SK5 Winch Mount print on page 223.

_____   **1.** The wall thickness of the tubing is _____″.

_____   **2.** The slot welds specified are _____″ wide.

_____   **3.** The strap is _____″ thick.

_____   **4.** The slot welds are ground to a(n) _____ contour.

_____   **5.** The fillet weld is _____″ long when completed.

_____   **6.** When finished to size, the slot welds are _____″ thick.

_____   **7.** The strap is bent to a(n) _____° angle.

_____   **8.** The slot welds are specified to be _____″ long.

_____   **9.** The overall length of the strap is _____″.

_____   **10.** Two _____ welds are specified to be ground flush with the tube.

_____   **11.** The drilled holes in the strap are_____″ in diameter.

T   F   **12.** The width of the strap is 2.50″.

T   F   **13.** The slot welds are welded last.

T   F   **14.** The distance between the slot welds is 2″.

_____   **15.** The radius of the strap at the bend is _____″.

_____   **16.** The tolerance for hundredths dimensions is ±_____″.

_____   **17.** All slot welds are made from the _____ side.

_____   **18.** The slot welds are centered _____″ from the left side.

_____   **19.** The strap material is Steel _____ HR.

_____   **20.** The holes drilled in the tube are _____″ in diameter.

Refer to the Beam Assembly B-43 print on page 224.

_____   **1.** The beam is attached to other structures using _____ connectors.

_____   **2.** The plug welds used to join the beam and connectors are spaced _____″ apart.

_____   **3.** Holes in the connectors are _____″ in diameter.

_____   **4.** Fillet welds with a(n) _____″ leg size are used to join the beam and connectors.

_____   **5.** All welds are to be made with _____ electrodes.

_____   **6.** Plug welds are specified to have a(n) _____ contour.

_____   **7.** The angle used for connectors has a $3\frac{1}{2}″ \times$ _____″ leg size.

_____   **8.** The structure is to be painted as specified with _____ coat(s).

_____   **9.** Fillet welds used to join the beam and connectors are _____ long.

_____   **10.** The plug welds are centered _____″ from the joining surface of the connectors.

_____   **11.** There are _____ holes that are not welded in each connector.

T   F   **12.** Both ends of the beam are recessed $1\frac{3}{4}″$.

T   F   **13.** The top plug weld is centered $3\frac{1}{2}″$ from the top of the beam.

T   F   **14.** The note specifies the beam is not to be painted in certain areas.

_____   **15.** The material used to construct the beam and connectors is ASTM _____.

_____   **16.** The scale of the drawing is _____.

_____   **17.** The beam weighs _____ lb per running foot.

T   F   **18.** The drawing was drawn on a $17″ \times 22″$ sheet.

T   F   **19.** The drawing was drawn by DRP.

_____   **20.** The assembly mark for the beam is _____.

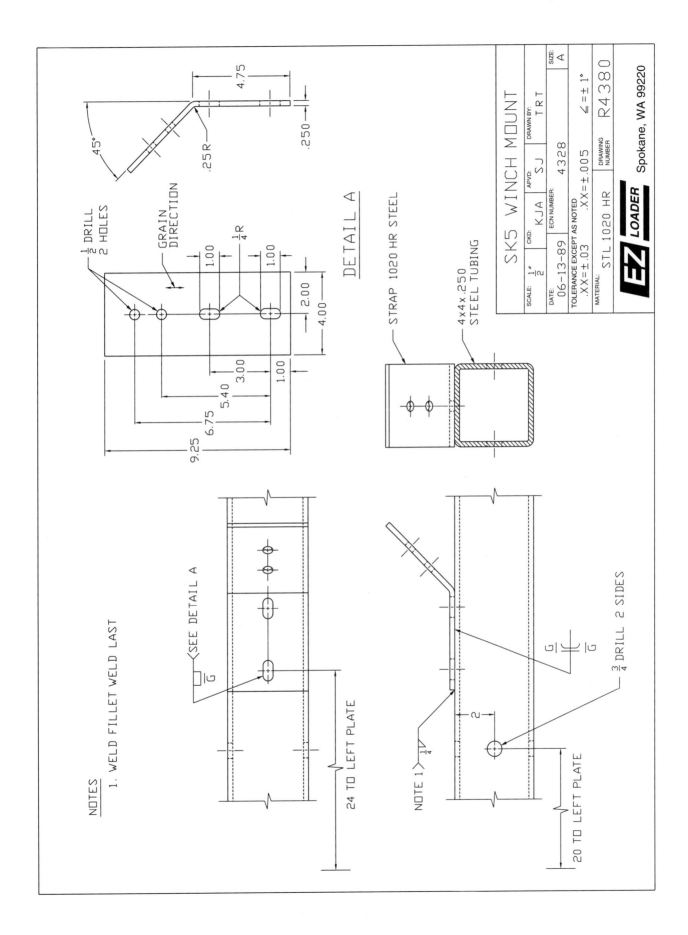

DETAIL A

STRAP 1020 HR STEEL

4×4×.250 STEEL TUBING

SK5 WINCH MOUNT

SCALE: 1/2"	CKD: KJA	APVD: SJ	DRAWN BY: T R T	SIZE: A
DATE: 06-13-89	ECN NUMBER: 4328	∠=± 1°
TOLERANCE EXCEPT AS NOTED	.XX=±.005
.XX=±.03	.XX=±.005
MATERIAL: STL 1020 HR	DRAWING NUMBER R4380

EZ LOADER Spokane, WA 99220

4.75

.250

45°

.25 R

1/2 DRILL 2 HOLES

GRAIN DIRECTION

1/4 R

1.00

1.00

2.00

4.00

3.00

1.00

5.40

6.75

9.25

NOTES

1. WELD FILLET WELD LAST

SEE DETAIL A

□/G

24 TO LEFT PLATE

NOTE 1 ⟩ 1/4

2

G/G

3/4 DRILL 2 SIDES

20 TO LEFT PLATE

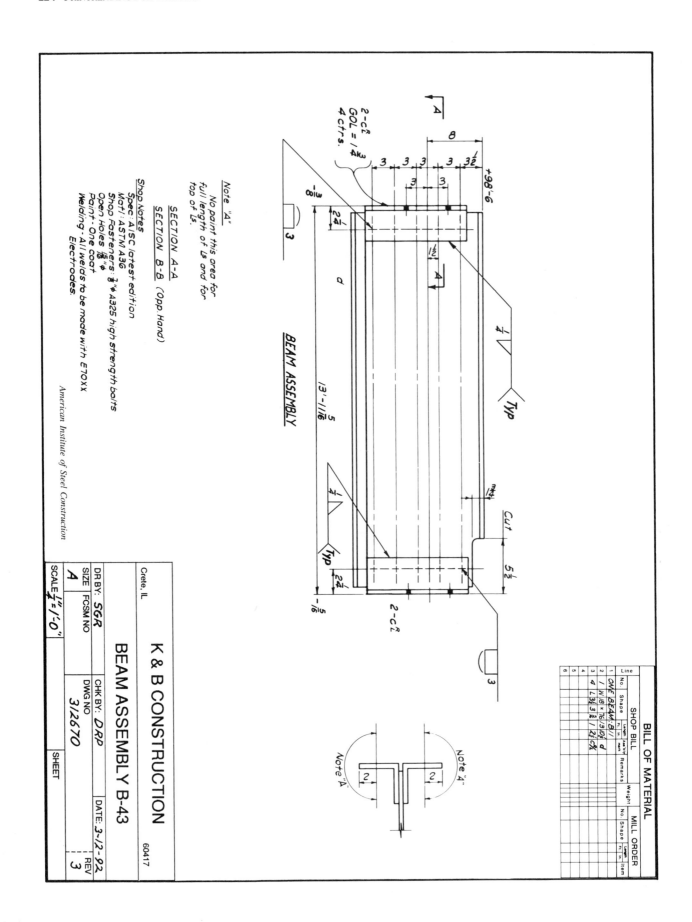

BEAM ASSEMBLY

Note "A"
No paint this area for
full length of Ls and for
top of Ls.

SECTION A-A

SECTION B-B (Opp. Hand)

Shop Notes
Spec: AISC latest edition
Mat'l: ASTM A36
Shop Fasteners: ⅞"⌀ A325 high strength bolts
Open Holes ⅝"⌀
Paint - One coat
Welding - All welds to be made with E70XX
Electrodes

American Institute of Steel Construction

		BILL OF MATERIAL					
		SHOP BILL			MILL ORDER		
Line No.	Shape	Length Ft. In.	No. Reqd.	Weight		No.	Shape	Length Ft. In.	Item

K & B CONSTRUCTION

Crete, IL   60417

BEAM ASSEMBLY B-43

DR BY: _SGR_	CHK BY: _DRP_	
SIZE	DWG NO	DATE: _3-12-92_
_A_ FCSM NO	312670	
SCALE ¼"=1'-0"	SHEET	REV 3

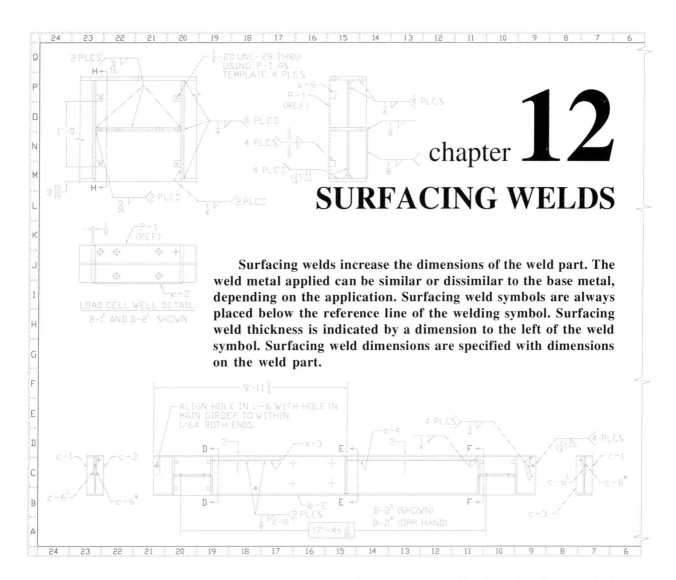

chapter 12

SURFACING WELDS

Surfacing welds increase the dimensions of the weld part. The weld metal applied can be similar or dissimilar to the base metal, depending on the application. Surfacing weld symbols are always placed below the reference line of the welding symbol. Surfacing weld thickness is indicated by a dimension to the left of the weld symbol. Surfacing weld dimensions are specified with dimensions on the weld part.

APPLICATIONS

Surfacing welds are weld beads deposited on a surface to increase the dimensions of the part or to add special properties to the weld part. Surfacing weld operations are classified as surfacing or hardfacing.

Surfacing is applying filler metals which have similar characteristics to the base metal. For example, worn parts such as shafts, plow points, bulldozer blades, etc. can be built up using surfacing welds and machined to the required dimensions. This saves on downtime, labor, and part replacement cost. Surfacing welds may be used to correct incorrect joint preparation or poor fit-up.

Surfacing welds can also be used in providing a layer of metal which is softer than the base metal. For example, shaft guides on a large pump are coated with bronze and machined to specifica-

tions. In this application, the shaft guide is designed to wear faster than the pump shaft. This protects the pump shaft from becoming damaged from normal wear. Building up and machining worn shaft guides is less expensive than the replacement cost of the pump shaft and reduces downtime. See Figure 12-1.

Hardfacing is applying filler metals, which provide a coating to protect the base metal from wear caused by impact, abrasion, erosion, or from other wear. Hardfacing allows the weld part to maintain the desired mechanical properties of the base metal, while adding new properties to specific areas of the weld part. For example, farm tools are commonly hardfaced to resist wear from dirt, rocks, and gravel. See Figure 12-2.

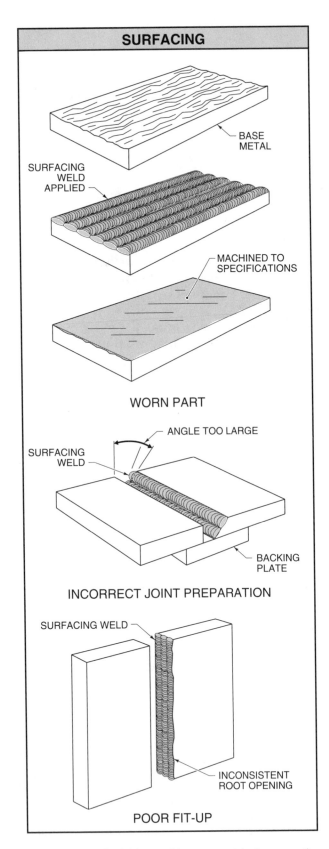

SURFACING

BASE METAL

SURFACING WELD APPLIED

MACHINED TO SPECIFICATIONS

WORN PART

ANGLE TOO LARGE

SURFACING WELD

BACKING PLATE

INCORRECT JOINT PREPARATION

SURFACING WELD

INCONSISTENT ROOT OPENING

POOR FIT-UP

Figure 12-1. Surfacing welds are used to increase the dimensions of parts.

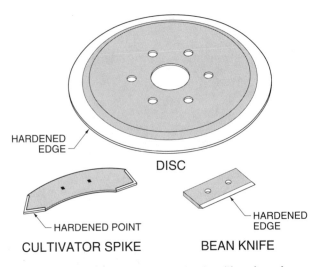

HARDENED EDGE

DISC

HARDENED POINT

CULTIVATOR SPIKE

HARDENED EDGE

BEAN KNIFE

Figure 12-2. Hardfacing extends the life of surfaces subjected to impact and abrasion.

SPECIFICATION

The surfacing weld symbol is always placed below the reference line of the welding symbol. There is no side significance, as the welding symbol arrow and dimensions indicate the weld area of the part. The thickness of the surfacing weld required above the base metal is indicated by a dimension to the left of the weld symbol. This dimension indicates thickness of weld metal applied to the base metal. See Figure 12-3.

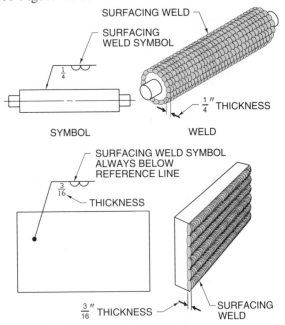

SURFACING WELD

SURFACING WELD SYMBOL

$\frac{1}{4}$

$\frac{1}{4}''$ THICKNESS

SYMBOL

WELD

SURFACING WELD SYMBOL ALWAYS BELOW REFERENCE LINE

$\frac{3}{16}$

THICKNESS

$\frac{3}{16}''$ THICKNESS

SURFACING WELD

Figure 12-3. The dimension on the left side of the surfacing weld symbol specifies the thickness of weld metal to be deposited.

If no specific thickness dimension is required, a thickness dimension is not included. Notes on the print or manufacturing standards may dictate surfacing weld dimensions. Single or multiple passes may be required to obtain the desired thickness.

The size and location dimensions of the surfacing weld are specified on the weld part in the drawing. If no dimensions are specified for the surfacing weld, the surfacing weld is applied to the entire surface of the weld part. See Figure 12-4.

Surfacing welds are used in combination with other weld types in specific applications. For example, a crack in a grader blade is repaired by grinding or gouging the crack for weld penetration. A groove weld joins the separated members and fills the remaining gap. The surface immediately surrounding the repaired area is then built up with a surfacing weld. The built-up area is then ground to match the contour of the blade.

Layer Orientation

Surfacing weld layers are specified for orientation by placing notes in the tail of the welding symbol. Multiple reference lines are used to indicate the order of welds. A surfacing weld which extends around a part is specified with the note CIRCUMFERENTIAL WELD in the tail.

Orientation is also specified with the notes LONGITUDINAL or LATERAL in separate welding symbol tails. Dimensions and welding processes of different layers are indicated using information in the tail and multiple reference lines of the welding symbol. See Figure 12-5.

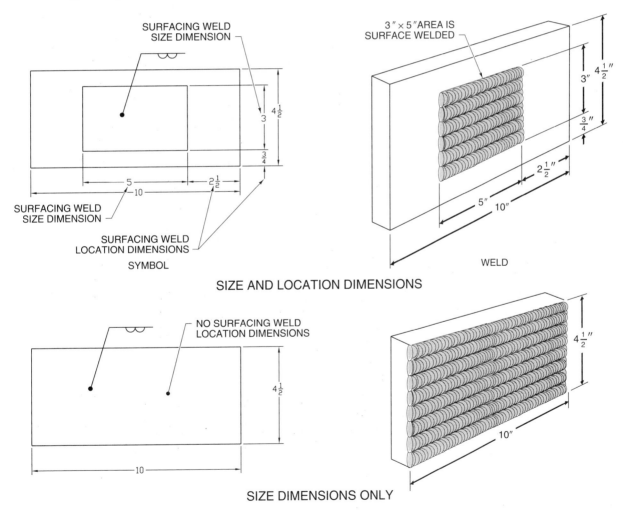

Figure 12-4. Print dimensions specify the area to be surface welded. If no dimensions are specified, the entire surface is surface welded.

LAYER ORIENTATION

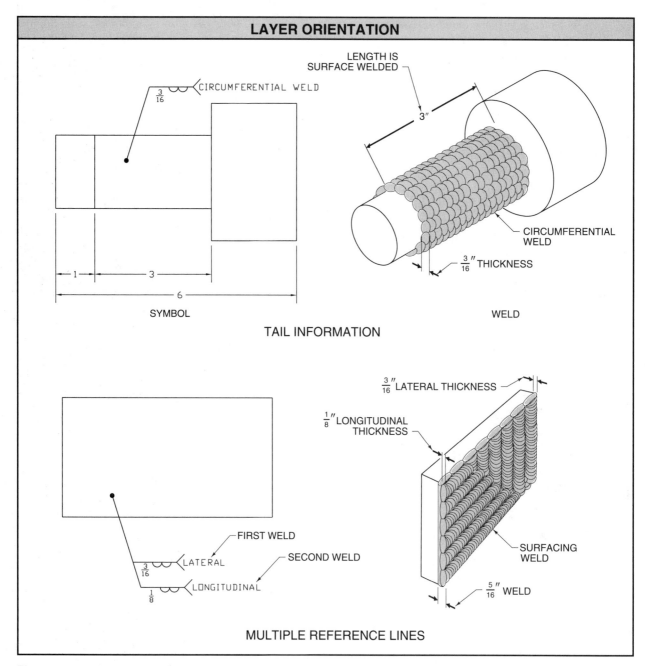

Figure 12-5. Layer orientation is specified in the tail or with multiple reference lines.

Review Questions

Name_____ Date _____

Completion

_____ 1. Surfacing welds increase the _____ of the weld part.

_____ 2. Surfacing weld symbols are always placed _____ the reference line of the welding symbol.

_____ 3. _____ is applying filler metals which provide a coating to protect the base metal from wear caused by impact or abrasion.

_____ 4. The thickness of the surfacing weld required above the base metal is indicated by a dimension placed to the _____ of the weld symbol.

_____ 5. _____ dimensions of the surfacing weld specify how much of the surface is to be surface welded.

_____ 6. _____ dimensions of the surfacing weld specify the part of the surface to be surface welded.

_____ 7. There is no _____ significance to the surfacing weld symbol, as the welding symbol arrow and dimensions indicate the weld area of the part.

_____ 8. Surfacing weld layers are specified for orientation by placing _____ in the tail of the welding symbol.

_____ 9. _____ reference lines are use to indicate the order of surfacing welds.

_____ 10. Single or multiple welding _____ may be required to obtain the desired thickness of a surface being surface welded.

True-False

T   F   1. Surfacing welds are weld beads deposited on a surface to increase the dimensions of the part.

T   F   2. The weld metal applied for a surfacing weld can be similar or dissimilar to the base metal.

T   F   3. Surfacing weld operations are classified as surfacing or spotfacing.

T   F   4. The surfacing weld metal can be softer than the base metal.

T   F   5. The surfacing weld symbol can be placed on either side of the reference line of the welding symbol.

T   F   6. Surfacing welds can be used in combination with other weld types.

T   F   **7.** Surfacing welds can be used to correct incorrect joint preparation.

T   F   **8.** If no specific thickness is given, the thickness of the surfacing weld is ⅛″.

T   F   **9.** Surfacing weld thickness is indicated by a dimension to the right of the weld symbol.

T   F   **10.** Worn parts can be surface welded and machined to avoid costly replacement.

T   F   **11.** The filler metal applied in a surfacing weld has similar characteristics to the base metal.

T   F   **12.** If no dimensions are specified, the surfacing weld is applied to the entire surface of the weld part.

T   F   **13.** Manufacturing standards may dictate surfacing weld dimensions.

T   F   **14.** Cracks in parts may be repaired by grinding before welding.

T   F   **15.** Location dimensions specify the size of the part to be welded.

Matching — Surfacing Weld Symbols

_____   **1.** First surfacing weld thickness

_____   **2.** Second surfacing weld thickness

_____   **3.** Note in welding symbol tail

_____   **4.** Weld part

_____   **5.** Depth of V-groove weld

_____   **6.** Surfacing weld symbol

_____   **7.** Welding symbol tail

_____   **8.** Groove angle

_____   **9.** Weld-part size dimension

_____   **10.** V-groove weld symbol

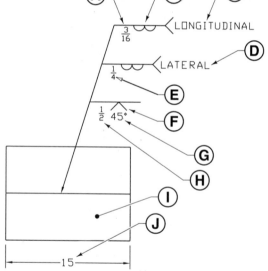

Trade Competency Test

Name_____ Date _____

Refer to the Shaft Details print on page 235.

_____ **1.** The maximum diameter dimension of Shaft A specified after build up and machining is _____″.

_____ **2.** The first surfacing weld completed on Shaft A is _____″ long.

_____ **3.** The _____ process is used to build up both shafts.

_____ **4.** The first surfacing weld of Shaft A deposits a minimum thickness of _____″.

_____ **5.** The end of Shaft A is chamfered _____°.
   A. 30
   B. 45
   C. 60
   D. none of the above

_____ **6.** Shaft A is preheated to _____°F before the axial weld is made.

_____ **7.** The minimum amount of weld metal applied before machining Shaft A is _____″.

_____ **8.** Shaft A is preheated to _____°F before the hardfacing electrode is used.

_____ **9.** The first weld applied specifies a(n) _____ electrode.

_____ **10.** The tolerance for all dimensions after machining is _____″.

_____ **11.** The overall diameter of Shaft A is _____″.
   A. 2.00
   B. 2.25
   C. 2.50
   D. 3.00

_____ **12.** The second surfacing weld on Shaft A deposits a minimum thickness of _____″.

_____ **13.** The center line shows that Shaft B is _____.

T   F   **14.** Shaft B is hollow.

_____ **15.** Shaft A is tapered for a total of _____″.
   A. ¼
   B. ⅜
   C. ½
   D. ⅝

_____ **16.** Shaft B is to be surface welded for a length of _____″.

_____ **17.** The first surfacing weld of Shaft B deposits a minimum thickness of _____″.

_____ **18.** The end of Shaft B is chamfered _____°.

_____ **19.** The overall diameter of Shaft B is _____″.
  A. 2.50
  B. 3.00
  C. 4.25
  D. none of the above

_____ **20.** The second surfacing weld of Shaft B deposits a minimum thickness of _____″.

_____ **21.** The repair work on the shafts was done by _____.

_____ **22.** The drawing number for the Shaft Details is _____.

_____ **23.** The shoulder of Shaft B should be broken.

T   F **24.** Both shafts connect to the Roller Assembly.

_____ **25.** The drawing was completed on a(n) _____ size sheet.

_____ **26.** This print represents the _____ revision of the Shaft Details.

T   F **27.** The axial weld of Shaft B is made before the circumferential weld.

T   F **28.** The print is not drawn to scale.

T   F **29.** The print was drawn by PAF.

T   F **30.** All shaft details are shown on two sheets.

_____ **31.** The print was checked by _____.

_____ **32.** The manufacturer of the shafts is located in _____.

T   F **33.** Both shafts are to be machined to specifications.

T   F **34.** The longitudinal surface weld on Shaft A is longer than the longitudinal surface weld on Shaft B.

T   F **35.** The overall diameter of Shaft B is larger than the overall diameter of Shaft A.

Refer to the Log Splitter Wedge Detail print on page 236.

_____ **1.** The wall thickness of the frame is _____″.
  A. ¼
  B. ⁵⁄₁₆
  C. ⅜
  D. ⁷⁄₁₆

_____ **2.** Weld parts forming the wedge are joined with _____ welds.

_____ **3.** Hardfacing material is applied using the _____ welding process.

_____ **4.** The top of the wedge is specified for a flat contour using the _____ finishing method.

_____ **5.** The inside width of the frame is _____″.

_____ **6.** The minimum thickness of hardfacing material applied is _____″.

_____ **7.** The hardfacing material specified is Class _____.
      A. 1A
      B. 2A
      C. 3B
      D. 3C

_____ **8.** A(n) _____ weld is specified to be welded-all-around.
      A. square-groove
      B. flare-bevel-groove
      C. axial
      D. none of the above

_____ **9.** The weld parts forming the wedge are _____″ thick.

_____ **10.** The welding symbol with Note 3 specifies a(n) _____ weld.

T   F   **11.** The fillet welds are completed last.

T   F   **12.** All square-groove welds must have a flat contour.

T   F   **13.** The hardfacing material is deposited $3/16″$ from the end of the wedge.

T   F   **14.** All welds on top of the wedge must have a concave contour.

T   F   **15.** The outside dimensions of the frame are $8″ \times 8″$.

_____ **16.** The length of the wedge is _____″.
      A. 6
      B. 8
      C. 10
      D. 12

_____ **17.** A(n) _____ drawing of the log splitter is shown.
      A. isometric
      B. oblique
      C. orthographic
      D. perspective

T   F   **18.** The wedge is centered on the width of the square tube.

T   F   **19.** The wedge is welded together before it is welded to the frame.

_____ **20.** The overall _____ dimension of the log splitter is not shown in the wedge detail.

_____   **21.** The wedge is manufactured by _____.

_____   **22.** The wedge is part of a(n) _____.

_____   **23.** The height of the wedge is _____″.
        A. 4
        B. 8
        C. 12
        D. 16

T   F   **24.** The manufacturer of the log splitter is located in Bessemer, Alaska.

T   F   **25.** The drawing is drawn to the scale of ¼″ = 1′- 0″.

_____   **26.** Section lines are shown only in the _____ view.

_____   **27.** A(n) _____ fillet weld joins the wedge to the tube.

_____   **28.** The tube has a wall thickness of _____″.

_____   **29.** Long _____ lines are used to indicate that the tube continues on both ends.

_____   **30.** The splitting portion of the wedge is ground to a(n) _____ contour.

T   F   **31.** The wedge is made from mild steel.

T   F   **32.** Hidden lines are used only in the front view.

_____   **33.** The wedge tapers from 0″ to _____″.

_____   **34.** The drawing was completed on _____.

_____   **35.** The drawing number of the log splitter is _____.

TO ROLLER ASSEMBLY

AXIAL WELD – SEE NOTE 4

CIRCUMFERENTIAL WELD – SEE NOTE 5

.067

.125

3.00

2.500

45°

.067

.067

2.00

SHAFT A

2.250 DIA

TO ROLLER ASSEMBLY

DO NOT BREAK SHOULDER

AXIAL WELD – SEE NOTE 4

CIRCUMFERENTIAL WELD – SEE NOTE 5

.045

.100

4.25

45°

.050

.250

SHAFT B

3.00 DIA

NOTES:

1. SMAW USED

2. MACHINE TO SPECIFICATIONS

3. ALL DIMENSIONS ±.002 AFTER MACHINING

4. E 7018 ELECTRODE PREHEAT TO 600°F

5. CLASS 4B HARDFACING ELECTRODE, PREHEAT TO 800°F

MEAD MACHINE CO.

318 S. WATERS

LOUISVILLE, KY 40223

SHAFT DETAILS

DR BY: PAF

CHK BY: RWA

DATE:

FCSM NO

DWG NO 1302-3

REV 4

SIZE A

SCALE NTS

SHEET 1 of 3

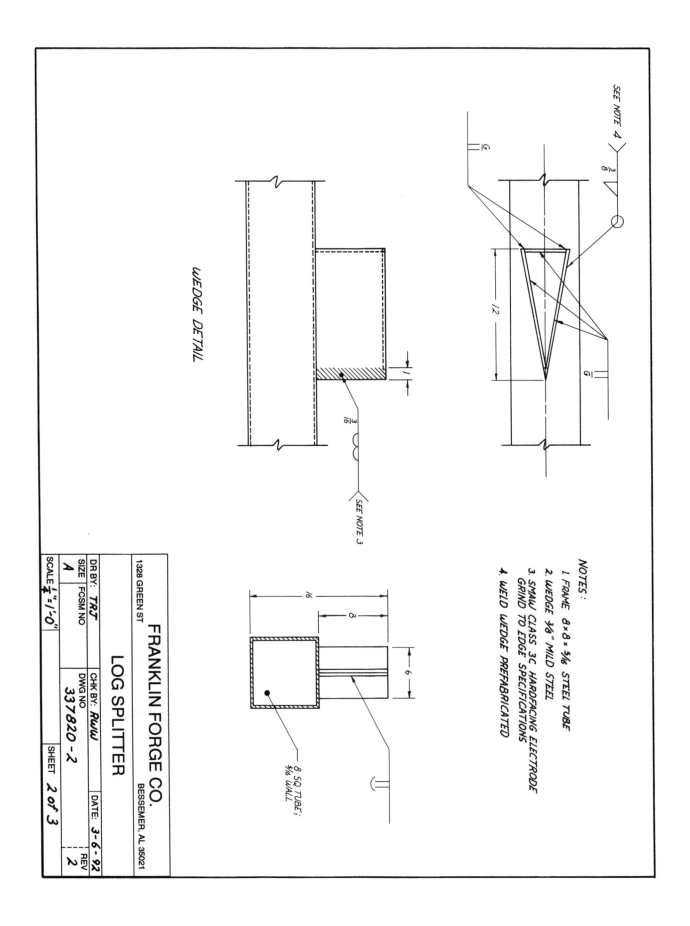

WEDGE DETAIL

SEE NOTE 3

SEE NOTE 4

NOTES :

1. FRAME 8×8×5⁄16 STEEL TUBE
2. WEDGE 3⁄8" MILD STEEL
3. SMAW CLASS 3C HARDFACING ELECTRODE
   GRIND TO EDGE SPECIFICATIONS
4. WELD WEDGE PREFABRICATED

8 SQ TUBE;
5⁄16 WALL

FRANKLIN FORGE CO.
BESSEMER, AL 35021

1328 GREEN ST

LOG SPLITTER

DR BY: *TRJ*	CHK BY: *RWW*	DATE: *3-6-92*
SIZE FCSM NO	DWG NO	
A	337820 - 2	REV 2
SCALE 1⁄4"=1'-0"		SHEET 2 of 3

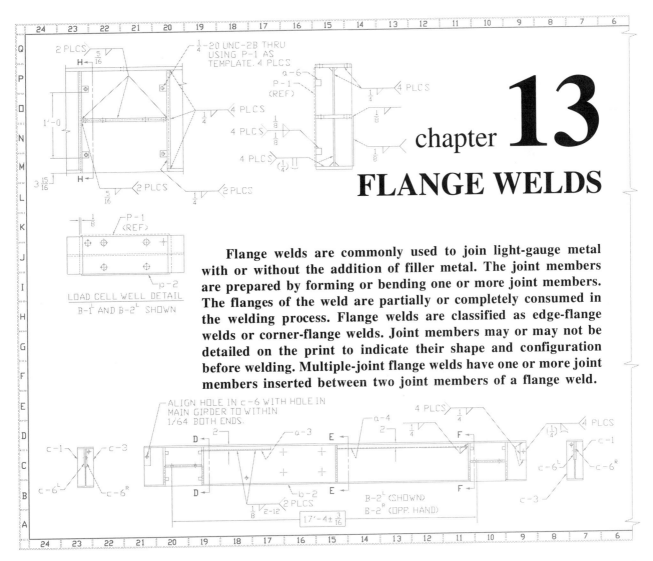

LOAD CELL WELL DETAIL
B-1^L AND B-2^L SHOWN

chapter 13

FLANGE WELDS

Flange welds are commonly used to join light-gauge metal with or without the addition of filler metal. The joint members are prepared by forming or bending one or more joint members. The flanges of the weld are partially or completely consumed in the welding process. Flange welds are classified as edge-flange welds or corner-flange welds. Joint members may or may not be detailed on the print to indicate their shape and configuration before welding. Multiple-joint flange welds have one or more joint members inserted between two joint members of a flange weld.

APPLICATIONS

A *flange weld* is a weld of light-gauge metal with one or both joint members bent at approximately 90°. Flange welds consist of two or more joint members of which at least one is flanged (bent). Edges are flanged by forming or bending one or both joint members at an approximate 90° angle.

Flange weld joint members may be melted together in the welding process to eliminate the need for addition of filler metal. Filler metal may be added to the flange weld if required for additional reinforcement.

Flange welds are used primarily on welds joining light-gauge (thin) metals. Sheets of light-gauge metal can be joined edge-to-edge to form larger sheets. Light-gauge metal can also be formed and welded to produce vessels or containers.

Flange welds are classified as edge-flange welds or corner-flange welds, depending on the configuration of the joint members. An *edge-flange weld* is a flange weld with both joint members flanged. A *corner-flange weld* is a flange weld with one joint member flanged.

SPECIFICATION

Flange welds are specified on prints with similar, but slightly different, weld symbols. Edge-flange welds are specified with a weld symbol consisting of two parallel lines, both of which are connected by radii to the reference line of the welding symbol. See Figure 13-1.

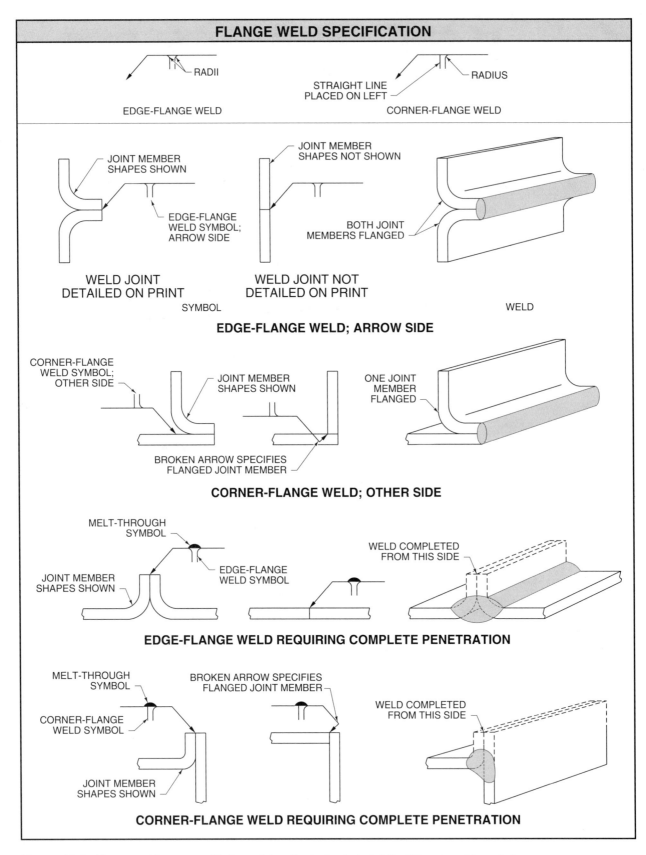

FLANGE WELD SPECIFICATION

EDGE-FLANGE WELD

RADII

STRAIGHT LINE
PLACED ON LEFT

CORNER-FLANGE WELD

RADIUS

JOINT MEMBER
SHAPES SHOWN

EDGE-FLANGE
WELD SYMBOL;
ARROW SIDE

WELD JOINT
DETAILED ON PRINT

JOINT MEMBER
SHAPES NOT SHOWN

BOTH JOINT
MEMBERS FLANGED

WELD JOINT NOT
DETAILED ON PRINT

SYMBOL

WELD

EDGE-FLANGE WELD; ARROW SIDE

CORNER-FLANGE
WELD SYMBOL;
OTHER SIDE

JOINT MEMBER
SHAPES SHOWN

ONE JOINT
MEMBER
FLANGED

BROKEN ARROW SPECIFIES
FLANGED JOINT MEMBER

CORNER-FLANGE WELD; OTHER SIDE

MELT-THROUGH
SYMBOL

EDGE-FLANGE
WELD SYMBOL

JOINT MEMBER
SHAPES SHOWN

WELD COMPLETED
FROM THIS SIDE

EDGE-FLANGE WELD REQUIRING COMPLETE PENETRATION

MELT-THROUGH
SYMBOL

BROKEN ARROW SPECIFIES
FLANGED JOINT MEMBER

CORNER-FLANGE
WELD SYMBOL

WELD COMPLETED
FROM THIS SIDE

JOINT MEMBER
SHAPES SHOWN

CORNER-FLANGE WELD REQUIRING COMPLETE PENETRATION

Figure 13-1. Flange welds are specified on prints with similar, but slightly different, weld symbols.

Corner-flange welds are specified with a weld symbol consisting of two parallel lines, one of which is connected by radius to the reference line. The straight line of the corner-flange weld symbol is connected directly to the reference line. It is always placed to the left of the line with the radius.

Weld joint configuration for a flange weld may or may not be detailed on the print. If the flange weld joint configuration is not detailed, the welding symbol arrow determines joint member preparation.

Both edge-flange welds and corner-flange welds are specified with the welding symbol arrow pointing to the location of the weld. The weld symbol is located on the arrow side or the other side to specify the weld location. Flange weld joints are never specified for both sides.

A break in the arrow of the welding symbol specifies the corner-flange weld joint member to be flanged. The corner-flange weld joint member to be flanged may also be shown by detailing its shape on the print.

Complete joint penetration is specified by placing a melt-through symbol opposite the edge-flange or corner-flange weld symbol. The melt-through weld symbol is shown as a solid half circle on the reference line of the welding symbol.

Flange Weld Dimensions

Flange weld dimensions are located to the left of the weld symbol. The dimensions included are flange radius, flange height, and flange weld thickness. The order in reading the dimensions is flange radius, flange height, and flange weld thickness. The flange radius and flange height are separated by a plus (+) sign. See Figure 13-2.

The root opening is not specified on the welding symbol of a flange weld. It is specified on the print using a detail with the required dimensions.

Flange Radius. *Flange radius* is the radius of the joint member(s) requiring edge preparation in a flange weld. The flange radius is specified with a dimension to the left of the plus sign.

Flange Height. *Flange height* is the distance from the point of tangency on the flange of a flange weld to the edge of the flange before welding. The point of tangency is the point at which the radius of the

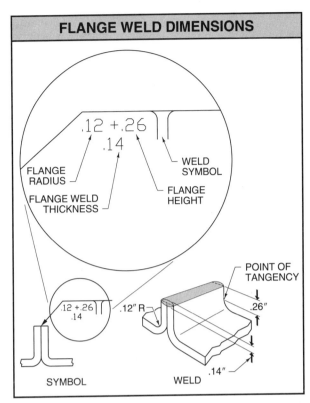

Figure 13-2. Flange weld dimensions are specified with three dimensions to the left of the weld symbol.

flange and the straight edge of the flange meet. Flange height is specified with a dimension to the right of the plus sign. The flange height specifies the flange dimension for both joint members in an edge-flange weld or one joint member in a corner-flange weld.

Flange Weld Thickness. *Flange weld thickness* is the cross-sectional distance of a flange weld from the weld face to the weld root. The flange weld thickness is specified with a dimension above or below the flange radius and flange height. The width of a flange weld depends upon the combined thickness and penetration of the joint members.

Multiple-Joint Flange Welds

Multiple-joint flange welds are specified when one or more joint members are inserted between two joint members of an edge-flange or corner-flange weld. The edge-flange or corner-flange weld symbol is used as required to specify the joint member configuration. See Figure 13-3.

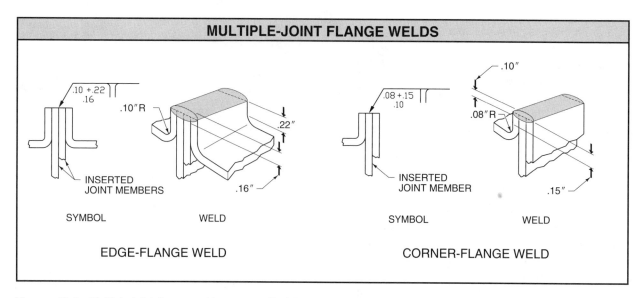

Figure 13-3. Multiple-joint flange welds are specified the same as other flange welds.

Review Questions

Name_____ Date _____

Completion

_____ 1. Flange welds are commonly used to join light-_____ metal.

_____ 2. One or both joint members of a flange weld may be bent at an approximate _____° angle.

_____ 3. _____ metal may be added to a flange weld if required for additional reinforcement.

_____ 4. A(n) _____-flange weld is a flange weld with both joint members flanged.

_____ 5. A(n) _____-flange weld is a flange weld with one joint member flanged.

_____ 6. The straight line of a corner-flange weld is always placed to the _____ of the line with the radius.

_____ 7. A(n) _____ in the arrow of the welding symbol is used to specify the corner-flange weld joint member to be flanged.

_____ 8. Flange weld dimensions are located to the _____ of the weld symbol.

_____ 9. The flange radius and flange height dimensions are separated by a(n) _____ sign.

_____ 10. Flange _____ is the distance from the point of tangency on the flange of a flange weld to the edge of the flange before welding.

True-False

T   F   1. The flange height specifies the flange dimension for both joint members of a corner-flange weld.

T   F   2. Flange weld thickness is the cross-sectional distance of a flange weld from the weld toe to the weld throat.

T   F   3. The flange weld thickness dimension may be placed above or below the flange radius and flange height dimensions.

T   F   4. The root opening of a flange weld is specified below the reference line of the welding symbol.

T   F   5. Flange welds always have at least two joint members.

T   F   6. Edge-flange welds are specified with a weld symbol consisting of two parallel lines connected by radii to the reference line of the welding symbol.

T   F   **7.** The weld joint configuration for a flange weld may or may not be detailed on a print.

T   F   **8.** The straight line of the corner-flange weld symbol is connected directly to the reference line of the welding symbol.

T   F   **9.** Complete joint penetration is specified by placing a melt-through symbol to the right of the flange weld symbol.

T   F   **10.** A multiple-joint flange weld has one or more joint members inserted between two joint members.

Matching — Flange Weld Symbols

_____   **1.** Edge-flange weld; other side

_____   **2.** Edge-flange weld joint; not detailed on print

_____   **3.** Edge-flange weld joint; detailed on print

_____   **4.** Edge-flange weld; arrow side

_____   **5.** Multiple-joint edge-flange weld joint; detailed on print

_____   **6.** Corner-flange weld; other side

_____   **7.** Corner-flange weld; arrow side

_____   **8.** Flanged joint member; A specified

_____   **9.** Flanged joint member; B specified

_____   **10.** Edge-flange weld; arrow side; complete penetration

Trade Competency Test

Name_____ Date _____

Refer to the Ductwork Shop Drawings print on page 245.

_____ **1.** The outlet duct length is _____.

_____ **2.** Hat channels are _____″ tall.

_____ **3.** Weld A specifies a fillet weld with a(n) _____″ leg size.

_____ **4.** Weld A specifies welds _____″ in length.

_____ **5.** The turning vane is joined to the outlet duct with a(n) _____ weld.

T   F **6.** Weld B specifies an edge-flange weld using the GTAW process.

_____ **7.** Hat channels are constructed of _____ gauge stainless steel.

_____ **8.** The flange is joined to the reducer duct with a(n) _____ weld.

_____ **9.** The total length of weld C is approximately _____″.

_____ **10.** Weld D specifies welds spaced on _____″ centers.

_____ **11.** The 24″ × 32″ section of the reducer duct is _____ long.

_____ **12.** Holes on the turning vane are located _____″ on center.

T   F **13.** The GTAW process is only specified for edge-flange welds.

_____ **14.** All duct is to be pressure tested to _____″ minimum water column.

_____ **15.** The total length of weld B is approximately _____″.

T   F **16.** The drawings were completed 11-91.

T   F **17.** The length of the reducer duct is 5′-10″.

T   F **18.** Duct 18″ to 29″ shall be reinforced with hat channel.

T   F **19.** The project number of the drawings is 536.

T   F **20.** The turning vane is attached to 24″ × 32″ duct.

T F   **21.** Flange reinforcing has equal legs measuring 1¾″ each.

_____   **22.** Z-bar reinforcing is joined to the duct with _____ welds.

T F   **23.** All welds are ground to remove excess slag.

T F   **24.** The scale for the drawings is ¼″ = 1′-0″.

T F   **25.** Duct larger than 30″ shall be reinforced with hat channel on 30″ centers.

_____   **26.** The hat channel is _____″, disregarding flange.

_____   **27.** The 16″ × 40″ piece of outlet duct is _____ long.

_____   **28.** The outlet duct has a(n) _____″ offset in 3′-0″.

_____   **29.** The diameter of the holes in the turning vain is _____″.

T F   **30.** The hat channel has ½″ flanges.

T F   **31.** The Z-bar has ½″ flanges.

T F   **32.** The reducer duct makes a transition from 24″ × 32″ to 34″ × 32″.

_____   **33.** The width of the Z-bar is _____″.

_____   **34.** All duct is constructed of 16 gauge _____ steel.

_____   **35.** The GTAW weld is made _____ the outlet duct.

T F   **36.** Hat channel is spaced on 24″ centers throughout the duct run.

T F   **37.** The height of the Z-bar is 1¾″.

T F   **38.** The reducer duct contains a flange on both ends.

_____   **39.** The stainless steel has a(n) _____ finish.

T F   **40.** The length of the hat channel specified is 14′-6″.

_____   **41.** All fillet welds have _____″ legs.

_____   **42.** The drawing was drawn by _____.

T F   **43.** All welds are welded-all-around.

T F   **44.** The offset section is 3′-0″ long.

T F   **45.** No drawing number is given for this print.

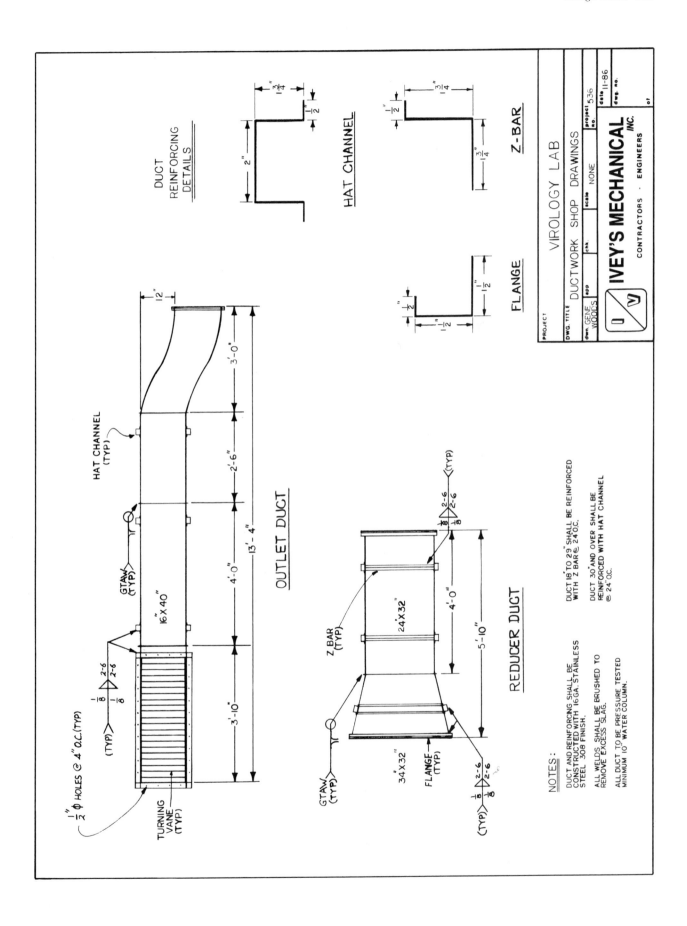

DUCT REINFORCING DETAILS

HAT CHANNEL

Z-BAR

FLANGE

OUTLET DUCT

HAT CHANNEL (TYP)

GTAW (TYP)

16" X 40"

12"

3'-0"

2'-6"

4'-0"

3'-10"

13'-4"

$\frac{1}{2}$" ∅ HOLES @ 4" O.C.(TYP)

TURNING VANE (TYP)

REDUCER DUCT

Z BAR (TYP)

GTAW (TYP)

24"X32"

34"X32"

FLANGE (TYP)

4'-0"

5'-10"

NOTES:

DUCT AND REINFORCING SHALL BE CONSTRUCTED WITH 16 GA. STAINLESS STEEL 308 FINISH.

ALL WELDS SHALL BE BRUSHED TO REMOVE EXCESS SLAG.

ALL DUCT TO BE PRESSURE TESTED MINIMUM 10" WATER COLUMN.

DUCT 18" TO 29" SHALL BE REINFORCED WITH Z BAR @ 24 O.C.

DUCT 30" AND OVER SHALL BE REINFORCED WITH HAT CHANNEL @ 24" O.C.

PROJECT VIROLOGY LAB

DWG. TITLE DUCTWORK SHOP DRAWINGS

IVEY'S MECHANICAL INC.

CONTRACTORS - ENGINEERS

dwn. GENE WOODS   app.   ckd.   scale NONE   project no. 536   date 11-86   dwg. no.   of

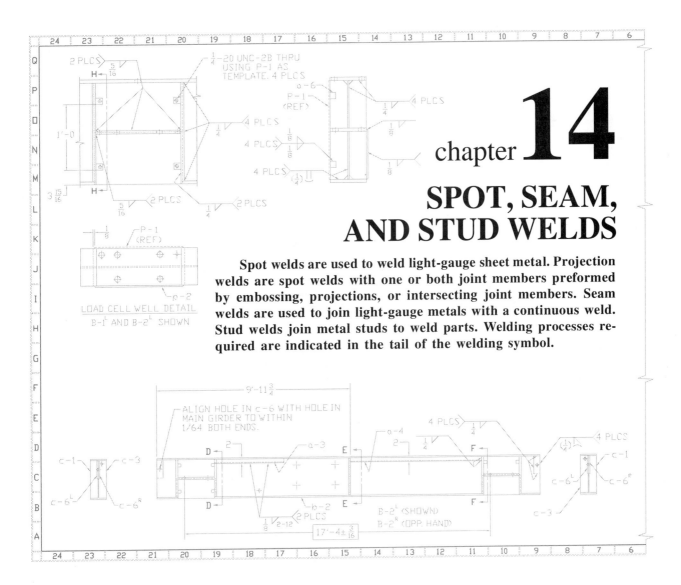

chapter 14

SPOT, SEAM, AND STUD WELDS

Spot welds are used to weld light-gauge sheet metal. Projection welds are spot welds with one or both joint members preformed by embossing, projections, or intersecting joint members. Seam welds are used to join light-gauge metals with a continuous weld. Stud welds join metal studs to weld parts. Welding processes required are indicated in the tail of the welding symbol.

SPOT WELDS

A *spot weld* is a weld type produced by confining the fusion of molten base metal using heat and pressure without preparation to the joint members. Spot welds join overlapping joint members with fusion on the faying surfaces which may extend to the outer surface of one of the joint members. Single or multiple spot welds can be made as required for strength. The cross-sectional shape of a spot weld is roughly circular.

The resistance welding process is the most common welding process used for spot welds. Other arc welding processes such as GTAW can also be used for spot welds.

Resistance welding heats joint members using heat generated by the flow of electricity through the joint members. The size and penetration of the weld is determined by electrode size, pressure applied to joint members, length of time current is passed through weld parts, and amount of current applied.

Spot welds are commonly used to join light-gauge metals. The process is widely used in the joining of sheet metal parts for the manufacture of automobiles, appliances, and electrical components. Spot welds can be easily adapted to automatic and robotics applications. Spot welds are indicated on prints using a weld symbol in the shape of an unfilled circle. See Figure 14-1.

Spot Weld Specification

The spot weld symbol may or may not have arrow side or other side significance. Spot welds that are

247

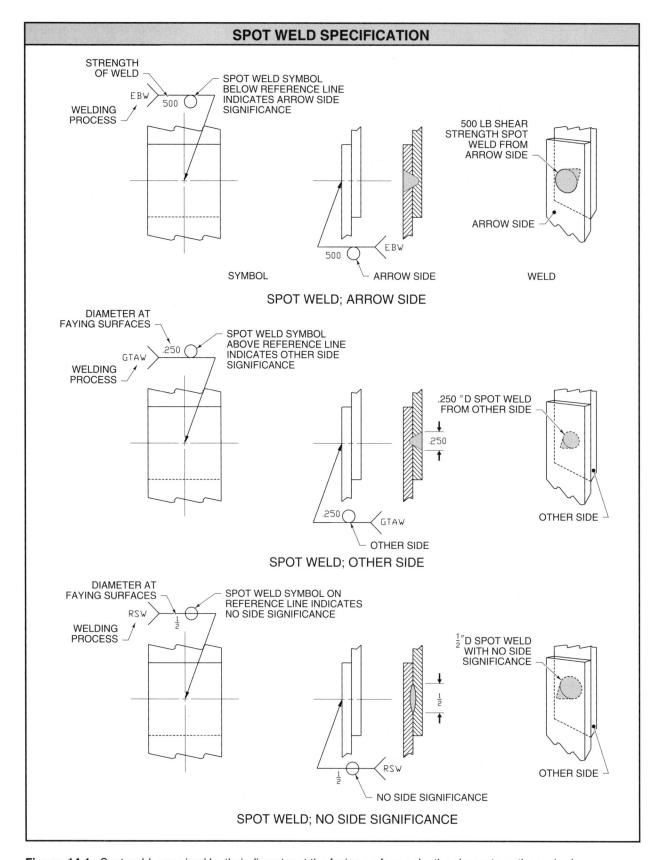

Figure 14-1. Spot welds are sized by their diameter at the faying surface or by the shear strength required.

completed from the arrow side are specified by placing the spot weld symbol below the reference line of the welding symbol. Welds completed from the other side are specified by placing the spot weld symbol above the reference line of the welding symbol. Spot welds with no side significance have the spot weld symbol centered on the reference line.

Size. All spot weld size dimensions (diameters) are placed on the same side of the reference line as the spot weld symbol. If the spot weld symbol is centered (indicating no side significance), the spot weld dimensions may be located on either side of the reference line. The welding process required is specified in the tail of the welding symbol. Spot welds are sized by the diameter of the spot weld at the faying surfaces of the joint members. The size of spot welds is specified by placing a dimension *to the left* of the spot weld symbol.

Spot weld strength is measured in shear strength. Shear strength is specified in pounds or newtons per spot. Spot weld strength required is specified by placing a number *to the left* of the spot weld symbol. See Figure 14-2.

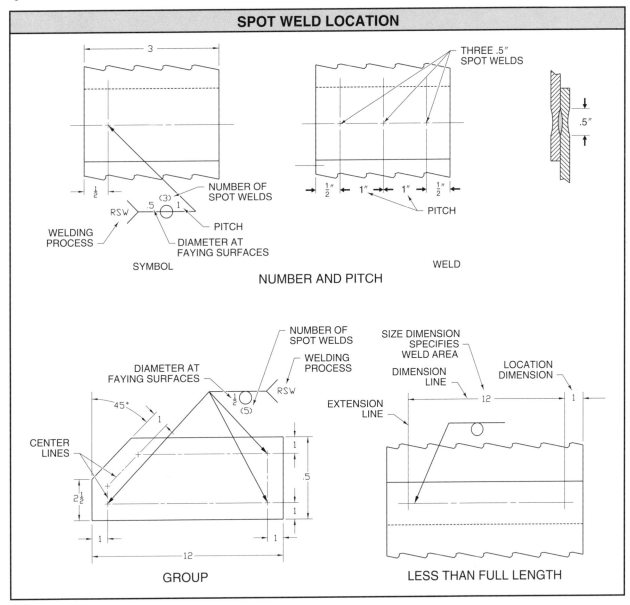

Figure 14-2. Spot weld location is specified by dimensions on the welding symbol and on the print.

Number. The number of spot welds is specified with a number in parentheses above or below the spot weld symbol, depending on its location on the reference line. The number is placed below the spot weld symbol when the weld is on the arrow side. It is placed above the spot weld symbol when the weld is on the other side. The number may be placed on either side when there is no side significance.

Pitch. Spot weld spacing in a straight line is specified with a pitch dimension placed to the right of the spot weld symbol. The pitch dimension may be placed on either side of the reference line when the spot weld symbol is centered on the reference line.

Group Spacing. Spot welds are specified as a group by using intersecting lines with dimensions on the print. Spot weld specification and dimensions included on the welding symbol supplement information included on the print.

Less Than Full Length. Spot welds that do not extend the full length of the weld part are dimensioned on the prints. Extension lines and dimension lines are used to indicate the area that is to be spot welded.

Projection Welds

A *projection weld* is a weld type produced by confining fusion of molten base metal using heat and pressure with a preformed dimple or projection in one of the joint members prior to welding. Projection welds use the resistance welding process in which resistance to the flow of current causes a buildup of heat in the parts to be joined. The heat is localized by the shape of the preformed projections. The preformed projections can be on one or both joint members.

Projection welds can also be completed at the intersection of joint members. Projection welds are used for the same applications as spot welds. Projection welding electrodes last longer than spot welding electrodes, as the heat generated from the welding process is confined to a smaller area of the projections.

Projection Weld Specification. Projection welds are specified on prints with the spot weld symbol located on either side of the reference line and with the abbreviation PW included in the tail of the welding symbol. The location of the weld symbol indicates which joint member has been preformed. See Figure 14-3.

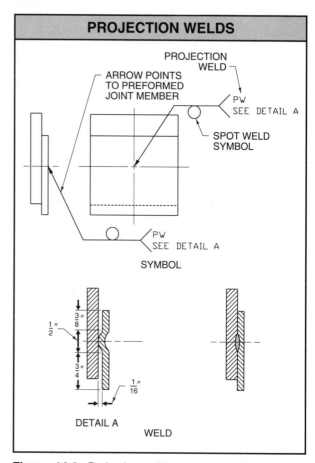

Figure 14-3. Projection welds are specified by placing a PW in the tail of the spot welding symbol.

SEAM WELDS

A *seam weld* is a weld type produced by confining fusion of molten base metal using heat and pressure for a series of continuous or overlapping successive spot welds on joint members. The continuous resistance weld is formed by equipment that moves the welding electrodes, the weld parts, or both.

Seam welds are commonly used in manufacturing light-gauge pipe or parts of light-gauge metal. Welding speed for seam welds is determined by the current applied and the thickness of the joint members. See Figure 14-4.

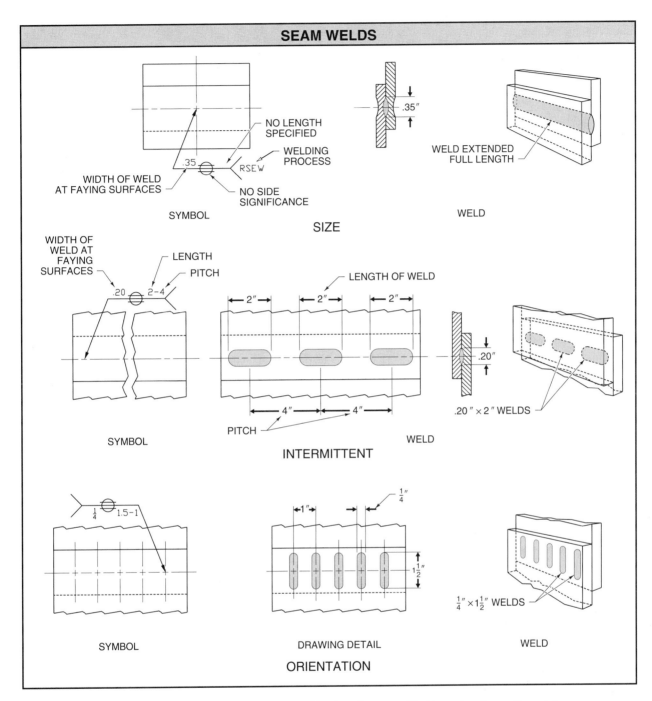

Figure 14-4. Seam welds are welded parallel to the weld axis unless specified by a drawing on the print.

Seam Weld Specification

Seam welds are indicated on prints using a weld symbol in the shape of a circle with two parallel, horizontal lines passing through. Like the spot weld, a seam weld may or may not have arrow or other side significance.

Seam welds completed from the arrow side are specified by placing the seam weld symbol below the reference line of the welding symbol. Seam welds completed from the other side are specified by placing the seam weld symbol above the reference line of the welding symbol. Seam welds with no side significance have the seam weld symbol centered on the reference line.

All seam weld dimensions are included on the same side of the reference line as the weld symbol. If the seam weld symbol is centered, the weld dimensions can be located on either side of the weld symbol. The welding process for seam welds is specified in the tail of the welding symbol.

Seam weld size is specified with a dimension to the left of the seam weld symbol which indicates the width of the weld at the faying surfaces of the weld. The strength is specified in pounds per linear inch or in newtons per millimeter. Length is specified by a dimension located to the right of the seam weld symbol. Seam welds that extend the entire length of the weld part do not require a length dimension. Seam weld lengths that do not extend the entire length of the weld part are dimensioned on the print.

Intermittent Seam Welds. Intermittent seam welds are specified on prints with length and pitch dimensions separated by a dash. This is similar to the specification for intermittent fillet welds. Length and pitch of intermittent seam welds are based on the weld located parallel to the axis of the weld. If the seam weld is not parallel to the axis of the weld, the orientation is indicated by a drawing on the print.

STUD WELDS

A *stud weld* is a weld type produced by joining threaded studs with other parts using heat and pressure. Stud welds are formed by joining a metal stud or a similar weld part to another weld part with arc, resistance, or friction welding, or with other welding processes. The welding process can be performed with or without the addition of a shielding gas.

Stud arc welding is the most common type of stud welding. It uses an arc formed between the stud and the workpiece with pressure to join the parts. Shielding gas or flux can be used for proper shielding protection and postweld cooling. The metal stud acts as an electrode and melts to provide molten filler metal. Pressure is applied to form the weld. See Figure 14-5.

Stud welding is commonly used in manufacturing products that require assembly and disassem-

bly of parts with studs and threaded fasteners. Stud welding also eliminates drilling requirements and provides fastening capability from one side of the joint parts.

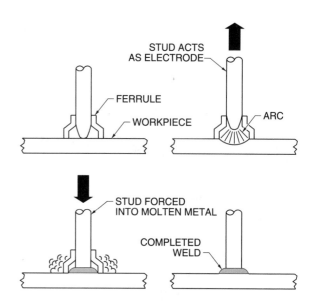

Figure 14-5. Stud welds join metal studs and fasteners to weld parts using heat generated from an arc.

Stud Weld Specification

Stud welds are indicated on prints using a weld symbol in the shape of a circle containing an X. The stud weld symbol is not used with any side significance. It is always placed below the reference line of the welding symbol.

All dimensions of stud welds are located on the same side of the reference line as the stud weld symbol. The size of the stud used in a stud weld is indicated to the left of the stud weld symbol.

More than one stud weld in a straight line is specified with a pitch dimension placed to the right of the stud weld symbol. Stud welds that are not located in a straight line are specified on the print with dimensions. More than one stud weld is specified with a number in parentheses below the weld symbol. The location of the first and last stud welds is indicated by dimensions on the prints. See Figure 14-6.

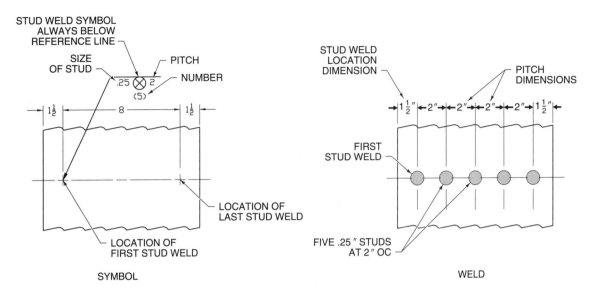

Figure 14-6. Stud welds are located on the weld part as specified on the welding symbol and print.

CONTOUR AND FINISH

Spot and seam weld contour is the shape of the completed weld face in sectional view. Both the contour and finish are specified on the welding symbol. The contour symbol is placed above or below the spot or seam weld symbol to indicate a flat or convex contour. Welds that require a flat, but not flush, contour require an additional note in the tail of the welding symbol.

Spot and seam welds requiring mechanical finishing after welding to obtain the desired contour are specified by a letter placed next to the weld contour symbol. The finishing methods that may be specified are C - chipping, H - hammering, G - grinding, M - machining, R - rolling, or U - unspecified. Contour and finish are specified in the welding symbol with the quality or degree of finish specified by the manufacturer.

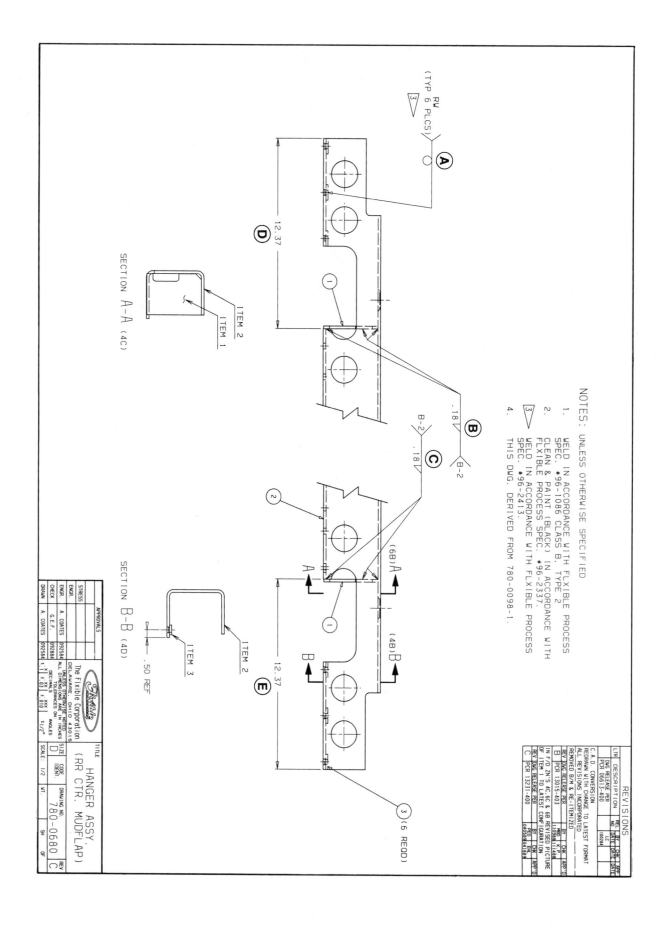

Review Questions

Name_____ Date _____

True-False

T   F   **1.** Spot welds are used to weld light-gauge sheet metal.

T   F   **2.** The GTAW process is the most common welding process used for spot welds.

T   F   **3.** Spot welds are specified on the welding symbol with an unfilled circle.

T   F   **4.** All spot weld size dimensions are placed on the same side of the reference line as the spot weld symbol.

T   F   **5.** Projection welds use the resistance welding process to join weld parts.

T   F   **6.** Seam welds are specified on the welding symbol by a solid circle.

T   F   **7.** The length of a seam weld is specified by a dimension located to the left of the seam weld symbol.

T   F   **8.** Seam welds are always parallel to the axis of the weld.

T   F   **9.** Stud arc welding is the most common type of stud welding.

T   F   **10.** The size of the stud used in a stud weld is indicated to the left of the stud weld symbol.

T   F   **11.** The cross-sectional shape of a spot weld is roughly circular.

T   F   **12.** Spot welds completed from the arrow side are specified by placing the spot weld symbol above the reference line of the welding symbol.

T   F   **13.** Projection welds are used for the same applications as spot welds.

T   F   **14.** The welding process used for seam welds is specified in the tail of the welding symbol.

T   F   **15.** The stud weld symbol is not used with any side significance.

Completion

_____   **1.** Spot welds join overlapping joint members with fusion at the _____ surfaces.

_____   **2.** Spot welds completed from the other side are specified by placing the spot weld symbol _____ the reference line of the welding symbol.

_____   **3.** Spot weld size is specified by placing a dimension to the _____ of a spot weld symbol.

_____ 4. The strength of a spot weld is specified by placing a number to the _____ of the spot weld symbol.

_____ 5. Spot weld spacing in a straight line is specified with a(n) _____ dimension placed to the right of the weld symbol.

_____ 6. Seam welds completed from the other side are specified by placing the seam weld symbol _____ the reference line of the welding symbol.

_____ 7. Intermittent seam welds are specified on the print with length and pitch dimensions separated by a(n) _____.

_____ 8. Length and pitch of intermittent seam welds are based on the weld located _____ to the axis of the weld.

_____ 9. More than one stud weld in a straight line is specified with a pitch dimension placed to the _____ of the stud weld symbol.

_____ 10. Spot and seam weld contour is the shape of the completed weld face in _____ view.

_____ 11. Spot welds with no side significance have the spot weld symbol _____ on the reference line.

_____ 12. The number of spot welds is specified with a number in parentheses placed _____ the spot weld symbol.

_____ 13. Stud welds are indicated on the welding symbol by a circle containing a(n) _____.

_____ 14. Spot and seam weld _____ is the shape of the completed weld face in sectional view.

_____ 15. The location of the first and last stud weld is indicated by _____.

Matching — Spot and Seam Welds

_____ 1. Three spot welds; 4″ OC; arrow side

_____ 2. Four spot welds; 3″ OC; arrow side

_____ 3. ¼″ spot weld; arrow side

_____ 4. ¼″ spot weld; other side

_____ 5. ¼″ spot weld; no side significance

_____ 6. ¼″ seam weld; no side significance

_____ 7. ¼″ seam weld; other side

_____ 8. Three spot welds; 4″ OC; other side

_____ 9. Four spot welds; 3″ OC; other side

_____ 10. ¼″ seam weld; arrow side

Trade Competency Test

Name_____ Date _____

Refer to the Hanger Ass'y. print on page 254.

_____   **1.** The drawing is drawn at _____ scale.

_____   **2.** Weld A specifies welding in _____ places.

_____   **3.** All dimensions are given in _____.

_____   **4.** Weld A is welded in accordance with Flxible Process Specification #_____.

T   F   **5.** Weld B specifies a weld leg size of .18″.

_____   **6.** Tolerance for decimals in hundredths is ±_____″.

_____   **7.** Weld A specifies a(n) _____ weld.

_____   **8.** Weld A specifies the weld to be completed using the _____ welding process.

T   F   **9.** The maximum leg size permitted for weld C is .21″.

_____   **10.** The print is a size _____ sheet.

_____   **11.** Weld C specifies a(n) _____ weld.

_____   **12.** The number of welds specified by weld B is _____.

_____   **13.** The number of the drawing is _____.

_____   **14.** The drawing was redrawn using the _____ process.

_____   **15.** Section B-B shows a(n) _____ weld in section view.

_____   **16.** The minimum size of the reference dimension in Section B-B is _____″.

_____   **17.** The drawing was checked by _____.

_____   **18.** The number of welds specified by weld C is _____.

_____   **19.** There are a total of _____ holes in the piece.

_____   **20.** The part is to be cleaned and _____ in accordance with Flxible Process Specification #96-2337.

_____ **21.** The maximum size for dimension D is 12.51″.

T   F   **22.** The engineer for the drawing was A. Coates.

T   F   **23.** The drawing derived from 780-0098-1.

T   F   **24.** The Flxible Corporation is located in Maryland.

T   F   **25.** Section A-A specifies a fillet weld before painting.

_____ **26.** The minimum size for dimension E is _____″.

_____ **27.** Long _____ lines are used to show that the hanger assembly continues in the center.

_____ **28.** A total of _____ parts labeled as Item 3 are required.

_____ **29.** A total of _____ parts labeled as Item 1 are required.

_____ **30.** This is Revision _____ of the print.

T   F   **31.** Any angle dimensions have a tolerance of $\pm\frac{1}{2}°$.

T   F   **32.** The drawing was completed by A. Coates.

T   F   **33.** The part was approved for stress by the engineer.

T   F   **34.** Item 1 parts are shown in detail in Section B-B.

T   F   **35.** Both cutting planes are taken in the same direction of sight.

_____ **36.** Weld B is to be welded in accordance with Flxible Process Specification #_____, Class B, Type 2.

_____ **37.** Weld B specifies a(n) _____ weld.

_____ **38.** All welds at weld C are made on the _____ side.

_____ **39.** All welds at weld B are made on the _____ side.

T   F   **40.** There is no side significance given for welds at Weld B.

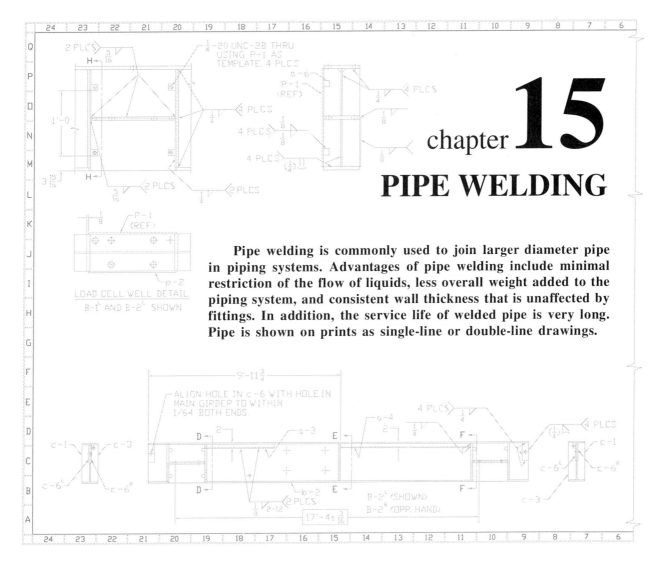

chapter 15
PIPE WELDING

Pipe welding is commonly used to join larger diameter pipe in piping systems. Advantages of pipe welding include minimal restriction of the flow of liquids, less overall weight added to the piping system, and consistent wall thickness that is unaffected by fittings. In addition, the service life of welded pipe is very long. Pipe is shown on prints as single-line or double-line drawings.

PIPE REPRESENTATION

Pipe is widely used in industry to transport liquids or gases in a system. Welding is the most common method of connecting pipe. Lines and symbols are used on prints to convey information regarding design, layout, materials, and operation of the system. Pipe is commercially available in a wide rage of diameters, wall thicknesses and length. Pipe is connected to increase its run.

Dimensions indicate location and size of pipe and fittings required. A *fitting* is a standard connection used to join two or more pieces of pipe. Pipe is represented on prints with single- or double-line drawings. Standard symbols are used to represent pipe, fittings, valves, etc. See Figure 15-1. See Appendix.

Single-Line Drawings

Single-line drawings use a single line to represent pipe and fittings. The single line acts as a center line for pipe and fittings throughout the piping system. Single-line drawings allow more information to be conveyed about piping systems in a smaller space than double-line drawings.

Orthographic, isometric, or oblique views are used with single-line drawings. Orthographic and isometric views are the most common. Pipe runs are shown in three planes in isometric drawings.

In single-line drawings, pipe and pipe fittings are dimensioned center to center. Center lines are used as extension and/or dimension lines to indicate location or a change of direction. The length of pipe, with or without fittings, is dimensioned from end to end.

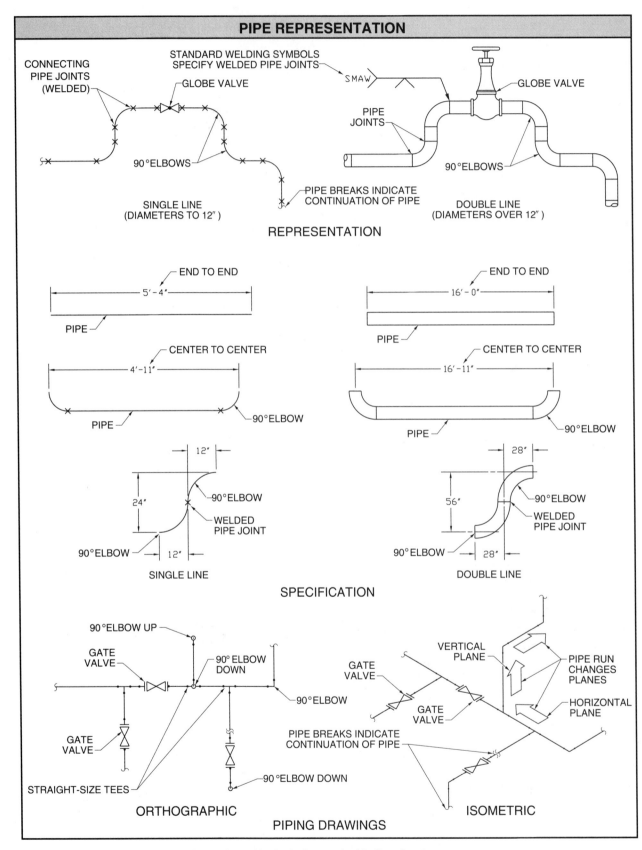

Figure 15-1. Pipe is represented on prints with single-line or double-line drawings.

Pipe breaks on single-line drawings are indicted by a double-curved line perpendicular to the pipe run. Single-line drawings are used to show pipe and fittings with nominal diameters up to 12″.

Double-Line Drawings

Double-line drawings use two lines to represent pipe and fittings in a pictorial form. Double-line drawings show fittings in greater detail than single-line drawings. Consequently, double-line drawings are more time-consuming than single-line drawings.

Standard welding symbols are used with double-line drawings to specify welding operations required. The center line of the pipe and fittings is used for location and size dimensions. Pipe break symbols are used to indicate that a section of the run has been removed to conserve space on the prints. Double-line drawings are commonly used to show pipe and fittings with nominal diameters over 12″.

PIPE CONNECTIONS

Pipes may be connected by welded joints, flanged joints, or screwed joints. The size of the pipe, material from which the pipe is manufactured, and use of the pipe dictate the connection method used. See Figure 15-2.

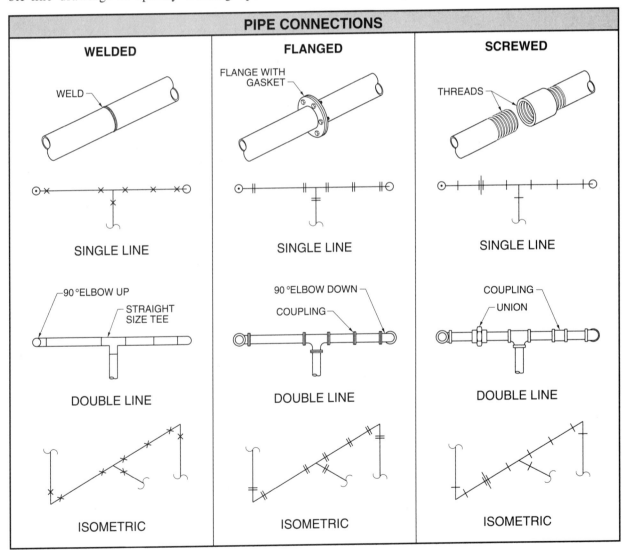

Figure 15-2. Pipes may be connected by welded joints, flanged joints, or screwed joints.

Welded Joints

Welding is the most common method of joining larger diameters of pipe. Pipe welding joins members without the addition of fittings. Welded pipe connections cause less restriction of the flow of materials in the pipe. When properly welded there is no gap, and joint strength is consistent with the surrounding sections of pipe.

Outside pipe dimensions are not greatly affected by the weld area. Welded joints are not designed to be disassembled. Repair or replacement of a line requires removal of a section by thermal or mechanical cutting.

Welded joints may be made with butt-welded fittings or socket fittings. See Figure 15-3. Butt welds require edge preparation of the pipe. A backing ring is recommended for butt-welded pipe having a wall thickness over ¾″.

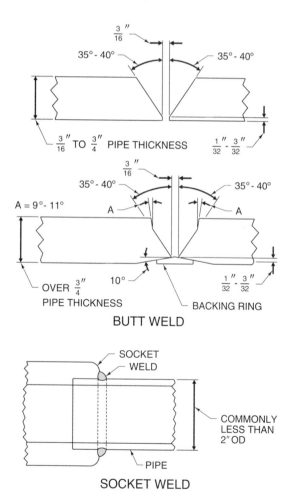

Figure 15-3. Welded joints may be made with butt-welded fittings or socket fittings.

Socket fittings join pipe and fittings with sleeves that are welded, brazed, or soldered. This joining method does not require any edge preparation. Socket fittings are commonly used on pipe less than 2″ OD.

Flanged Joints

Flanged pipe fittings allow removal of pipe and fittings with less labor costs than other joining methods. Bolts and nuts are used to secure flanges together. A gasket is placed between the flanges prior to assembly. The flanges may be cast as an integral part of the fitting or welded to the fitting.

Screwed Joints

Screwed pipe fittings use threaded pipe and fittings to join members in the piping system. Fittings and pipe are joined in sequence by tightening clockwise. Union fittings are located throughout the run to facilitate efficient disassembly. A *union* is a fitting consisting of three parts having threads and flanges which draw together when tightened. Unions can be installed with little disturbance to the position of the pipes.

PIPE CLASSIFICATION

Pipe is dimensioned using nominal inside diameter and nominal pipe size (NPS). For pipe up to 12″ in diameter, the NPS is the same as the inside diameter. Pipe wall thickness is specified using one of two standards. ANSI classifies pipe thicknesses using schedule numbers (Schedule 40, 60, 80, etc.). ASTM and ASME classify pipe wall thickness using nominal inside diameter in the three different weights available as required by load requirements: standard (STD), extra-strong (XS), and double extra-strong (XXS).

The nominal inside diameter is determined using standard weight pipe. Extra-strong pipe and double extra-strong pipe have a reduced inside diameter as the wall thickness is increased. Outside wall thickness remains constant in the three weight classifications.

For example, 3″ nominal inside diameter pipe has an outside diameter of 3.500″. In the standard weight, the pipe has an inside diameter of 3.068″. In the extra-strong weight, the pipe has an inside diameter of 2.900″. In the double extra-strong weight, the pipe has an inside diameter of 2.300″. See Appendix.

In pipe welding, pipe can be classified as thick wall or thin wall. Thick-wall pipe is pipe with a wall thickness greater than ⁵⁄₁₆″. Thin-wall pipe is pipe with a wall thickness of ⅛″ to ⁵⁄₁₆″. The wall thickness of the pipe to be welded determines the joint preparation required. For example, thin-wall pipe with a wall thickness of ⅛″ commonly does not require edge preparation. Pipe with wall thicknesses greater than ⅛″ usually requires edge preparation. See Figure 15-4.

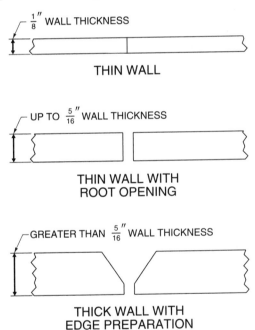

Figure 15-4. Thick-wall pipe is pipe with a wall thickness greater than ⁵⁄₁₆″.

Pipe in most common applications is made from wrought steel, low-carbon steel, or low-alloy steel. In special applications, chromium-molybendum, nickel steel, stainless steel, copper, aluminum, or brass piping is used. The selection of pipe materials is based on the pressures and the material controlled in the pipe. For example, steam lines in a nuclear power plant must be strong enough to withstand high pressures without the possibility of failure caused by defective welds or corrosion.

PIPE WELDING PROCESS

Pipe welding processes vary, depending on the pipe material, size, and function of the piping system. The composition of the pipe dictates the weld filler metal and welding process used. For example, welding stainless steel pipe with a ⅜″ wall thickness requires deep penetration. Pipe in a critical application may be purged with shielding gas. The GTAW process is used for weld purity. The pipe welding process includes edge preparation, fit-up, and welding.

Edge Preparation

Pipe weld joints requiring edge preparation are prepared as V-groove welds. Pipe weld V-groove parts include the root opening, root face, groove face, bevel angle, and included angle. For example, pipe weld specifications commonly used on ⁵⁄₁₆″ thick-wall pipe call for a 75° included angle (37½° bevel angle). See Figure 15-5. The root opening is approximately ³⁄₃₂″ to ⅛″. The root face is approximately ³⁄₃₂″ to ⅛″.

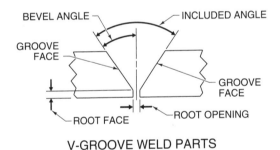

V-GROOVE WELD PARTS

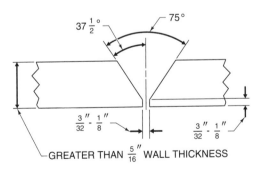

STANDARD THICK WALL
EDGE PREPARATION DIMENSIONS

Figure 15-5. Pipe weld joints that require edge preparation are prepared as V-groove welds.

Root opening and included angle increase as required for pipe having greater wall thicknesses. In addition, variations of edge preparation are required to assure proper penetration with the welding process used. The groove face can be altered for access and to limit filler metal required without compromising weld strength.

Pipe is commonly purchased with the bevel already machined on the ends. A bevel can be cut or ground in the field. However, this method is time-consuming and is less accurate than machine-beveled pipe. Edge preparation is determined by pipe wall thickness, pipe composition, and the welding process used. Pipe welding techniques are also affected by pipe dimensions, location, requirements of the pipe and the weld, and welding equipment available.

Fit-Up

Fit-up is the positioning of pipe with other pipe or fittings before welding. Fit-up of pipe includes the alignment of the pipe ends and consistent proper root opening. Surfaces affected in the welding process must be clean and free of foreign matter before welding. The pipe is aligned with pipe jigs, spacers, or consumable inserts.

A *pipe jig* is a device which holds sections of pipe or fittings before tack welding. A *tack weld* is a weld that joins the joint members at random points to keep the joint members from moving out of their required positions. Tack welds are welds used to hold joint members temporarily in place. See Figure 15-6.

Spacers provide a gap between the joint members. The joint members are tack welded, and the spacers are removed before welding. Backing rings are commonly used in the GTAW process. Backing rings have spacers attached to a ring which fits in the pipe before welding.

Consumable inserts are melted in the welding process and become part of the filler metal added to the weld joint. A consumable insert provides the proper opening of the weld joint and becomes part of the filler metal in the welding process. Information pertaining to the consumable insert is included in the tail of the welding symbol. The classification numbers 1, 2, 3, 4, and 5 refer to AWS classes of consumable inserts.

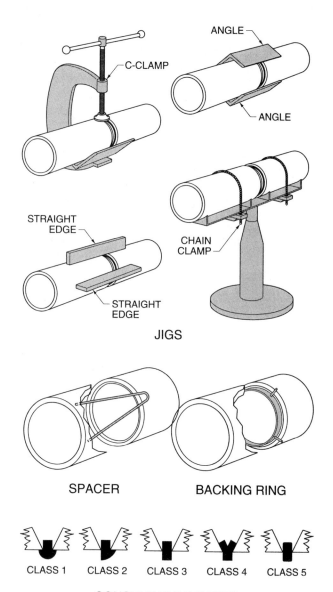

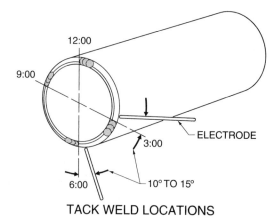

Figure 15-6. Jigs are used for pipe fit-up during tack welding.

Welding

Thin-wall pipe is commonly welded using the downhill welding technique. In the downhill welding technique, the weld is started at the twelve o'clock position and ends at the six o'clock position. The other side is started at the twelve o'clock position to meet with the previous weld at the six o'clock position.

The downhill welding technique uses gravity as molten weld metal forms the weld bead. This allows faster welding speed with adequate deposition without an excess amount of penetration. See Figure 15-7.

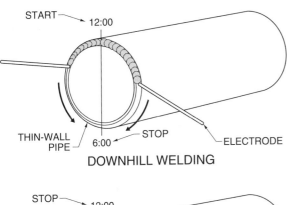

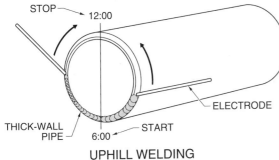

Figure 15-7. Thin-wall pipe is commonly welded using the downhill welding technique. Thick-wall pipe is commonly welded using the uphill welding technique.

Thick-wall pipe is commonly welded using the uphill welding technique. In the uphill welding technique, the weld is started at the six o'clock position and ends at the twelve o'clock position. The other side is started at the six o'clock position to meet with the previous weld at the twelve o'clock position. During the welding process, gravity pulls the metal back into the weld root and toward the weld bead. This results in greater amounts of weld metal applied.

Welding thick-wall pipe requires multiple weld passes. A *weld pass* is the single progression of welding along a joint. A *weld bead* is the weld that results from a weld pass. Multiple weld passes for thick-wall pipe include the root pass, hot pass, filler passes (if required), and cover pass.

Root Pass. A *root pass* is the initial weld pass that provides complete penetration through the thickness of the joint member. Amperage is set to provide maximum penetration without burn-through or excessive weld metal deposited inside the pipe. The root pass deposits weld metal in the root of the weld as a "keyhole" formed by the penetration of the weld pass. An improper root pass will cause the entire weld to be rejected. Subsequent weld passes cannot compensate for a defective root pass. See Figure 15-8.

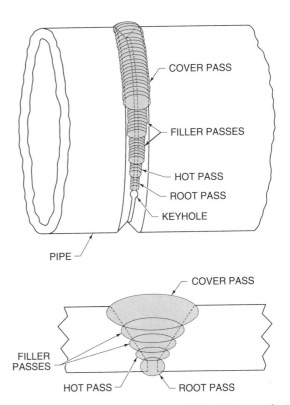

Figure 15-8. A root pass is the initial weld pass that provides complete penetration through the thickness of the joint member.

Hot Pass. A *hot pass* is the weld pass that penetrates deeply into the root pass and the root face of the joint. This eliminates the possibility of slag inclusions or porosity left from the root pass. Any undercutting at the toes of the root pass is eliminated by the hot pass.

Filler Pass. A *filler pass* is the weld pass that fills the remaining portion of the weld after the root pass and hot pass. The number of filler passes required depends on the wall thickness, the included angle, the size of the electrode, and the welding process used. Each of the filler passes penetrates completely into the previous weld bead.

Cover Pass. A *cover pass* is the final weld pass deposited. The cover pass is usually weaved to provide a complete cover of the weld and present a neat appearance. The cover pass also provides the weld reinforcement required for strength and protection.

PIPE WELDING STANDARDS

Pipe welding standards assure pipe welding quality. Pipe welding standards have been established by the American Petroleum Institute (API), the American Society of Mechanical Engineers, and the American Welding Society for specifying material requirements, preparation, welder proficiency, and weld testing. In some cases, other agencies adopt these standards for specific applications. For example, the U.S. Department of Defense has adopted several standards published by the American Welding Society.

Welder Certification

Information regarding test specifications and procedures are detailed in the AWS publication "Standard for Welding Procedure and Performance Qualification" (ANSI/AWS B2.1). As part of the procedure for qualification, forms are completed which specify all welding directives and requirements ("Welding Procedure Specifications"), and specific welding variables and procedures used to complete an acceptable test weld ("Procedure Qualification Record"). See Appendix.

Certification of welders is based on the proficiency of the welder making welds in specific positions. Pipe weld positions are based on AWS test positions 1G, 2G, 5G, 6G, and 6GR. G stands for groove welds. See Figure 15-9.

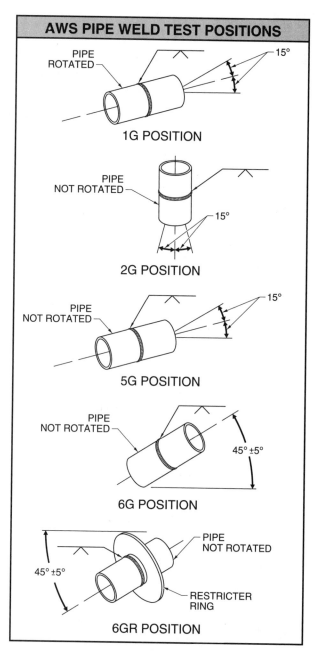

Figure 15-9. Pipe weld positions are based on AWS test positions 1G, 2G, 5G, 6G, and 6GR.

1G Position. Test position 1G is positioning the axis of the pipe in a horizontal position or within 15° above or 15° below horizontal. The weld is completed in the flat position as the pipe is rotated.

2G Position. Test position 2G is positioning the axis of the pipe in a vertical position or within 15°

on any side of vertical. The weld is completed in the horizontal position. The pipe is fixed in position and is not rotated during welding.

5G Position. Test position 5G is positioning the pipe in a horizontal position or within 15° above or below horizontal. The weld is completed in the flat, vertical, and overhead fixed positions. The pipe is fixed in position and is not rotated during welding.

6G Position. Test position 6G is positioning the axis of the pipe at a 45° angle, plus or minus 5°. The pipe is fixed in position and is not rotated during welding.

6GR Position. Test position 6GR is positioning the axis of the pipe at a 45° angle, plus or minus 5°. The pipe is fixed in position and is not rotated during welding. This position requires a restricter ring around the pipe before welding.

Pipe Weld Testing

Pipe weld testing can be conducted by using destructive and nondestructive examinations. Destructive testing is used primarily in the qualification of welder performance. In destructive testing, a test specimen is removed from a weld and analyzed using tensile or guided bend testing. See Figure 15-10.

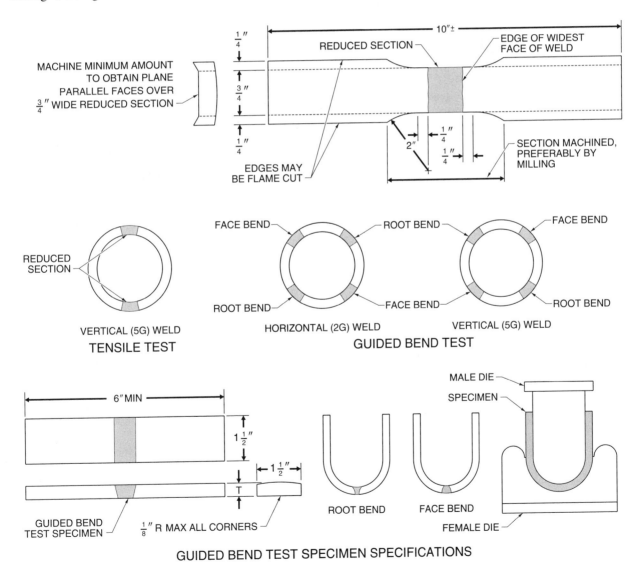

Figure 15-10. Tensile and guided bend tests are used in destructive testing.

In the tensile test, the test specimen has an area reduced which includes the weld area. The test specimen is subjected to force in opposite directions. The tensile strength achieved is compared with weld strength requirements.

The guided bend test uses a test specimen in a guided bend tester to identify points of failure when the test specimen is subjected to bending force. The guided bend test requires two test specimens, one for the face bend test and one for the root bend test.

In the face bend test, the test specimen is checked for proper weld fusion, porosity, inclusions, or other weld defects. The test specimen is placed in the guided bend tester with the face side down.

In the root bend test, the test specimen is tested for penetration. The test specimen is placed in the guided bend tester with the root side down. After bending, the test specimen is inspected for cracks.

Nondestructive testing is used to determine weld quality without adversely affecting the performance of the weld. In nondestructive testing of pipe welds, the most common methods used are penetrant, proof, radiographic, ultrasonic, and visual testing. Nondestructive test requirements are specified on prints using nondestructive examination symbols. See Figure 15-11.

NONDESTRUCTIVE EXAMINATION	
METHOD	LETTER DESIGNATION
Acoustic emmision	AET
Electromagnetic	ET
Leak	LT
Magnetic particle	MT
Neutron radiographic	NRT
Penetrant	PT
Proof	PRT
Radiographic	RT
Ultrasonic	UT
Visual	VT

METHODS USED FOR TESTING PIPE WELDS

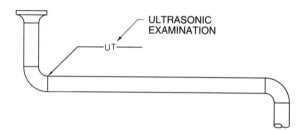

Figure 15-11. Nondestructive testing does not adversely affect the performance of the weld.

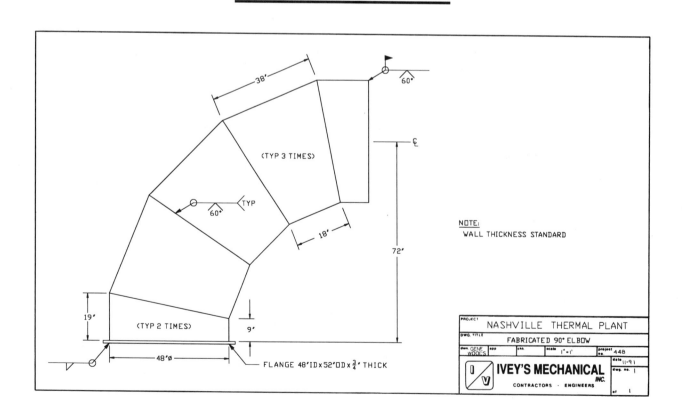

Review Questions

Name_____ Date _____

True-False

T   F   **1.** The single line of a single-line drawing acts as a center line for pipe and fittings.

T   F   **2.** Single-line drawings show fittings in greater detail than double-line drawings.

T   F   **3.** A backing ring is recommended for butt-welded pipe having a wall thickness over $\frac{5}{16}''$.

T   F   **4.** Extra-strong pipe and double extra-strong pipe have a reduced inside diameter as the wall thickness is increased.

T   F   **5.** The root opening and included angle decrease as required for pipe having greater wall thicknesses.

T   F   **6.** Spacers can be used to provide a gap between joint members while tack welding.

T   F   **7.** The classification numbers 1, 2, 3, 4, and 5 refer to ASME classes of consumable inserts.

T   F   **8.** The weld is started at the six o'clock position when welding thin-wall pipe.

T   F   **9.** The filler pass is the final weld pass deposited.

T   F   **10.** In the 6G position, the pipe is rotated during welding.

T   F   **11.** In single-line drawings, pipe and pipe fittings are dimensioned center to center.

T   F   **12.** Flanged pipe fittings cannot be removed.

T   F   **13.** Pipe is dimensioned using nominal inside diameter and nominal inside pipe size.

T   F   **14.** Thin-wall pipe with a wall thickness of $\frac{1}{8}''$ commonly does not require edge preparation.

T   F   **15.** Pipe is commonly purchased with the bevel already machined on the ends.

T   F   **16.** In the uphill welding technique, the weld is started at the twelve o'clock position.

T   F   **17.** In the 2G position, the pipe is rotated during welding.

T   F   **18.** The guided bend test requires two test specimens.

T   F   **19.** Radiographic and ultrasonic testing are not destructive to the test specimen.

T   F   **20.** Welding thick-wall pipe may require multiple weld passes.

Completion

_____ **1.** A(n) _____ is a standard connection used to join two or more pieces of pipe.

_____ **2.** Pipe runs are shown in _____ planes in isometric drawings.

_____ **3.** Single-line drawings are used to show pipe and fittings with nominal diameters up to _____".

_____ **4.** A pipe _____ symbol indicates that a section of the pipe run has been removed to conserve space on the prints.

_____ **5.** _____ is the most common method of joining larger diameters of pipe.

_____ **6.** Socket fittings are commonly used on welded pipe less than _____" OD.

_____ **7.** A(n) _____ is a fitting consisting of three parts having threads and flanges which draw together when tightened.

_____ **8.** ANSI classifies pipe thicknesses using _____ numbers.

_____ **9.** Thick-wall pipe is pipe with a wall thickness greater than _____".

_____ **10.** _____ is the positioning of pipe with other pipe or fittings before welding.

_____ **11.** The _____ pass is the initial weld pass that provides complete penetration through the thickness of the joint member.

_____ **12.** Thick-wall pipe is commonly welded using the _____ welding technique.

_____ **13.** In test position 1G, the pipe is positioned in a(n) _____ position.

_____ **14.** In the root bend test, the specimen is tested for _____.

_____ **15.** _____ testing is used to determine weld quality without adversely affecting the performance of the weld.

_____ **16.** In single-line drawings, the length of the pipe is dimensioned from _____.

_____ **17.** A pipe _____ is a device which holds sections of pipe on fittings before tack welding.

_____ **18.** A(n) _____ weld is a temporary weld that joins joint members at random points to keep the joint members from moving out of their required positions.

_____ **19.** Information pertaining to consumable inserts is included in the _____ of the welding symbol.

_____ **20.** The downhill welding technique uses _____ as molten metal forms the weld bead.

_____ **21.** Test position 2G is completed in the _____ position.

_____ **22.** In the face bend test, the test specimen is placed in the guided bend tester with the face side _____.

_____ 23. A(n) _____ pass is the weld pass that penetrates deeply into the root pass and root face of the joint.

_____ 24. Test Position 6GR requires a(n) _____ ring around the pipe before welding.

_____ 25. _____ testing is used primarily in the qualification of welder performance.

_____ 26. _____ pipe dimensions are not greatly affected by the weld area.

_____ 27. Double extra-strong pipe is abbreviated as _____.

_____ 28. The _____ angle of a V-groove weld contains one-half the degrees of the included angle.

_____ 29. Pipe breaks on single-line drawings are indicated by a double-curved line _____ to the pipe run.

_____ 30. In the _____ position, the pipe is not rotated during welding.

Matching — Pipe Symbols

_____ 1. Connecting pipe joint; flanged

_____ 2. Straight-size tee; screwed

_____ 3. 45° elbow; welded

_____ 4. Sleeve; welded

_____ 5. Straight-size tee; welded

_____ 6. Sleeve; flanged

_____ 7. Connecting pipe joint; welded

_____ 8. 45° elbow; screwed

_____ 9. Straight-size tee; flanged

_____ 10. Sleeve; screwed

Matching — V-Groove Weld Parts

_____ **1.** Root opening

_____ **2.** Included angle

_____ **3.** Bevel angle

_____ **4.** Root face

_____ **5.** Groove face

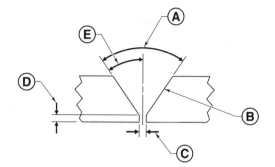

Matching — Welded Pipe Symbols

_____ **1.** 45° elbow

_____ **2.** Connecting pipe joint

_____ **3.** 90° elbow; turned up

_____ **4.** Concentric reducer

_____ **5.** 90° elbow

_____ **6.** Break symbol

_____ **7.** Union

_____ **8.** Straight-size tee

_____ **9.** Lateral

_____ **10.** Gate valve

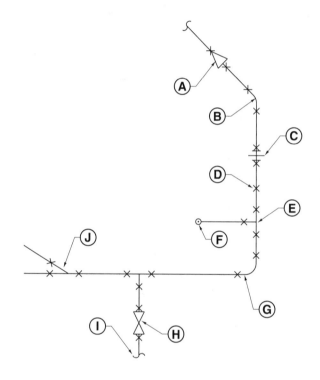

Matching — Test Positions

_____ **1.** 1G position

_____ **2.** 2G position

_____ **3.** 5G position

_____ **4.** 6G position

_____ **5.** 6GR position

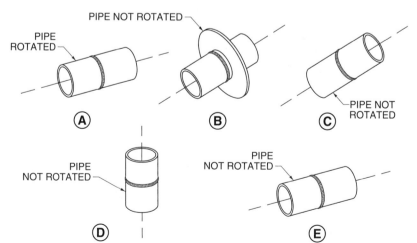

Trade Competency Test

Name＿＿＿＿＿＿＿＿＿＿＿＿＿＿＿＿＿＿＿＿＿＿＿＿＿＿＿＿＿ Date ＿＿＿＿＿＿＿＿＿＿＿＿

Refer to the Fabricated 90° Elbow print on page 268.

＿＿＿＿＿＿＿＿ 1. The elbow contains a total of ＿＿＿＿＿＿ sections and a flange.

＿＿＿＿＿＿＿＿ 2. The flange is ＿＿＿＿＿＿″ thick.

＿＿＿＿＿＿＿＿ 3. A(n) ＿＿＿＿＿＿ weld is specified to join the flange to the elbow sections.

T　　F　　4. The inside diameter of the elbow sections is 52″.

＿＿＿＿＿＿＿＿ 5. A(n) ＿＿＿＿＿＿ weld is specified to join elbow sections.

＿＿＿＿＿＿＿＿ 6. The outside diameter of the elbow sections is ＿＿＿＿＿＿″.

＿＿＿＿＿＿＿＿ 7. The end of the elbow without the flange is welded in the ＿＿＿＿＿＿.

T　　F　　8. The flange specifies sixteen ¾″ drilled holes for mounting the elbow to a base.

＿＿＿＿＿＿＿＿ 9. The drawing is drawn to the scale of 1″ = ＿＿＿＿＿＿′.

＿＿＿＿＿＿＿＿ 10. The circumference of the elbow sections is ＿＿＿＿＿＿″.
　　　　　　　A. 18
　　　　　　　B. 56
　　　　　　　C. 150.72
　　　　　　　D. 172.78

＿＿＿＿＿＿＿＿ 11. The project number for the elbow is ＿＿＿＿＿＿.

＿＿＿＿＿＿＿＿ 12. The flange has an outside diameter of ＿＿＿＿＿＿″.
　　　　　　　A. 9
　　　　　　　B. 19
　　　　　　　C. 48
　　　　　　　D. 52

T　　F　　13. All welding on the elbow is completed in the field.

＿＿＿＿＿＿＿＿ 14. Welds specified to join elbow sections have a(n) ＿＿＿＿＿＿° groove angle.

＿＿＿＿＿＿＿＿ 15. The flange has an inside diameter of ＿＿＿＿＿＿″.
　　　　　　　A. 18
　　　　　　　B. 38
　　　　　　　C. 48
　　　　　　　D. 52

T   F   **16.** The drawing was drawn by _____.

T   F   **17.** All 60° bevel welds are welded-all-around.

_____   **18.** Two sections have a minor height of _____".
          A. 9
          B. 18
          C. 19
          D. 38

_____   **19.** Three sections have a major height of _____".
          A. 9
          B. 18
          C. 19
          D. 38

_____   **20.** The horizontal center line of the uppermost elbow section is _____" above the flange.

_____   **21.** The wall thickness of the pipe is _____.

          A. determined by the installer
          B. $1\frac{1}{4}$"
          C. standard
          D. not given

_____   **22.** The drawing was completed on _____.

_____   **23.** The total shortest distance of the elbow sections from horizontal to vertical is _____.

_____   **24.** The total longest distance of the elbow sections from horizontal to vertical is _____.
          A. 108
          B. 144
          C. 152
          D. 168

T   F   **25.** Ivey's Mechanical, Inc. performs engineering and contracting work.

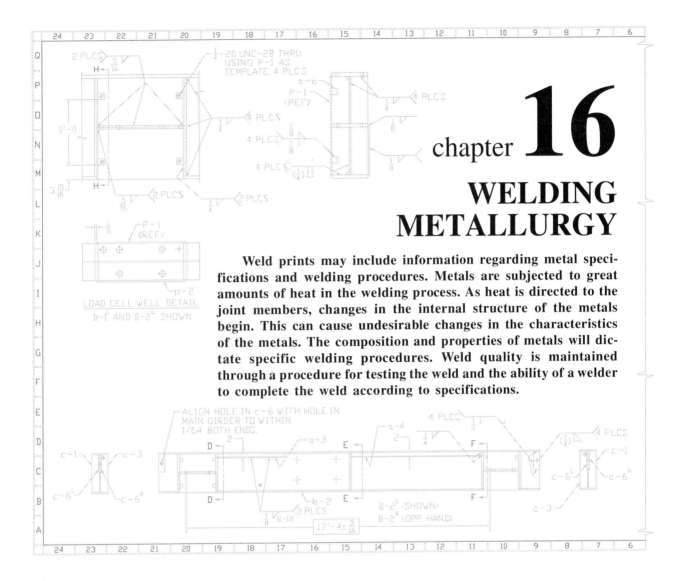

chapter **16**

WELDING METALLURGY

Weld prints may include information regarding metal specifications and welding procedures. Metals are subjected to great amounts of heat in the welding process. As heat is directed to the joint members, changes in the internal structure of the metals begin. This can cause undesirable changes in the characteristics of the metals. The composition and properties of metals will dictate specific welding procedures. Weld quality is maintained through a procedure for testing the weld and the ability of a welder to complete the weld according to specifications.

METALS

A *metal* is a material consisting of one or more chemical elements having crystalline structure, high thermal and electrical conductivity, the ability to be deformed when heated, and high reflectivity. A metal is either a pure metal or an alloy.

A *pure metal* is a metal that consists of one chemical element. Pure metals are usually soft and have relatively low strength. This limits their applications. An *alloy* is a metal that consists of more than one chemical element, with at least one of the elements being a pure metal. For example, iron is a pure metal that is very soft and ductile. With the alloying additions of carbon and other chemical elements, carbon steel

(an alloy) is produced, increasing the hardness and the tensile strength.

All weld joints must maintain mechanical and physical properties of the weld metal similar to the base metal. The chemical composition of the weld metal is commonly the same as the base metal. Weld metal different from the chemical composition of the base metal may be used, however, provided the properties meet engineering specifications.

Metal Structure

Metals have no arrangement of atoms or specific structure when heated to a liquid state. The atoms have no boundaries and move freely. As the metal

275

begins to cool, the reduction of heat energy in the atoms limits their movement. The atoms no longer move freely and begin to form space lattices. A *space lattice* is the uniform pattern produced by lines connected through the atoms.

Different metals have different space lattices, resulting in different pattern formation. For example, iron has a body-centered space lattice. Aluminum has a face-centered space lattice. Magnesium has a close-packed hexagonal space lattice. See Figure 16-1.

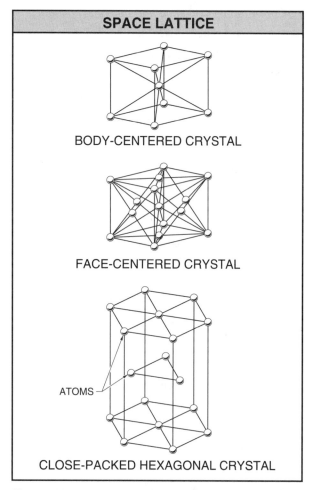

SPACE LATTICE

BODY-CENTERED CRYSTAL

FACE-CENTERED CRYSTAL

ATOMS

CLOSE-PACKED HEXAGONAL CRYSTAL

Figure 16-1. A space lattice is the uniform pattern produced by lines connected through the atoms.

In a solid state, the space lattice is known as a crystal. A *crystal* is a solid composed of atoms arranged in a pattern that is repetitive in three dimensions. All metals are composed of crystalline solids formed from atoms that have a specific space lattice.

Crystals grow as a metal continues to cool from its liquid state. The crystals grow freely until com-

ing in contact with other crystals. The intersecting of crystals with other crystals further shapes the crystals into grains. A *grain* is an individual crystal in a metal that has multiple crystals.

The size of crystals, grains, and grain structure is dependent on the length of time the atoms can move freely. Large crystals tend to form when a metal is cooled slowly. When the metal is cooled rapidly, the crystal size tends to be smaller.

Each metal has a unique microstructure of space lattice, grain structure, and grain boundaries. A *microstructure* is the microscopic arrangement of the components within a metal. Most metals used in industry are alloys which have more than one microstructure. The microstructures overlap and interface with one another to provide certain properties. The microstructure is affected by the heat input and the cooling rate resulting from the welding process. Metals are broadly classified into two groups, ferrous and nonferrous metals.

Ferrous Metals. A *ferrous metal* is a metal with iron as a major alloying element. Ferrous metals are magnetic. Pure iron is very soft, extremely ductile, and melts at a low temperature. Iron is commonly alloyed with carbon and other elements to form carbon steel. The addition of carbon increases the tensile strength of iron. Carbon steel is a ferrous metal and is the most common metal used in fabrication and manufacturing. Carbon steel can be grouped into low-, medium-, and high-carbon steel classifications, depending on the percent of carbon. See Figure 16-2.

Other ferrous metals include stainless steel and cast iron. Stainless steel contains steel alloyed with chromium or chromium and nickel. This provides high corrosion resistance. Cast iron is a metal consisting of 2% to 4% carbon. Cast iron is commonly formed into shapes by casting into molds. Cast iron cannot be shaped by bending, forging, or other forming methods. Cast iron is classified as white, gray, malleable, and ductile (nodular).

CARBON STEEL	
CLASSIFICATION	% CARBON
Low	0.10 – 0.25
Medium	0.26 – 0.50
High	.051 – 1.03

Figure 16-2. Carbon steel is the most common metal used in fabrication and manufacturing.

Nonferrous Metals. A *nonferrous metal* is a pure metal, other than iron or metals with iron as a major alloying element. Nonferrous metals do not contain iron, are not magnetic, and have distinctive color differences. Compared to ferrous metals, nonferrous metals are relatively soft. In addition, nonferrous metals as a whole have lower melting temperatures than ferrous metals. Nonferrous metals commonly used in manufacturing include aluminum, copper, brass, and bronze.

Aluminum is silver in color, has good electrical and thermal conductivity, and is relatively light weight. Pure aluminum is very ductile and is often alloyed with silicon, copper, manganese, zinc, and other elements. Pure aluminum has approximately $\frac{1}{5}$ the tensile strength of alloyed aluminum and weighs approximately $\frac{1}{3}$ less than iron. Aluminum has a coefficient of thermal expansion 50% greater than copper, and twice that of steel and cast iron. Aluminum will not rust. It is used in sheet form and for electrical conductors, castings, machine and motor parts.

Copper is an element which easily forms chemical bonds with other elements. Copper is reddish brown in color and has excellent electrical and thermal conductivity characteristics. It is used for electrical conductors, coins, cooking utensils, and as an alloy in brass and other nonferrous metals.

Brass is the most common alloy of copper. It is yellow in color and consists of 1% to 50% zinc, depending on the application. Brass is very soft, but may be hardened by adding tin. Brass is used for brazing light materials and cast iron. It is also used for making ornaments, jewelry, and clocks.

Bronze is reddish yellow in color and is an alloy of copper that consists of copper with elements other than zinc, such as tin, silicon, aluminum, and beryllium. Bronze is used for making castings, coins, ornaments, and bearings for machines.

Metal Identification

Weld prints often list metals to be used in the fabrication of the print part. The metals are listed in the Bill of Materials or as a note on the prints. See Figure 16-3.

Metals not labelled are best identified by qualified personnel using the proper instruments and equipment. In maintenance and repair, welders

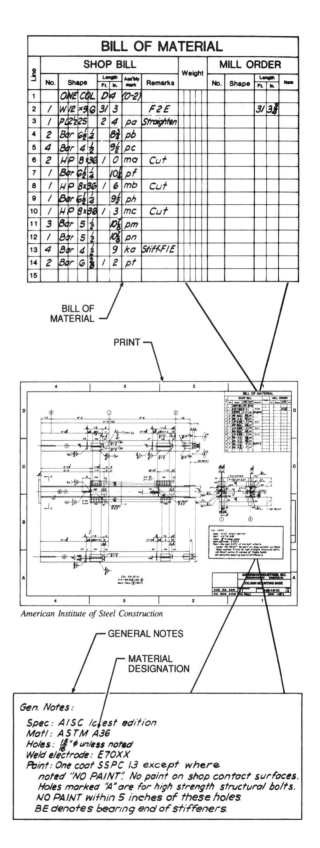

American Institute of Steel Construction

Figure 16-3. Weld prints specify materials required using applicable material designations.

must often identify the type of metal without the benefit of other means of identification. Not all methods of identification work on all metals. However, some quick observations can simplify the identification process. For example, if the metal appears to have rusted or is magnetic, it contains iron. If the metal is nonmagnetic and reddish in color, it is copper or an alloy of copper.

Methods commonly used to identify metals include the color, spark, chemical, magnetic, file, torch, chip, and fracture tests. When using any of these tests, the welder must consider the potential damage to the sample or weld part.

Color Test. The *color test* is a metal identification test that identifies metals by their color. In the color test, the color of the metal is examined. For example, gray usually indicates a type of carbon steel. Zinc and lead are bluish white, tin is silvery white, etc.

Spark Test. The *spark test* is a metal identification test that identifies metals by the shape, length, and color of spark emitted from contact with a grinding wheel. For example, a small red spark emitted from a metal sample on a grinding wheel indicates white or gray cast iron.

Chemical Test. The *chemical test* is a metal identification test that uses chemicals which react when placed on certain types of metals. Chemical tests can also be used on metal shavings immersed in chemicals, which change color to indicate composition.

Magnetic Test. The *magnetic test* is a metal identification test that checks for the presence of iron in a metal. A magnet is placed on the sample to determine magnetism.

File Test. The *file test* is a metal identification test in which a file is used to indicate the hardness of steel compared with that of the file. For example, if the metal can be marked with a file only after great effort, the steel is likely to be high-carbon steel or harder. If the file marks the sample easily, it is likely to be mild steel.

Torch Test. The *torch test* is a metal identification test that can be used to identify a metal by its color change with the application of heat, its melting point, and its behavior in the molten state. For example, if there is no color change before melting occurs and a great deal of heat is required to obtain a puddle, the metal is probably aluminum. Additionally, if the molten puddle is clear and free from floating impurities or pores, it will weld easily.

Chip Test. The *chip test* is a metal identification test that identifies metal by the shape of its chips. A chisel is used to remove small chips of the sample for examination. Long and curled chips are taken from mild steel and soft metals, such as aluminum. Short, broken chips are taken from cast steel. Sample chips are very difficult to obtain from high-carbon steel.

Fracture Test. The *fracture test* is a metal identification test that breaks the metal sample to check for ductility and grain size. For example, mild steel bends before fracturing.

HEAT

Welding introduces a tremendous amount of heat to specific areas of a weld part. Heat is directed to the weld area. As metal is heated, it expands. As metal is cooled, it contracts. An entire weld part that is heated expands and contracts in all directions. If unrestricted, the weld part returns to the same size and original shape after cooling. If restricted, the weld part expands in greater amounts in the unrestricted axis or axes. See Figure 16-4.

The weld part is restricted in most welding situations. If a portion of a weld part is heated, the heated portion is affected differently by the expansion caused from heat and contraction from cooling. Different rates of heating and cooling in a weld part result in residual stress and strain.

Properties of metals such as melting point, thermal conductivity and coefficient of thermal expansion, and dimensions determine how heat from welding affects the metal. The welding process used, dimensions, and properties of the base metal affect the metal when welding. Heat applied during the welding process should be minimized to reduce distortion and weld defects.

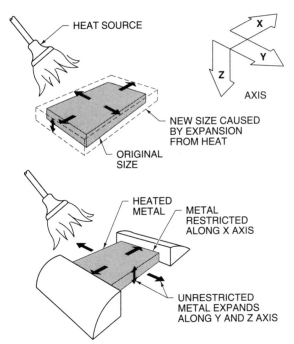

Figure 16-4. Metal expands when heated and contracts when cooled.

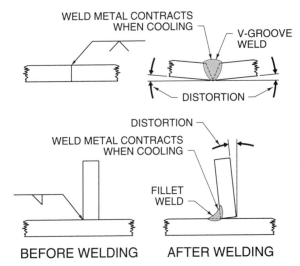

Figure 16-5. Distortion is a change in the original shape of the metal.

Distortion

Distortion is a change in the original shape of the metal as metal expands when heated and contracts when cooled. Generally, thinner metals are more susceptible to distortion because of the amount of heat conducted. Joint members in the correct position before welding contract as the weld and heat-affected zone cool. See Figure 16-5. For example, in a V-groove weld, a large amount of weld metal is deposited to the sides of the joint members. The molten weld metal meets the weld metal at the weld interface. As the weld part cools, more contraction occurs where the metal has been affected by the welding heat. This results in contraction and distortion. The same distortion occurs on a single fillet T-joint.

Distortion Control. Distortion control is accomplished by joint member control, joint member position, heat sinks, or welding technique. Movement of joint members can be controlled using fixtures or tack welds to maintain the proper position prior to, during, and after welding. Joint member position can be controlled by allowing for expansion and contraction of the weld area to minimize the effects of distortion caused from the weld. The an-

gle of offset (X°) required for joint members depends on their dimensions.

Chill blocks can be used as heat sinks to draw heat away from the weld area. The chill blocks are allowed to cool before reuse.

Welding techniques, such as backstep welding, intermittent welding, consistent time between welding passes on opposite sides, and minimizing the number of passes, reduces the distortion caused from welding. *Backstep welding* is welding in which individual passes are made in the opposite direction of the weld. See Figure 16-6.

Heat-Affected Zone

The *heat-affected zone* is the area of base metal in which the mechanical properties and structure are affected by the welding process. See Figure 16-7. Atoms in metal become very active when the metal is heated. Bonds that were present in the metal at a lower temperature are lost. This occurs at the melting point of the metal. The weld metal, the weld interface, and the metal surrounding the weld area assume certain properties, depending on the amount of heat applied and the rate at which cooling occurs.

The size of the heat-affected zone is determined by the type and size of metal, amount of heat applied, welding process, and the type of weld joint. The weld part of metals with high thermal conductivity is affected to a higher degree than metals with low thermal conductivity.

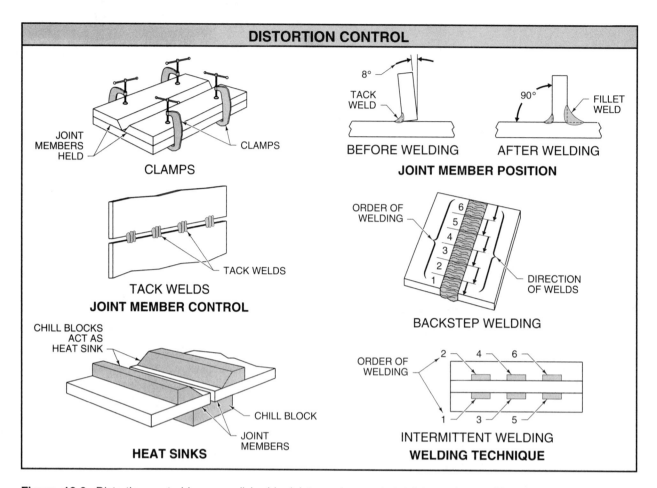

Figure 16-6. Distortion control is accomplished by joint member control, joint member position, heat sinks, or welding technique.

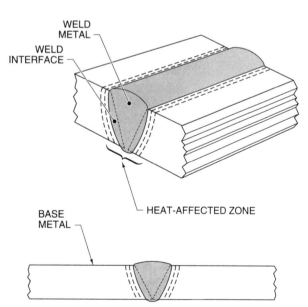

Figure 16-7. The heat-affected zone is the area of base metal in which the mechanical properties and structure are affected by the welding process.

Thinner metals have less mass and heat more quickly than thicker metals. The GTAW process provides a more concentrated heat source than the OAW process. This results in less heat applied to the entire weld part in the GTAW process than the OAW process. A V-groove joint has a larger weld area than a square-groove joint and has less heat input during welding. Effects from heat can be controlled by preheating and postheating the weld and/or weld part.

Preheating

Preheating is the application of heat to the base metal before welding to reduce the temperature difference between the weld metal and the surrounding base metal. Depending on the shape of the weld parts, heat is applied using an oxyfuel torch, electric resistance heaters, or gas burners to preheat the weld parts to the required temperature.

The temperature required for preheating varies and is specified by the engineer or by applicable codes. Temperature indicating crayons can be used to indicate when the desired temperature is achieved. On small assemblies, the weld parts can be placed in a furnace to provide more accurate control over the temperature of the weld parts.

Interpass temperature is the temperature of the weld area between passes of a multiple-pass weld. The maximum or minimum interpass temperature may be specified. If there is a concern regarding overheating, the maximum interpass temperature is specified. If there is a concern regarding a lack of heat, the minimum interpass temperature is specified. The time and temperature required for interpass temperature varies and is specified by the engineer or by applicable codes.

Postheating

Postheating is the application of heat to the weld part after welding to facilitate a controlled cooling rate. Postheating is primarily concerned with stress relieving of the weld joint and joint members. This reduces the internal residual stresses and minimizes changes in dimensions resulting from the welding process.

Postheating methods are the same as preheating methods. The time and temperature required for postheating varies and is specified by the engineer or by applicable codes. Temperature can be monitored using a temperature indicating crayon. Stress relieving can also be accomplished by peening. Peening is performed by striking the weld area with a hammer to relieve stress.

WELD QUALITY

Quality control of welds is maintained by minimizing weld defects through weld testing and welder certification. Weld quality is maximized by using prescribed weld specifications and welding techniques and procedures in accordance with the applicable federal, state, and local codes. Additional ASME, API, and AWS standards and codes may also apply. Other agencies, such as ASTM, AISI, and AISC, may also specify weld require-

ments. AISI is the American Iron and Steel Institute. See Appendix.

Weld Defects

Weld defects are undesirable characteristics of a weld which may cause the weld to be rejected. The cause for rejection is determined by the inability of the weld part to meet standards or codes applied to the specific weld. Common weld defects include incomplete fusion, weld cracks, undercutting, overlapping, porosity, slag inclusions, and segregation. See Figure 16-8.

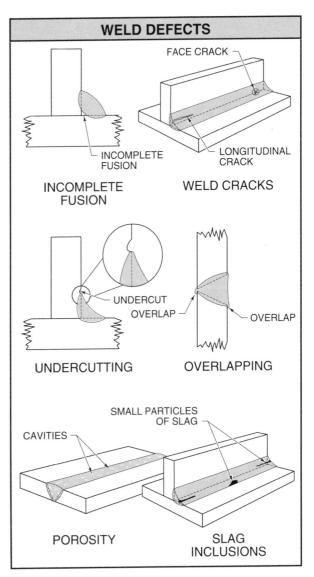

Figure 16-8. Weld defects reduce joint strength and may cause the weld to be rejected.

Incomplete fusion is the discontinuance of a weld where complete fusion does not occur between the weld metal and the fusion faces of the joint members. *Weld cracks* are linear discontinuities that occur in the base metal, weld interface, or the weld metal. Weld cracks may appear as face or longitudinal cracks.

Undercutting is creating a groove that is not completely filled by weld metal in the base metal during the welding process. *Overlapping* is extending weld metal beyond the weld toes or weld root. *Porosity* is a cavity or cavities in the weld metal or weld interface caused by trapped gas. *Slag inclusions* are small particles of slag (cooled flux) trapped in the weld metal which prevent complete penetration.

Segregation is the separation of elements comprising the base metal. Segregation is commonly caused by improper preheat and postheat treatment. For example, stainless steel overheated during welding loses some corrosion resistance as the nickel becomes separated from the steel.

Weld Testing

Weld testing is required to verify the strength of a given weld. Weld testing is conducted in accordance with Welding Procedure Specifications (WPS) established for the weld.

The WPS specifies the base metal, preheating and postheating specifications, and filler metal for a given weld. The Procedure Qualification Record (PQR) documents the specific welding variables and procedures used to complete an acceptable test weld and the results of the required weld tests. Weld testing can be conducted using destructive and nondestructive testing methods.

Destructive Testing. *Destructive testing* is any type of testing that damages the test part (specimen). It is used primarily in the qualification of welders. In destructive testing, a test specimen is removed from a weld and analyzed. The weld cannot be used after destructive testing.

Destructive tests using mechanical methods include bend, break, tension, fracture, and shear tests. See Figure 16-9. These tests are detailed in the publication "Standard Methods for Mechanical Testing of Welds" (ANSI/AWS B4.0). Destructive tests using chemicals and instrument analysis may also be used as required.

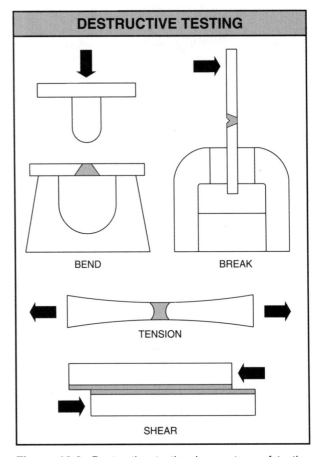

Figure 16-9. Destructive testing is any type of testing that damages the specimen.

Nondestructive Testing. *Nondestructive testing* is any type of testing that leaves the test part undamaged. It is used to determine weld quality without adversely affecting the performance of the weld.

The most common nondestructive testing methods used are visual, penetrant, radiographic, ultrasonic, and magnetic particle. Nondestructive test requirements are specified on prints using nondestructive examination symbols (NDE). See Figure 16-10.

The method of examination required can be specified on a separate reference line of the welding symbol or as a separate NDE symbol.

When used as a separate reference line, the order of operation is the same as multiple welding operations. The reference line furthest from the arrowhead indicates the last operation to be performed. When used separately, NDE symbols

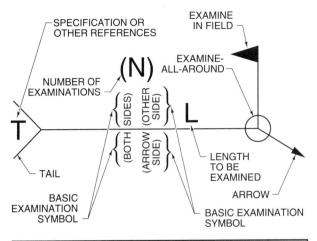

EXAMINATION METHODS	
METHOD	LETTER DESIGNATION
Acoustic emmision	AET
Electromagnetic	ET
Leak	LT
Magnetic particle	MT
Neutron radiographic	NRT
Penetrant	PT
Proof	PRT
Radiographic	RT
Ultrasonic	UT
Visual	VT

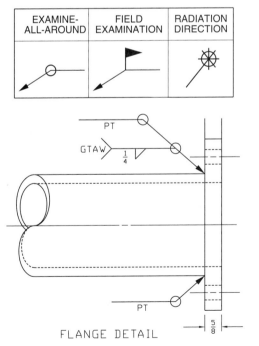

Figure 16-10. Nondestructive testing is any type of testing that leaves the test part undamaged.

include an arrow, reference line, examination letter designation, dimensions, areas, number of examinations, supplementary symbols, tail, and specifications or other references.

The location of the basic examination symbol is treated like a welding symbol. Examination required on the arrow side of the welded part is indicated by having the basic examination symbol located on the same side as the arrowhead.

Examination required on the other side of the welded part is indicated by locating the basic examination symbol on the opposite side of the arrowhead. Examinations required on both sides are indicated by having the symbol on both sides. When side has no significance, the basic examination symbol is located in the center of the reference line.

The arrow of the NDE symbol can be used to indicate direction of radiographic examination. The direction of radiographic examination can also be specified with a separate line in the required direction and by including the required angle on the print for clarity.

Welder Certification

Welder certification is not governed by a central agency. Efforts have been made to standardize a nationwide certification program. However, because of the many requirements of the welding industry, these efforts have not been successful. Individual companies and agencies have preferred to certify welders in their own controlled environments based on the applicable WPS. WPS's and PQR's must be filed for each weld procedure and welder qualified. This allows testing on specific fabrications produced and/or regulated by the agency.

The WPS specifies welds to be completed in a specific test weld position. Test weld positions are based on AWS groove, fillet, and stud weld test positions. In groove and fillet weld tests, 1 is flat, 2 is horizontal, 3 is vertical, and 4 is overhead. See Figure 16-11.

Information regarding test specifications and procedures are detailed in the AWS publication "Standard for Welding Procedure and Performance Qualification" (AWS B2.1). This publication also includes WPS and PQR forms. See Appendix.

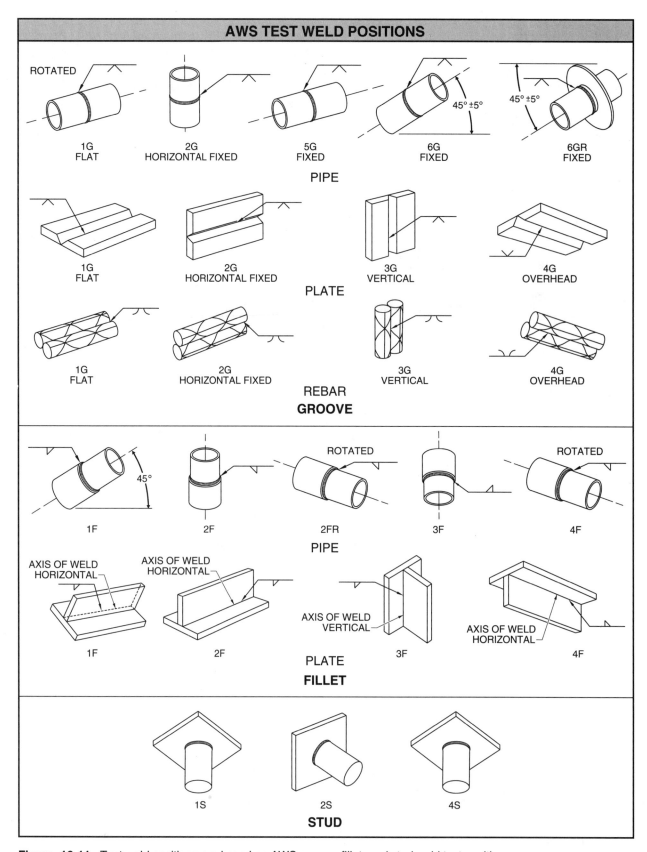

Figure 16-11. Test weld positions are based on AWS groove, fillet, and stud weld test positions.

Review Questions

Name_____ Date _____

Completion

_____ **1.** A(n) _____ metal is a material that consists of one chemical element.

_____ **2.** A(n) _____ is a metal that consists of more than one chemical element with at least one of the elements being a pure metal.

_____ **3.** A space lattice is the uniform _____ produced by lines connected through atoms.

_____ **4.** A(n) _____ is an individual crystal in a metal that has multiple crystals.

_____ **5.** A(n) _____ is the microscopic arrangement of the components within a metal.

_____ **6.** A(n) _____ metal is a metal with iron as a major alloying element.

_____ **7.** Cast iron is a metal consisting of 2% to 4% _____.

_____ **8.** A(n) _____ metal is a pure metal other than iron or metals with iron as a major alloying element.

_____ **9.** _____ is the most common alloy of copper.

_____ **10.** The _____ test is a metal identification test that breaks the metal sample to check for ductility and grain size.

_____ **11.** _____ is a change in the original shape of metal as it heats or cools.

_____ **12.** _____ temperature is the temperature of the weld area between passes of a multiple pass weld.

_____ **13.** _____ testing is any type of testing that leaves the test part undamaged.

_____ **14.** _____ is extending weld metal beyond the weld toes or weld root.

_____ **15.** _____ is the application of heat to the weld part after welding to facilitate a controlled cooling rate.

True-False

T   F   **1.** The chemical composition of the weld metal is commonly the same as the base metal.

T   F   **2.** Metals have no arrangement of atoms or particular structure when heated to a liquid state.

T   F   **3.** A space lattice is also known as a diamond.

T   F   **4.** Heat has no effect on the microstructure of a metal.

T   F   **5.** Crystals grow as metal cools from its liquid state.

T   F   **6.** Carbon steel is a ferrous metal.

T   F   **7.** Carbon steel is the most common metal used in fabrication and manufacturing.

T   F   **8.** Ferrous metals have no magnetic properties.

T   F   **9.** Ferrous metals are generally softer than nonferrous metals.

T   F   **10.** Cast iron can be shaped by bending or forging.

T   F   **11.** Copper is reddish brown in color.

T   F   **12.** Porosity is a cavity in the weld metal caused by trapped gas.

T   F   **13.** The NDE method can be specified on a separate reference line of the welding symbol.

T   F   **14.** Welder certification is governed by a central agency.

T   F   **15.** Postheating methods are the same as preheating methods.

Matching — Test Weld Positions

_____   **1.** Pipe; 1G

_____   **2.** Rebar; 1G

_____   **3.** Plate; 4G

_____   **4.** Pipe; 2F

_____   **5.** Plate; 1G

_____   **6.** Plate; 3F

_____   **7.** Plate; 3G

_____   **8.** Plate; 4F

_____   **9.** Pipe; 2G

_____   **10.** Rebar; 4G

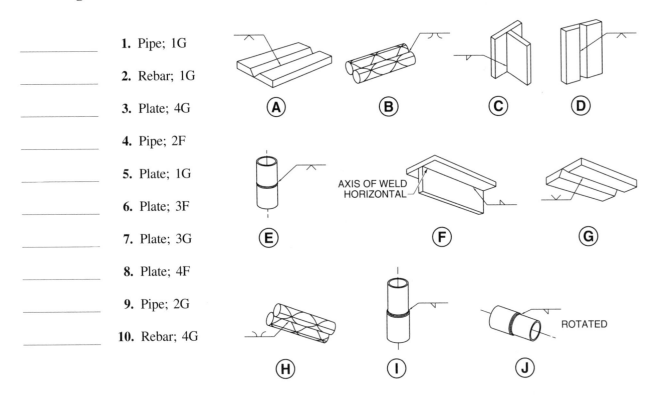

Trade Competency Test

Name _____   Date _____

WELDING PROCEDURE SPECIFICATION (WPS)

Date __2/4/92__   Identification __302__
Company name __National Fabricators, Inc., Atlanta, GA__   Revision __NA__
Supporting PQR no.(s) __1__   Type - Manual ( ) Semi-Automatic (**X**)
Welding proces (es) __GMAW__   Machine ( )   Automatic ( )
Backing: Yes ( ) No (**X**) __NA__
Backing material (type) __NA__
Material number __M-1__ Group __M-1__ To material number __M-1__ Group __M-1__
Material spec. type and grade __ASTM A36__ To material spec. type and grade __ASTM A36__
Base metal thickness range: Groove __½ inch__ Fillet __NA__
Deposited weld metal thickness range __As required__
Filler metal F no. __8__   A no. __1__
Spec. no. (AWS) __A5.18 and A5.28__   Flux tradename __NA__
Electrode-flux (Class) __E70S-5__   Type __NA__
Consumable insert: Yes ( ) No (**X**)   Classifications __NA__
                                            Shape __NA__
Position(s) of joint __Flat 1G__   Size __NA__
Welding progression: Up (—) Down (—)   Ferrite number (when reqd.) __NA__
PREHEAT: **GAS:**
  Preheat temp., min __400°F__   Shielding gas(es) __Argon/Oxygen__
  Interpass temp., max __NA__   Percent composition __95/5__
  (continuous or special heating, where   Flow rate __40-50 CFH__
  applicable, should be recorded)

WELDING PROCEDURE SPECIFICATION (WPS)

Refer to the Welding Procedure Specification (WPS) form on page 287.

_____   **1.** The welding position required is _____.

_____   **2.** The welding process specified is GMAW which stands for _____ welding.

_____   **3.** The electrode specified is _____.

_____   **4.** The material specified is ASTM _____.

_____   **5.** Filler metal specified is AWS filler metal F No. _____.

T   F   **6.** The weld is heated to 400°F after welding.

_____   **7.** The weld type specified is a(n) _____ weld.

_____   **8.** The shielding gas flow is _____ cubic feet per hour.

_____   **9.** The shielding gas specified is 95% _____.

_____   **10.** A 400°F _____ temperature is specified.

Welding progression: Up (—) Down (—) Ferrite number (when reqd.) _____ **NA**

PREHEAT:

Preheat temp., min _____ **400°F**

Interpass temp., max _____ **NA**
(continuous or special heating, where
applicable, should be recorded)

POSTWELD HEAT TREATMENT:

Temperature range _____ **NA**

Time range _____ **NA**

Tungsten electrode, type and size _____ **NA**

Mode of metal transfer for GMAW: Short-circuiting ( ) Globular ( ) Spray (✗)

Electrode wire feed speed range: _____ **105–110 IPM**

Stringer bead ( ) Weave bead ( ) Peening: Yes ( ) No (✗)

Oscillation _____ **As required**

Standoff distance **⅝ – ⅞**

Multiple ( ) or single electrode (✗)

Other _____

GAS:

Shielding gas(es) **Argon/Oxygen**

Percent composition **95/5**

Flow rate **40–50 CFH**

Root shielding gas **NA**

Trailing gas composition **NA**

Trailing gas flow rate **NA**

| Filler metal | | | | | Current | | |
Weld layer(s)	Process	Class	Dia.	Type & polarity	Amp range	Volt range	Travel speed range
ALL	GMAW	E70S-5	3/32"	DCEP	400-425	26-27	30 IPM

10° PUSH ANGLE

Approved for Production by _____
Employer

Note: Those items that are not applicable should be marked N.A.

WELDING PROCEDURE SPECIFICATION (WPS), CONTINUED

Refer to the Welding Procedure Specification (WPS) form, continued, on page 288.

_____ 1. The postweld heat treatment is _____.

_____ 2. The diameter of the electrode is _____″.

_____ 3. The wire feed speed is _____ inches per minute.

_____ 4. The electrode polarity is _____.

_____ 5. The metal transfer is by the _____ method.

_____ 6. The travel speed is _____ inches per minute.

_____ 7. A(n) _____° push angle is specified.

T   F   8. The tungsten electrode required is 3/32″ in diameter.

T   F   9. A standoff distance of 3/4″ meets specifications.

T   F   10. The work is grounded to the negative lead from the welding machine.

Trade Test 1

Name_____ Date _____

Refer to the Checking Bracket — Left Model 7260 print on page 309.

_____ **1.** Item No. 2 weighs _____ lb.

_____ **2.** The finish on the bracket weldment is black _____.

T   F   **3.** Location dimensions for the four holes are shown in Detail 1.

_____ **4.** The bracket weldment is a total of _____″ high.

_____ **5.** The overall dimensions of the plate are _____.

_____ **6.** The drawing was completed by _____.

_____ **7.** Item No. 2 is located _____″ from the front edge of the plate.

_____ **8.** The 2½″ × 5″ × 6″ bar weighs _____ lb.

T   F   **9.** Detail 1 shows the front view of the plate.

_____ **10.** A ½″ × 45° _____ is ground on three edges of Item No. 2.

_____ **11.** Item No. 4 contains _____ cu in.

_____ **12.** Item No. 3 is _____″ thick.

_____ **13.** Each hole of the bracket weldment has a diameter of _____″.

_____ **14.** Item No. 4 is located _____″ from the left edge of the plate.

T   F   **15.** Center lines are shown in both detail views.

_____ **16.** The front and top views are drawn as _____ projections.

T   F   **17.** Four bars of various sizes are required for the bracket weldment.

T   F   **18.** The scale of the drawing is ¼″ = 1′-0″.

_____ **19.** The manufacturer of the bracket is _____.

_____ **20.** _____ welds are used exclusively on the bracket weldment.

_____ **21.** Item No. 3 is _____″ in length.

T   F   **22.** The Bill of Materials contains four items.

T   F   **23.** The four holes are on $6\frac{1}{2}″$ and 12″ centers.

T   F   **24.** All holes are centered $1\frac{1}{2}″$ from the nearest two edges of the plate.

_____ **25.** Item No. 2 is _____″ in its longest dimension.

_____ **26.** The ID number of Item No. 4 is _____.

_____ **27.** Item No. _____ is the heaviest part.

_____ **28.** The overall dimensions of Item No. 3 are _____.

_____ **29.** The edges of the holes are represented by _____ lines in the front view.

T   F   **30.** No revisions have been made to this drawing.

_____ **31.** A(n) _____ fillet weld joins Item No. 4 to Item No. 1.

_____ **32.** The top of Item No. 3 must be positioned to make a(n) _____° angle with the left side of Item No. 4.

_____ **33.** The date the drawing was completed is _____.

T   F   **34.** All welding on the bracket must conform to AWS specifications.

_____ **35.** The overall dimensions of Item No. 4 are _____.

T   F   **36.** Three pieces of bar stock are required for fabrication of the bracket.

T   F   **37.** The Checking Bracket is shown on drawing number TB200609.

T   F   **38.** The tolerance for fractional dimensions is $\pm\frac{1}{32}″$.

T   F   **39.** The bracket weldment must be preheated before welding.

_____ **40.** The front edge of Item No. 4 is _____″ from the center of the nearest hole.

_____ **41.** Item No. 3 is located _____″ from the right edge of the plate.

_____ **42.** The height of Item No. 3 is _____″.

_____ **43.** Item No. 3 extends _____″ above the top of Item No. 2.

_____ **44.** Item No. 4 extends _____″ closer to the front edge of the plate than Item No. 2.

T   F   **45.** Section lines are used in Detail 2 to show the chamfer.

Trade Test 2

Name_____ Date _____

Refer to the Upright print on page 310.

_____ 1. All tubes for the Upright have a wall thickness of _____″.

_____ 2. The plate is _____″ thick.

_____ 3. The sides of the horizontal tube are welded to the top of the vertical tube with a(n) _____ weld.

_____ 4. The straps are spaced apart _____″ on center.

_____ 5. A total of _____ straps are required for one assembly.

_____ 6. The diameter of the drilled holes in the strap is _____″.

_____ 7. The diameter of the hole in the vertical tube is _____″.

T   F   8. The holes in the vertical tube are centered from end to end.

_____ 9. A total of _____ hole(s) is/are required for the horizontal tube.

T   F   10. The manufacturing operation for producing the length and width of the plate is not given.

T   F   11. Both tubes are polished after assembly.

_____ 12. The plates are welded to the vertical tube with a(n) _____ weld.

T   F   13. All fillet welds are arrow side.

T   F   14. All fillet welds have $\frac{3}{16}$″ legs.

_____ 15. The vertical tube is _____″ in length.

_____ 16. The strap is _____″ in length.

_____ 17. The plate is welded-_____ with a fillet weld to the tubes.

_____ 18. The centers of the holes in the horizontal tube are located _____″ above the bottom of the plates.

_____ 19. The plate measures $\frac{1}{4}$ × _____ × $4\frac{5}{8}$.

_____ 20. The vertical and horizontal tubes are cut from _____ steel.

_____ **21.** The hole in the plate is centered _____″ from the front edge.

_____ **22.** The holes in the horizontal tube are centered _____″ from the far end.

_____ **23.** The material for the plate is _____ steel.

_____ **24.** The Upright was drawn on a(n) _____ size sheet of paper.

_____ **25.** A total of _____ assemblies are required per unit.

_____ **26.** Holes in the spacers are spaced on _____″ centers.

T   F   **27.** The strap is sawed, punched, and sandblasted.

T   F   **28.** The strap is $\frac{3}{8}″$ thick.

T   F   **29.** The center of the hole in the vertical tube is $\frac{5}{8}″$ from the near end.

T   F   **30.** The complete assembly is painted with Sherwin Williams 2801-99993, semi-gloss, white latex paint.

_____ **31.** The _____ is centered on the tubes.

_____ **32.** The drawing is drawn to the scale of _____.

_____ **33.** The ends of the plates extend _____″ on both sides of the vertical tube.

_____ **34.** The maximum angle of the horizontal tube to the vertical tube is _____.

_____ **35.** A total of _____ coats of paint are required.

_____ **36.** The diameter of the holes in the horizontal tube is _____″.

_____ **37.** The total number of pieces required for one assembly is _____.

T   F   **38.** The centerpoint for the 2^R on the straps is offset $\frac{1}{2}″$ from the center line between holes.

T   F   **39.** The horizontal tube is $10\frac{1}{8}″$ in length.

T   F   **40.** The drawing was approved by HJR.

_____ **41.** The edges of the strap extend _____″ on either side of the vertical tube.

T   F   **42.** The two notes on the print are specific notes.

T   F   **43.** The drawing was approved by TRH.

T   F   **44.** The drawing was completed on 2-19-92.

T   F   **45.** Square-groove welds join the strap to the vertical tube.

Trade Test 3

Name_____ Date _____

Refer to the Spring Tower Coulter Mount print on page 311.

_____ **1.** Part _____ is a round tube.

_____ **2.** Weld D specifies a(n) _____ weld.

_____ **3.** The Spring Tower Coulter Mount requires _____ drawing sheet(s).

_____ **4.** The width of part 20-425-984 is _____″.

_____ **5.** Holes in part 20-425-985 are located _____″ from center.

_____ **6.** Weld H is _____″ long.

_____ **7.** The distance from the center of part 6-136-630 to the end of part 20-425-988 is _____″.

_____ **8.** Weld E has a(n) _____″ leg size.

_____ **9.** The drawing was completed on _____.

_____ **10.** The length of part 20-425-987 is _____″.

T   F   **11.** Weld E specifies that two welds be made.

T   F   **12.** The drawing specifies that two 20-425-986 parts are required.

T   F   **13.** Part 20-425-984 has the same width and length dimensions.

T   F   **14.** The length of Weld I is 6″.

_____ **15.** The length of part 20-425-984 is _____″.

_____ **16.** The drawing number is _____.

_____ **17.** Weld F is specified to be welded _____.

_____ **18.** Part 20-425-986 is specified to be positioned at a(n) _____° angle from the horizontal center line.

T   F   **19.** Part 20-425-988 is 5″ long.

T   F   **20.** Weld I specifies a fillet weld.

_____ **21.** Part 20-425-985 requires a total of _____ holes.

_____ **22.** Weld G specifies a(n) _____″ leg size.

_____ **23.** Weld D specifies that welds are to be made in a total of _____ locations.

_____ **24.** Weld E specifies a weld length of _____″.

_____ **25.** Parts 6-136-630 and 20-425-986 are joined with _____ welds.

_____ **26.** The hole diameters of part 20-425-985 are _____″.

_____ **27.** Part _____ is a rectangular-shaped tube.

_____ **28.** Weld B specifies a weld _____″ long.

_____ **29.** Weld F specifies a(n) _____″ leg size.

_____ **30.** Part 20-425-986 requires a total of _____ hole(s).

_____ **31.** Parts 20-425-988 and 20-425-986 are joined with welds having a(n) _____″ leg size.

T   F **32.** The drawing was drawn by AL.

T   F **33.** Tolerances are specified as $\pm\frac{1}{16}$″ for all dimensions.

T   F **34.** Weld F specifies a weld 3″ long.

T   F **35.** Weld D specifies a concave weld contour.

_____ **36.** Holes on part 20-425-985 are located _____″ from center to the edge of the part.

_____ **37.** Part 20-425-984 requires a total of _____ hole(s).

_____ **38.** Calkins Manufacturing Company is located in the city of _____, WA.

_____ **39.** Weld J specifies a weld _____″ long.

_____ **40.** Weld A joins part 20-425-985 and part _____.

_____ **41.** Parts 20-425-987 and 20-425-986 are joined welds having a(n) _____″ leg size.

_____ **42.** Weld C specifies a weld _____″ long.

T   F **43.** Part 6-136-630 is centered on part 20-425-985.

T   F **44.** Part 20-425-987 is less than 6″ long.

T   F **45.** Part 20-425-986 is less than 21″ long.

Trade Test 4

Name_____ Date _____

Refer to the Front Hinge Mount print on page 312.

_____   **1.** Weld H specifies a(n) _____″ leg size.

_____   **2.** Weld G joins parts 20-410-969 and _____.

_____   **3.** Part 10-425-997 is _____″ wide.

_____   **4.** Part 10-425-936 requires holes _____″ in diameter.

_____   **5.** Part 10-425-700 is _____″ thick.

_____   **6.** Part 10-425-700 and 10-425-701 are located _____″ apart.

_____   **7.** Weld B was revised on _____.

_____   **8.** Weld E is _____″ long.

_____   **9.** Weld G specifies the _____ welding process.

T   F   **10.** There are three sheets to the drawing.

T   F   **11.** Part 10-425-936 requires four holes.

T   F   **12.** Weld L specifies $\frac{3}{8}$″ depth of penetration.

_____   **13.** The hole in part 10-425-997 is _____″ in diameter.

_____   **14.** Parts 10-425-699 and 10-425-700 are joined with a(n) _____ weld.

_____   **15.** Holes in part 10-425-701 are located _____″ from center.

_____   **16.** Part 10-425-997 is _____″ long.

_____   **17.** Parts 10-425-700 and 10-425-997 are joined with a fillet leg size of _____″.

_____   **18.** Weld I is _____″ long.

T   F   **19.** The Front Hinge Mount is part number 425-002.

T   F   **20.** The latest revision on the drawing was made on 7-18-91.

T F   **21.** Holes in part 10-425-700 are ½″ in diameter.

_____ **22.** Holes in parts 20-410-969 and 10-425-997 are located _____″ apart vertically.

_____ **23.** The overall height of the Front Hinge Mount is _____″.

_____ **24.** Weld D specifies _____″ depth of preparation.

_____ **25.** Weld B specifies a(n) _____″ fillet leg size.

_____ **26.** The drawing was drawn by _____.

_____ **27.** Weld L specifies a(n) _____ weld on the arrow side.

_____ **28.** Weld A specifies a(n) _____″ leg size.

_____ **29.** Part 10-425-997 is _____″ thick.

_____ **30.** Weld F is _____″ in length on each side.

_____ **31.** Part 20-410-969 is _____″ long.

_____ **32.** Weld J specifies a(n) _____ weld.

_____ **33.** Weld E specifies a(n) _____″ fillet weld leg size.

_____ **34.** Part 10-425-701 is _____″ thick.

_____ **35.** Part 10-425-936 is _____″ long.

_____ **36.** Weld F was revised on _____.

_____ **37.** The drawing was approved by _____.

_____ **38.** Weld C specifies a(n) _____″ fillet weld leg size.

_____ **39.** Weld D specifies a(n) _____ weld on the other side.

_____ **40.** Weld K specifies a fillet weld on the _____ side.

T F   **41.** Weld G specifies a fillet weld-all-around.

T F   **42.** Part 20-410-969 is round.

T F   **43.** Part 10-425-936 is 2″ wide.

_____ **44.** Part 10-425-701 requires _____ holes.

_____ **45.** The overall width of the Front Hinge Mount is _____.

Trade Test 5

Name_____ Date _____

Refer to the Boom Arm Ass'y RH print on page 313.

_____ 1. Weld A joins part 2542731 to part _____.

_____ 2. Weld H specifies a leg size of _____ cm.

_____ 3. Weld G specifies a(n) _____ weld.

_____ 4. Weld I specifies a(n) _____ weld.

_____ 5. Weld A specifies a leg size of _____".
    A. .25
    B. .50
    C. .75
    D. none of the above

T   F 6. View A-A is drawn to the same scale as all other views of the print.

T   F 7. The center line to center line distance from the left hole to the center hole is 2010.

T   F 8. Part 2542738 is round in shape.

_____ 9. Part 12578872 is centered _____ cm from the center of the left hole.

_____ 10. Weld F specifies a .25" _____.

_____ 11. The center hole is centered _____" below the center line of the left hole.

_____ 12. Weld C joins part 2551520 to part _____.

_____ 13. Weld J joins part 12578872 to part _____.

_____ 14. Part 2553624 is joined on its concave surface by weld _____.

_____ 15. Weld B specifies fillet welds in a total of _____ locations.

_____ 16. Part 2542733 is _____ in shape.
    A. square
    B. rectangular
    C. round
    D. none of the above

T   F 17. Part 2542731 contains two holes.

T F **18.** Weld J specifies a fillet weld that is welded-all-around.

_____ **19.** The position angle of part 2553624 is shown in view _____.

_____ **20.** The metric equivalent of $\frac{1}{2}''$ is _____ cm.

_____ **21.** The drawing number is _____.

_____ **22.** Unless otherwise specified, the tolerance for all linear dimensions is _____ cm.
   A. .5
   B. 1,0
   C. 1,5
   D. 2,0

_____ **23.** Weld D joins parts 2555736 to part _____.

_____ **24.** Parts 2555736 are positioned at _____° off center.

_____ **25.** The drawing was drawn by _____.

_____ **26.** A taper in parts 2555736 is shown where weld _____ is specified.

_____ **27.** The drawing is drawn at a(n) _____ scale.

_____ **28.** Weld E shows the weld using graphic representation in place of a(n) _____.

_____ **29.** Weld D specifies a(n) _____ weld.

_____ **30.** The tolerance for angles is specified at _____°.

T F **31.** The drawing was completed July 24, 1984.

T F **32.** Weld D specifies leg sizes of 12.7 cm.

T F **33.** All burrs and sharp edges are to be removed.

T F **34.** Part 2551520 is symmetrical.

_____ **35.** The drawing was checked by _____.

_____ **36.** Welds at Weld E are located _____° off the center line of part 2555736.

_____ **37.** The drawing is sheet 1 of _____.

_____ **38.** Weld D specifies a leg size of _____″.
   A. .375
   B. .50
   C. .625
   D. .750

T F **39.** The length of Weld D is specified on the print.

T F **40.** No welds are permitted between the welds specified at Weld E.

Trade Test 6

Name_____ Date _____

Refer to the Left Plate Ass'y — Side print on page 314.

_____ **1.** The overall height of the left plate assembly is _____″.

_____ **2.** Weld F joins item _____ to items 7 and 3.

_____ **3.** Weld H specifies a(n) _____ weld in two locations.

_____ **4.** Weld A joins items 2 and _____.

_____ **5.** Weld C specifies a(n) _____″ leg size.

_____ **6.** Section A-A is drawn at _____ scale.

_____ **7.** Weld F specifies a(n) _____ contour.

_____ **8.** Weld _____ specifies a fillet weld to be welded-all-around.

_____ **9.** Weld G specifies a(n) _____″ leg size.

_____ **10.** The distance between the roots of welds G and H is _____″.

T   F **11.** Weld F specifies weld penetration to be a minimum of ¼″.

T   F **12.** Weld C specifies a total of two fillet welds.

T   F **13.** Weld A specifies the same weld as weld G.

_____ **14.** Weld B specifies a(n) _____ weld, where required.

_____ **15.** Weld E specifies a(n) _____″ leg size.

_____ **16.** Weld B specifies a(n) _____ weld in two locations.

_____ **17.** Weld F specifies a(n) _____ weld.

_____ **18.** The distance between the roots of the welds specified by weld H is _____″.

_____ **19.** Items 3 and 1 are joined by weld _____.

_____ **20.** Weld B specifies a ⅜″ _____.

299

Refer to the Left Plate Ass'y — Side print on page 315.

_____ **1.** The drawing was completed on _____.

_____ **2.** All dimensions are given in _____.

_____ **3.** Weld C joins part 12575098 to part _____.

_____ **4.** Weld D specifies a groove weld where required in two _____.

_____ **5.** The drawing is drawn to a(n) _____ scale.

_____ **6.** Weld _____ does not specify weld-all-around.

T   F **7.** The name or initials of the checker are not on the print.

T   F **8.** Weld A specifies a $5/16''$ leg size.

T   F **9.** The drawing number is 12583490.

_____ **10.** Item 4 is parallel to item _____.

_____ **11.** Weld D specifies a(n) _____ weld.

_____ **12.** The drawing is sheet 1 of _____.

_____ **13.** A total of _____ part(s) 12575121 is/are required.

_____ **14.** All welds are to be completed with an E-_____XX electrode.

_____ **15.** Weld C specifies a(n) _____'' leg size.

_____ **16.** Item 5 is part number _____.

_____ **17.** A total of _____ items are listed on the print.

_____ **18.** Weld B joints item _____ to items 2, 3, and 6.

_____ **19.** Item 4 is spaced _____'' from the bottom of item 1.

_____ **20.** The last section letter on the print is _____.

_____ **21.** Unless otherwise specified, identical welds are used on similar _____.

_____ **22.** A total of _____ part(s) 12575088 is/are required.

_____ **23.** The drawing was drawn by _____.

_____ **24.** Weld D joins item _____ to items 1, 2, and 3.

_____ **25.** A total of _____ frame(s) is/are required to show all print parts.

Final Exam 1

Name_____ Date _____

Refer to the Body Ass'y Welded print on page 316.

_____ **1.** View D is drawn at _____ size.

_____ **2.** The inside to inside dimension of the side plates is _____″.

_____ **3.** Weld C specifies a(n) _____ weld to be welded first.

_____ **4.** Weld E specifies a fillet weld with a(n) _____″ leg size.

_____ **5.** Weld B joins item numbers 5 and _____.

_____ **6.** The outside to outside dimension of the body assembly is _____″.

_____ **7.** Weld A specifies a(n) _____ weld.

_____ **8.** View D specifies how item number _____ is joined to item numbers 1 and 4.

T   F   **9.** Weld B specifies that a fillet weld be welded-all-around.

T   F   **10.** Weld C joins item numbers 1 and 4.

T   F   **11.** The distance from the center to the inside of the side plate is 79″.

_____ **12.** Item numbers 5 and 3 are joined by a fillet weld with a(n) _____″ leg size.

T   F   **13.** Item number 1 is shown in view D with hidden lines.

_____ **14.** Weld C specifies a fillet weld leg size of _____″ for item number 4.

_____ **15.** Weld D joins item number 4 to item numbers 2, 3, and 1 with a(n) _____ weld.

_____ **16.** Item number 15 is located _____″ above item number 1.

_____ **17.** Weld C specifies that a(n) _____ weld is to be made last.

_____ **18.** Item number 1 is to be located _____″ above the bottom edge of item number 16.

T   F   **19.** Weld D specifies that a fillet weld is to be made all-around.

T   F   **20.** The first hole from the center on item number 7 is located 32″ from the center of the assembly.

Refer to the Body Ass'y Welded print on page 317.

_____ 1. Weld F specifies a leg size of _____".

_____ 2. The overall height of the body is _____".

_____ 3. The distance from the intersection of the floor and front plate to the top of the body is _____".

_____ 4. Weld C joins item number _____ to item numbers 2 and 3.

T   F   5. Weld B specifies the same weld as weld C.

T   F   6. Weld F specifies a bevel-groove weld.

T   F   7. Section H-H is drawn at half size.

_____ 8. Weld D specifies a(n) _____ weld.

_____ 9. Weld E specifies a $\frac{1}{4}$" leg size on the _____ side of item number 4.

_____ 10. Weld B specifies a fillet weld with a(n) _____" leg size.

_____ 11. Item numbers 1 and 8 are joined by weld _____.

_____ 12. Weld E specifies a fillet weld on _____ side(s).

T   F   13. Section H-H specifies body pivot dimensions.

T   F   14. Weld A specifies a fillet weld, other side.

T   F   15. Weld D specifies a $\frac{3}{8}$" root opening.

_____ 16. Weld A specifies a fillet weld with a(n) _____" leg size.

_____ 17. Weld D specifies that the _____ weld is to be completed first.

_____ 18. Weld D joins item numbers 1 and _____.

_____ 19. Weld B specifies a fillet weld on the _____ side.

_____ 20. Section H-H shows the thickness of item numbers 2 and _____.

Refer to the Body Ass'y Welded print on page 318.

_____ 1. The rear axle center is _____" from the body pivot center.

_____ 2. Weld D joins item number _____ to item numbers 2 and 3.

_____ 3. Weld A specifies a(n) _____ contour.

T F **4.** Section E-E is drawn to quarter size.

T F **5.** Weld F specifies a fillet weld 3″ in lengtʰ on the other side.

T F **6.** Section E-E specifies that item number₁ 10 and 12 are to be joined.

T F **7.** Weld D specifies that a fillet weld is to be made all-around.

_____ **8.** Weld A specifies a ⅛″ minimum _____.

_____ **9.** Weld F specifies ³⁄₁₆″ fillet weld leg sizes on welds centered on _____″ intervals.

_____ **10.** Weld _____ specifies the same weld as welds B, C, and D.

_____ **11.** Section E-E shows item number _____ with hidden lines.

_____ **12.** Weld F specifies ³⁄₁₆″ fillet weld leg sizes on welds _____″ in length.

_____ **13.** Weld A specifies a(n) _____ weld.

_____ **14.** Dimensions shown in boxes are basic engineering, non-toleranced _____.

_____ **15.** Weld E specifies a(n) _____ weld.

_____ **16.** View J is drawn to the scale of _____.

_____ **17.** Weld C specifies a fillet weld with a(n) _____″ leg size.

T F **18.** Weld E joins item numbers 1 and 6.

T F **19.** Weld F specifies a double-fillet weld.

T F **20.** Weld E specifies a flush contour.

Refer to the Body Ass'y Welded print on page 319.

_____ **1.** The scale of the drawing is _____.

_____ **2.** A total of _____ corner caps are required.

_____ **3.** Unless otherwise specified, dimensions are given in _____.

_____ **4.** Weld B specifies a fillet weld with a(n) _____″ leg size.

_____ **5.** Section B-B is drawn at _____ size.

_____ **6.** Weld E specifies a(n) _____ weld.

_____ **7.** Weld E specifies ³⁄₁₆″ depth of _____.

_____ **8.** Item number 14 is specified to be located _____″ above the top of the floor plate.

_____ **9.** Weld C specifies that a(n) _____ weld is to be completed first.

_____ **10.** Weld E has a(n) _____ contour.

T   F **11.** View A-A is drawn at quarter size.

T   F **12.** Weld E specifies a weld more than 13″ in length.

T   F **13.** The drawing number is 12575679.

_____ **14.** Section B-B is located in view _____.

_____ **15.** A total of _____ bars are required.

_____ **16.** All sharp edges are to be broken to _____″ maximum.

_____ **17.** The drawing was completed on _____.

_____ **18.** Weld B specifies that the weld is to be made _____.

T   F **19.** All weld sizes specified are minimum as measured.

T   F **20.** Plate-mud is listed as item number 15.

T   F **21.** Weld D specifies a fillet weld with equal leg sizes.

T   F **22.** Weld C specifies a weld length of ⅜″.

_____ **23.** The drawing was drawn by _____.

_____ **24.** Weld A joins part number _____ to item numbers 12, 10, and 8.

_____ **25.** Weld C specifies that a(n) _____ weld is to be completed last.

T   F **26.** The Body Ass'y Welded was released to production on 3-6-89.

_____ **27.** Unless otherwise specified, all welds are to be made with E-_____ electrodes.

T   F **28.** Item numbers 8 and 9 are welded to item 4.

_____ **29.** View F-F is drawn at _____ size.

_____ **30.** The corner plate left is part number _____.

Final Exam 2

Name_____ Date _____

Refer to the Fabrication Dwg. A-21 print (Front View, Section Views) on page 320.

_____ **1.** Parts 1 and 12 are joined with a(n) _____ weld.

_____ **2.** The horizontal divider and back are joined with 4″ fillet welds on _____″ centers.
  A. 4
  B. 6
  C. 8
  D. 10

_____ **3.** Parts 3 and 10 are joined with a(n) _____ weld.

_____ **4.** The weld joining the side and the top has a flush contour using the _____ finishing method.

_____ **5.** Parts 5 and 1 are joined with fillet welds with a(n) _____″ leg size.

_____ **6.** Parts 1 and 10 are joined with a(n) _____ weld on the arrow side.

_____ **7.** The width of the safe as shown in the front view is _____″.
  A. 15
  B. 15½
  C. 32
  D. none of the above

_____ **8.** The height of the safe as shown in the front view is _____″.

_____ **9.** The sides and horizontal divider are joined with 4″ welds on _____″ centers.

_____ **10.** Parts 27 and 22 are joined with fillet welds with a(n) _____″ leg size.

_____ **11.** Part 6 is located _____″ below part 3.

_____ **12.** Part 4 is joined to part 1 with a(n) _____ weld.

_____ **13.** The weld joining the sides and the back is specified for a(n) _____″ leg size.

_____ **14.** Part 4 is recessed _____″ from the back of the safe.
  A. ⅛
  B. ¼
  C. ½
  D. 1

_____ 15. Parts 12 and 4 are joined with a weld _____″ long.

_____ 16. Parts 29 and 22 are joined with a weld _____″ long.

_____ 17. Part 6 is located _____″ from the front of the safe.

T   F   18. The mounting angles are welded to the sides with 4″ long welds.

T   F   19. All groove welds are specified for a flush contour.

_____ 20. The horizontal divider is joined to the back with welds _____″ in length.
   A. ½
   B. ¾
   C. 1
   D. 1½

_____ 21. Mounting angles are welded _____″ from the front of the safe.

T   F   22. Butted surfaces on the front of the safe are welded with square-groove welds and ground to a flush contour.

_____ 23. The top right door is _____″ in height.

_____ 24. Section B-B shows the section view of the _____ side of the safe.

T   F   25. The vertical divider is centered in the width of the safe.

_____ 26. The bottom is welded to the sides with the welds starting behind the mounting _____.

_____ 27. Section A-A shows the section view of the _____ of the safe.

_____ 28. The bottom is _____″ above the floor line.
   A. ¾
   B. 6⅜
   C. 18
   D. 39⅜

T   F   29. The side is 19½″ wide.

_____ 30. The angle is added to the bottom with 2″ fillet welds on _____″ centers.

Refer to the Fabrication Dwg. A-21 (Details, Doors) print on page 321.

_____ 1. The height of the bottom left door is _____″.

_____ 2. The height of the top left door is _____″.
   A. 9⁹⁄₁₆
   B. 13⁷⁄₁₆
   C. 17⁵⁄₁₆
   D. 21¹³⁄₁₆

          3. Parts 22 and 27 are joined with a(n) _____ weld.

          4. The weld joining the sides and horizontal divider (3) starts behind door _____.

          5. Part 22 is located _____″ above the bottom of part 29.

          6. The width of the bottom right door is _____″.
             A. $14\frac{5}{16}$
             B. $14\frac{7}{16}$
             C. $14\frac{15}{16}$
             D. none of the above

          7. The interlock (25) is located _____″ from the top of the bottom right door.

          8. The bottom hinge detail specifies fillet welds with a(n) _____″ leg size.

T   F   **9.** The left and right doors are the same width.

T   F   **10.** The detail of the locking lug specifies hinge hole sizes.

T   F   **11.** Fillet welds with $\frac{1}{4}$″ legs are used to join top hinge weld parts.

          12. The detail of the locking lug specifies _____″ fillet welds.

          13. A total of _____ doors are required for the safe.

T   F   **14.** All welds made on the doors are fillet welds.

          15. The left doors are _____″ wider than the right doors.

Refer to Fabrication Dwg. A-21 print (Welding Symbol Notes, Bill of Materials, and Title Block) on page 322.

          1. Part 18 is $\frac{1}{4}$″ × 2″ × _____″ angle.

          2. Part 8 is _____″ thick.
             A. $\frac{1}{8}$
             B. $\frac{1}{4}$
             C. $\frac{1}{2}$
             D. $\frac{3}{4}$

T   F   **3.** Tolerances are not specified for the parts on this print.

          4. Part 5 is _____″ long.

          5. The print specifies a quantity of _____ $\frac{1}{2}$″ × $\frac{1}{2}$″ × $9\frac{9}{16}$″ hinge bars.

          6. Part 26 is _____″ thick.

          7. The horizontal divider is made from $\frac{1}{2}$″ _____.

T F   **8.** The scale for the drawing is $\frac{1}{4}'' = 1'\text{-}0''$.

_____   **9.** The side and the hinge bar are specified to be welded at the same location in all four _____.

_____   **10.** The interlock (25) is _____'' long.
     A. 10
     B. 15
     C. 20
     D. 25

T F   **11.** Eight hinge hasps are specified.

_____   **12.** The drawing number is _____.

_____   **13.** The $2'' \times 2''$ angle is _____'' long.
     A. $21^{13}\!/_{16}$
     B. $25^{5}\!/_{16}$
     C. $30^{7}\!/_{8}$
     D. none of the above

_____   **14.** The print is drawn on a(n) _____ size sheet.

T F   **15.** There are eight gussets required.

T F   **16.** Allied/Gary Safe & Vault Co., Inc. is located in Spokane, WA.

_____   **17.** Part 10 is _____'' long.

T F   **18.** The drawing was drawn by K.A.

T F   **19.** The drawing was completed 8/31/91.

_____   **20.** The back is _____'' thick.

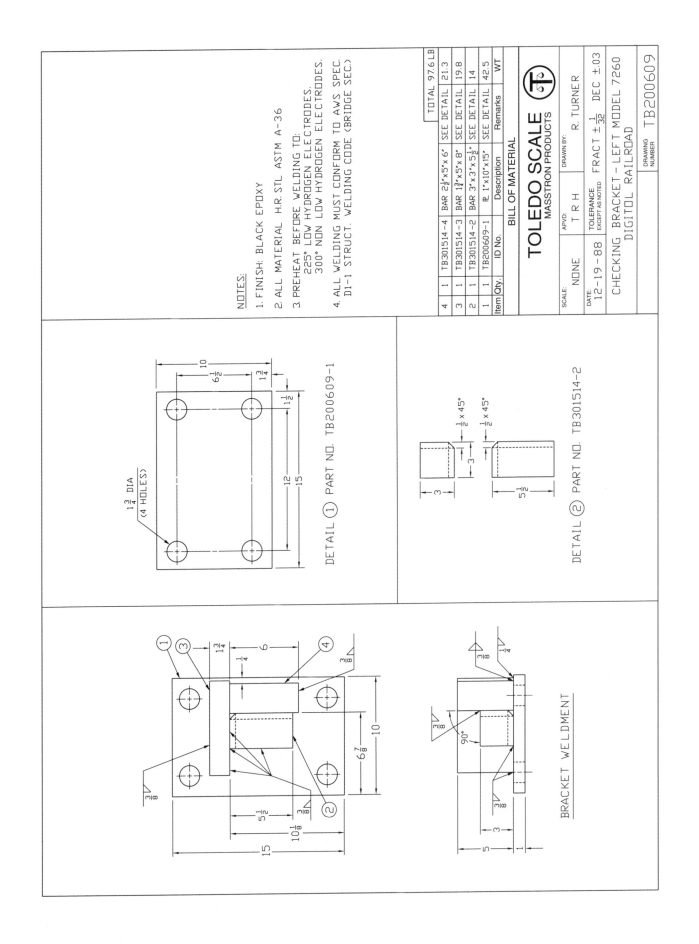

NOTES:

1. FINISH: BLACK EPOXY

2. ALL MATERIAL H.R. STL ASTM A-36

3. PREHEAT BEFORE WELDING TO:
   225° LOW HYDROGEN ELECTRODES.
   300° NON LOW HYDROGEN ELECTRODES.

4. ALL WELDING MUST CONFORM TO AWS SPEC.
   D1-1 STRUCT. WELDING CODE (BRIDGE SEC.)

DETAIL ① PART NO. TB200609-1

$1\frac{3}{4}$ DIA (4 HOLES)

DETAIL ② PART NO. TB301514-2

$\frac{1}{2} \times 45°$

BRACKET WELDMENT

BILL OF MATERIAL

Item	Qty.	ID No.	Description	Remarks
4	1	TB301514-4	BAR $2\frac{1}{2}$"x5"x6"	SEE DETAIL
3	1	TB301514-3	BAR $1\frac{3}{4}$"x5"x8"	SEE DETAIL
2	1	TB301514-2	BAR 3"x3"x$5\frac{1}{2}$"	SEE DETAIL
1	1	TB200609-1	Pl. 1"x10"x15"	SEE DETAIL

	WT
TOTAL	97.6 LB
	21.3
	19.8
	14
	42.5

TOLEDO SCALE Ⓣ
MASSTRON PRODUCTS

SCALE: NONE	APVD: T R H	DRAWN BY: R. TURNER
DATE: 12-19-88	TOLERANCE EXCEPT AS NOTED	FRACT $\pm\frac{1}{32}$ DEC $\pm.03$

CHECKING BRACKET - LEFT MODEL 7260
DIGITOL RAILROAD

DRAWING NUMBER: **TB200609**

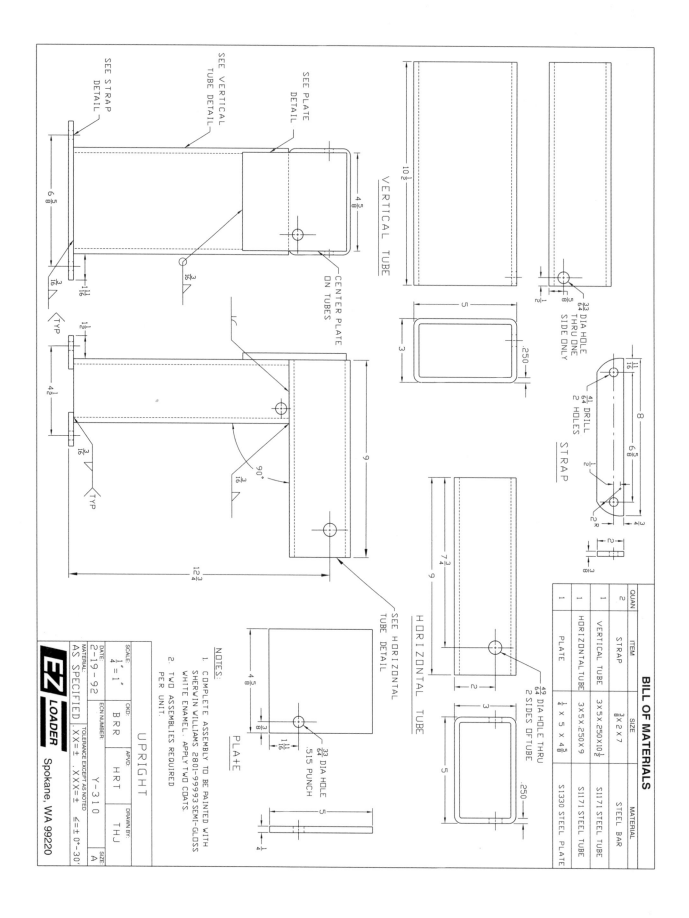

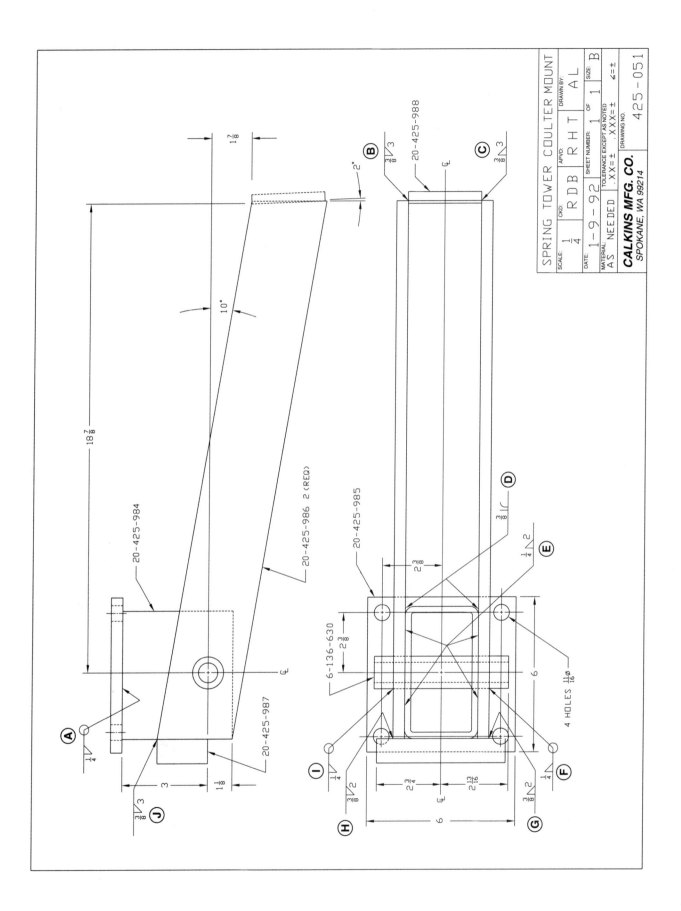

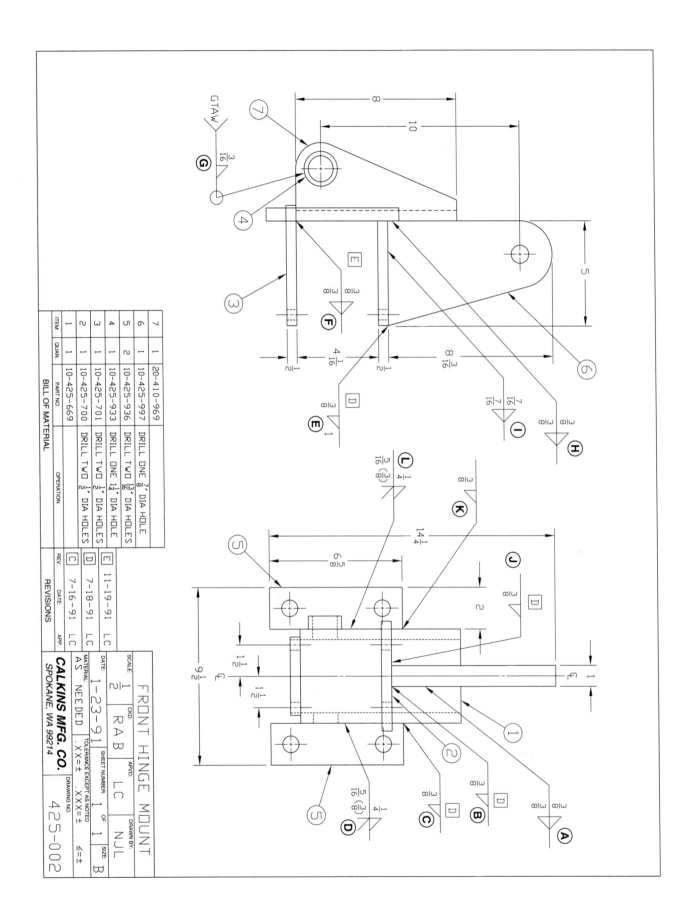

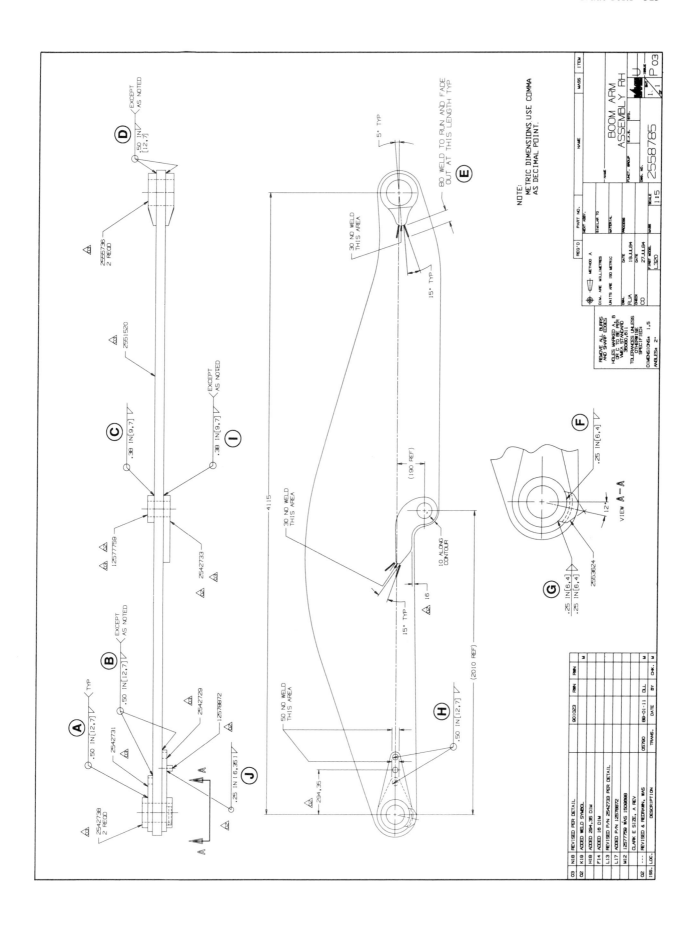

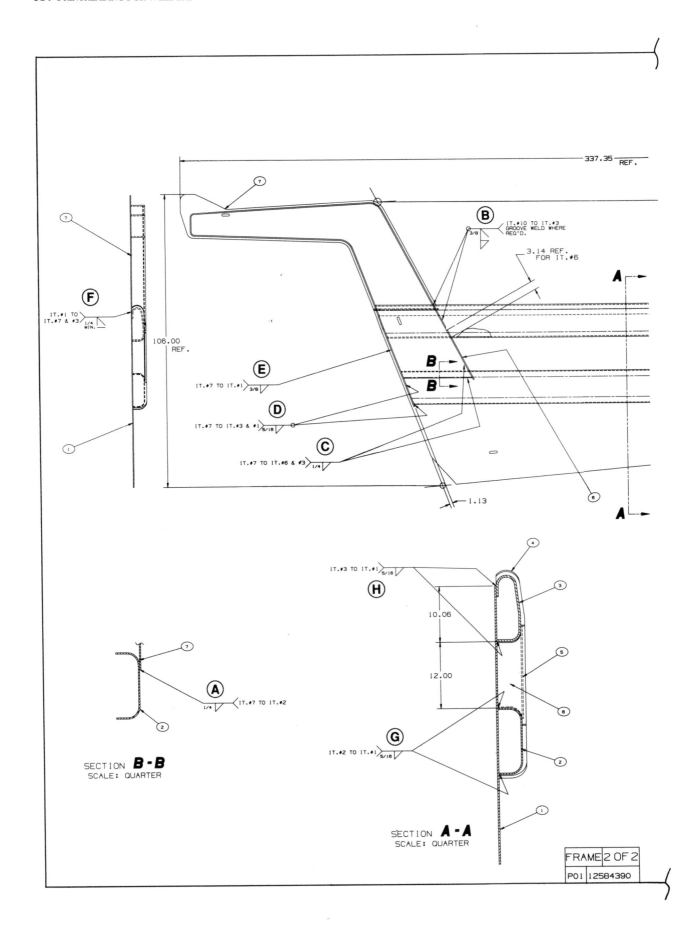

337.35 REF.

IT.#10 TO IT.#3
GROOVE WELD WHERE
REQ'D.

3.14 REF.
FOR IT.#6

IT.#1 TO
IT.#7 & #3
1/4
MIN.

106.00
REF.

IT.#7 TO IT.#1
3/8

IT.#7 TO IT.#3 & #1
5/16

IT.#7 TO IT.#6 & #3
1/4

1.13

IT.#3 TO IT.#1
5/16

10.06

12.00

IT.#2 TO IT.#1
5/16

IT.#7 TO IT.#2
1/4

SECTION **B-B**
SCALE: QUARTER

SECTION **A-A**
SCALE: QUARTER

FRAME 2 OF 2
P01 12584390

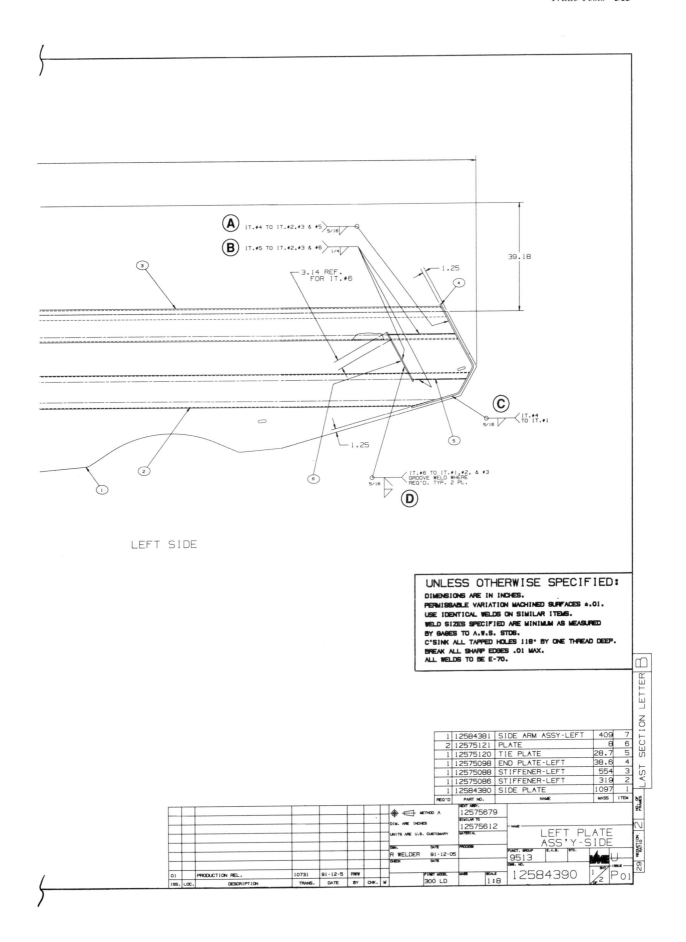

A — IT.#4 TO IT.#2,#3 & #5 [5/16]

B — IT.#5 TO IT.#2,#3 & #6 [1/4]

3.14 REF.
FOR IT.#6

1.25

39.18

3

4

C — IT.#4
TO IT.#1 [5/16]

2

5

1.25

1

6

IT.#6 TO IT.#1,#2, & #3
GROOVE WELD WHERE
REQ'D. TYP. 2 PL. [5/16]

D

LEFT SIDE

UNLESS OTHERWISE SPECIFIED:

DIMENSIONS ARE IN INCHES.
PERMISSABLE VARIATION MACHINED SURFACES ±.01.
USE IDENTICAL WELDS ON SIMILAR ITEMS.
WELD SIZES SPECIFIED ARE MINIMUM AS MEASURED
BY GAGES TO A.W.S. STDS.
C'SINK ALL TAPPED HOLES 118° BY ONE THREAD DEEP.
BREAK ALL SHARP EDGES .01 MAX.
ALL WELDS TO BE E-70.

REQ'D	PART NO.	NAME	MASS	ITEM
1	12584381	SIDE ARM ASS'Y-LEFT	409	7
2	12575121	PLATE	8	6
1	12575120	TIE PLATE	28.7	5
1	12575098	END PLATE-LEFT	38.6	4
1	12575088	STIFFENER-LEFT	554	3
1	12575086	STIFFENER-LEFT	319	2
1	12584380	SIDE PLATE	1097	1

LAST SECTION LETTER B

METHOD A
DIM. ARE INCHES
UNITS ARE U.S. CUSTOMARY

NEXT ASSY.
12575679
SIMILAR TO
12575612
MATERIAL
PROCESS

LEFT PLATE
ASS'Y-SIDE

DSN. DATE
R WELDER 91-12-05
CHECK DATE

FUNCT. GROUP
9513 E.A.S. STD.

01	PRODUCTION REL.	10731	91-12-5	RWW		
ISS.	LOC.	DESCRIPTION	TRANS.	DATE	BY	CHK.

FIRST MODEL
300 LD

SCALE
1:8

DWG. NO.
12584390

SHT
1/2
OF 2

ISSUE
P01

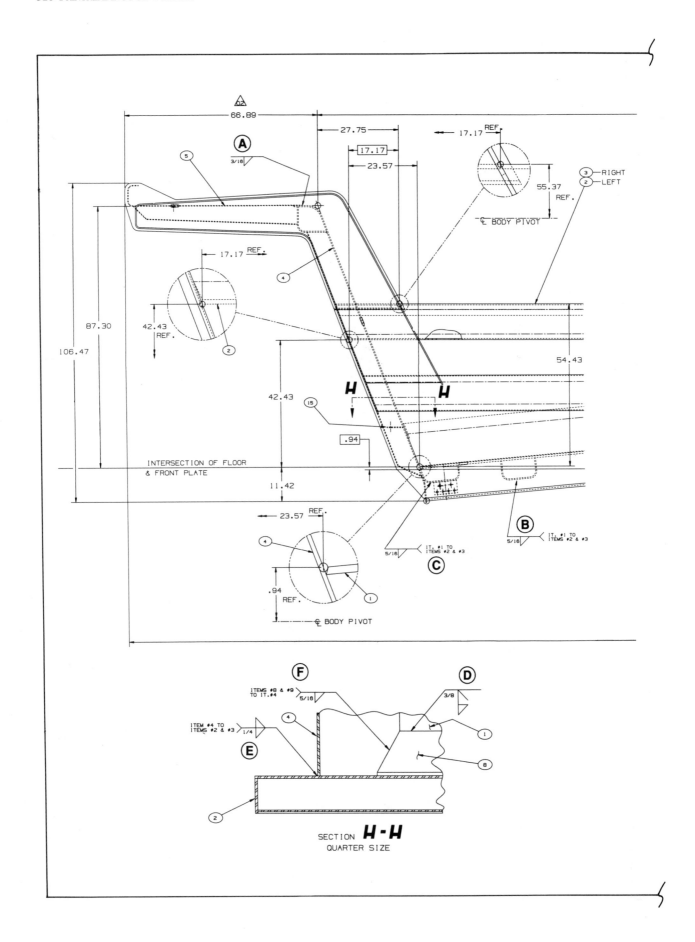

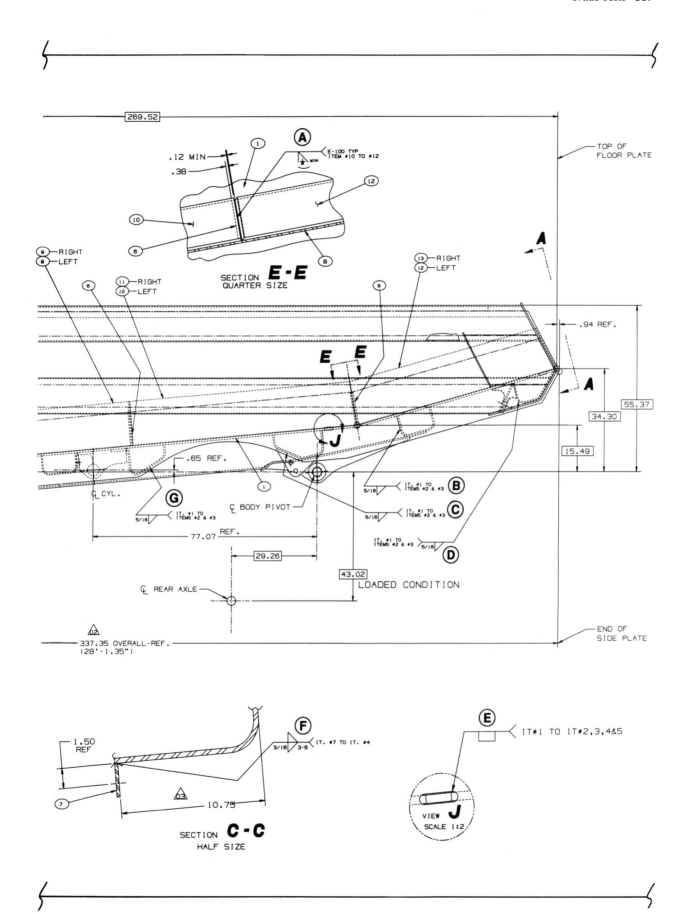

TOP OF
FLOOR PLATE

269.52

.12 MIN

.38

.04 MIN

E-100 TYP
ITEM #10 TO #12

SECTION **E-E**
QUARTER SIZE

9 — RIGHT
8 — LEFT

11 — RIGHT
10 — LEFT

13 — RIGHT
12 — LEFT

.94 REF.

55.37

34.30

15.49

.65 REF.

C CYL.

C BODY PIVOT

G

5/16 IT. #1 TO
ITEMS #2 & #3

5/16 IT. #1 TO
ITEMS #2 & #3   B

5/16 IT. #1 TO
ITEMS #2 & #3   C

IT. #1 TO
ITEMS #2 & #3 5/16   D

43.02
LOADED CONDITION

77.07   REF.

29.26

C REAR AXLE

337.35 OVERALL-REF.
(28'-1.35")

END OF
SIDE PLATE

1.50
REF

7

10.75

SECTION **C-C**
HALF SIZE

F
3/18 IT. #7 TO IT. #4
3-6

E
IT#1 TO IT#2,3,4&5

VIEW **J**
SCALE 1:2

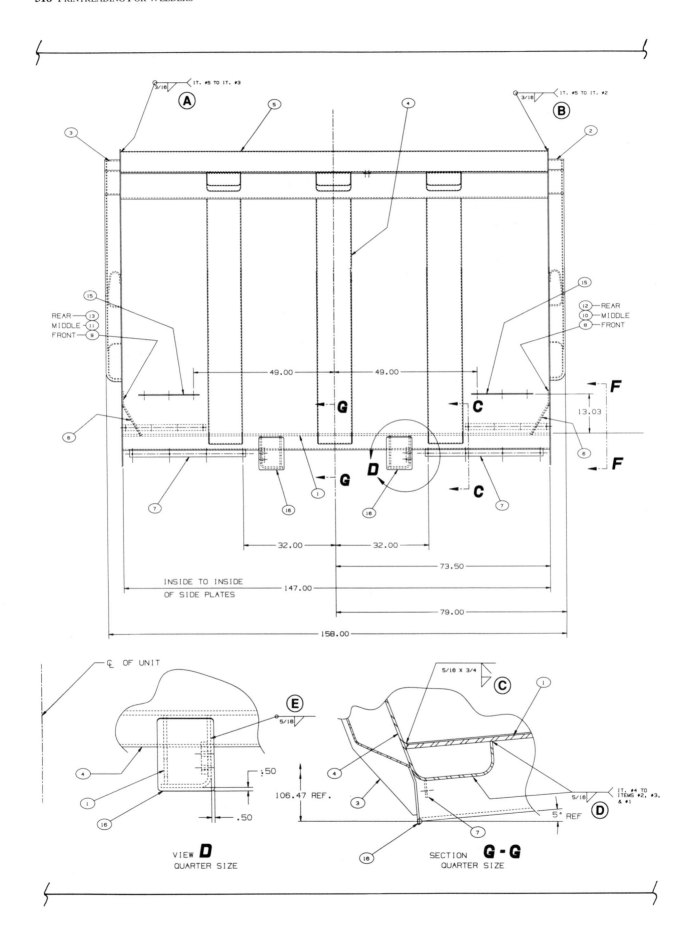

IT. #5 TO IT. #3

(A)

IT. #5 TO IT. #2

(B)

5° REF

INSIDE TO INSIDE
OF SIDE PLATES

147.00

158.00

49.00   49.00

13.03

32.00   32.00

73.50

79.00

REAR — 13
MIDDLE — 11
FRONT — 9

12 — REAR
10 — MIDDLE
8 — FRONT

CL OF UNIT

5/16 X 3/4

(E)

106.47 REF.

.50

.50

VIEW **D**
QUARTER SIZE

IT. #4 TO
ITEMS #2, #3,
& #1

5/16

(D)

SECTION **G-G**
QUARTER SIZE

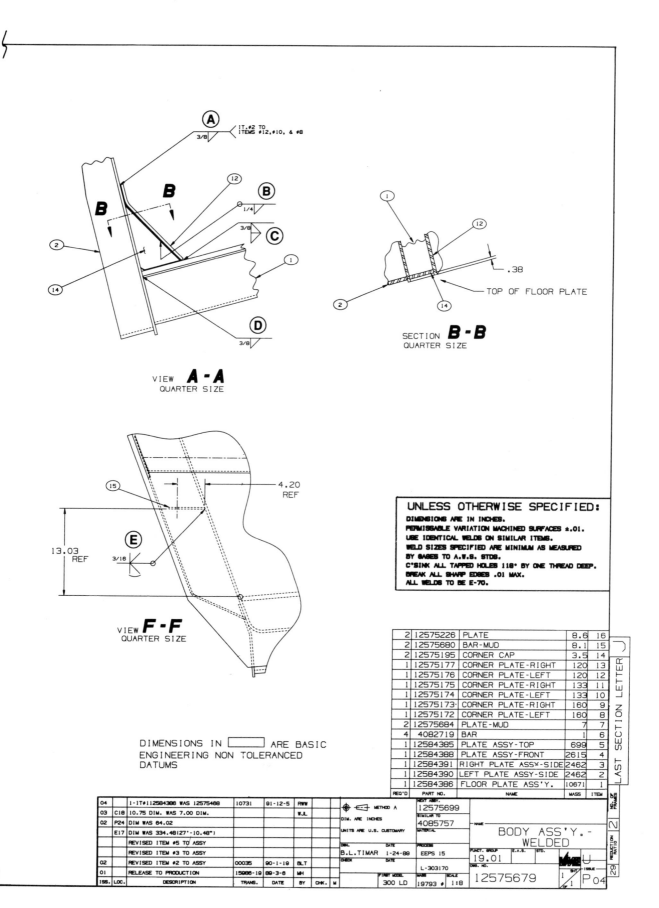

Ⓐ
IT.#2 TO
ITEMS #12,#10, & #8
3/8

VIEW **A - A**
QUARTER SIZE

SECTION **B - B**
QUARTER SIZE

.38

TOP OF FLOOR PLATE

4.20
REF

13.03
REF

3/16

VIEW **F - F**
QUARTER SIZE

DIMENSIONS IN ☐ ARE BASIC
ENGINEERING NON TOLERANCED
DATUMS

UNLESS OTHERWISE SPECIFIED:

DIMENSIONS ARE IN INCHES.
PERMISSABLE VARIATION MACHINED SURFACES ±.01.
USE IDENTICAL WELDS ON SIMILAR ITEMS.
WELD SIZES SPECIFIED ARE MINIMUM AS MEASURED
BY GAGES TO A.W.S. STDS.
C'SINK ALL TAPPED HOLES 118° BY ONE THREAD DEEP.
BREAK ALL SHARP EDGES .01 MAX.
ALL WELDS TO BE E-70.

REQ'D	PART NO.	NAME	MASS	ITEM
2	12575226	PLATE	8.6	16
2	12575680	BAR-MUD	8.1	15
2	12575195	CORNER CAP	3.5	14
1	12575177	CORNER PLATE-RIGHT	120	13
1	12575176	CORNER PLATE-LEFT	120	12
1	12575175	CORNER PLATE-RIGHT	133	11
1	12575174	CORNER PLATE-LEFT	133	10
1	12575173	CORNER PLATE-RIGHT	160	9
1	12575172	CORNER PLATE-LEFT	160	8
2	12575684	PLATE-MUD	7	7
4	4082719	BAR	1	6
1	12584385	PLATE ASSY-TOP	699	5
1	12584388	PLATE ASSY-FRONT	2615	4
1	12584391	RIGHT PLATE ASSY-SIDE	2462	3
1	12584390	LEFT PLATE ASSY-SIDE	2462	2
1	12584386	FLOOR PLATE ASS'Y.	10671	1

LAST SECTION LETTER J

ISS.	LOC.	DESCRIPTION	TRANS.	DATE	BY	CHK.	M
04		1-IT#112584388 WAS 12575488	10731	91-12-5	RWW		
03	C18	10.75 DIM. WAS 7.00 DIM.			W.L		
02	P24	DIM WAS 64.02					
	E17	DIM WAS 334.48(27'-10.48")					
		REVISED ITEM #5 TO ASSY					
		REVISED ITEM #3 TO ASSY					
02		REVISED ITEM #2 TO ASSY	00035	90-1-19	BLT		
01		RELEASE TO PRODUCTION	15988-19	89-3-8	MH		

METHOD A

DIM. ARE INCHES
UNITS ARE U.S. CUSTOMARY

B.L.TIMAR 1-24-89
CHECK   DATE

NEXT ASSY.
12575699
SIMILAR TO
4085757
MATERIAL
PROCESS
EEPS 15

L-303170

NAME
BODY ASS'Y.-
WELDED
FUNCT. GROUP
19.01

FIRST MODEL
300 LD
MASS
19793 #
SCALE
1:8

12575679

1/1

U
P04

29

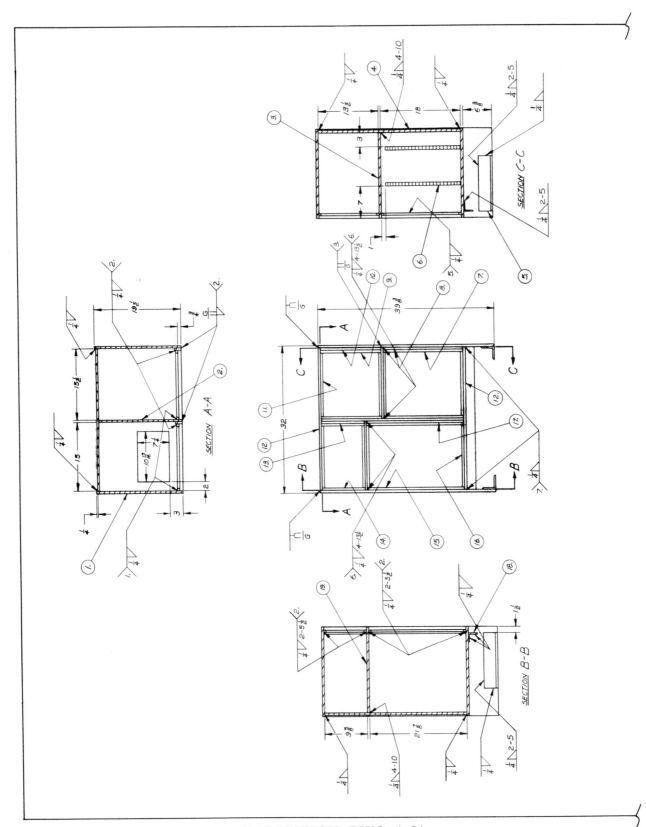

FABRICATION DWG. A-21
(Front View, Section Views)

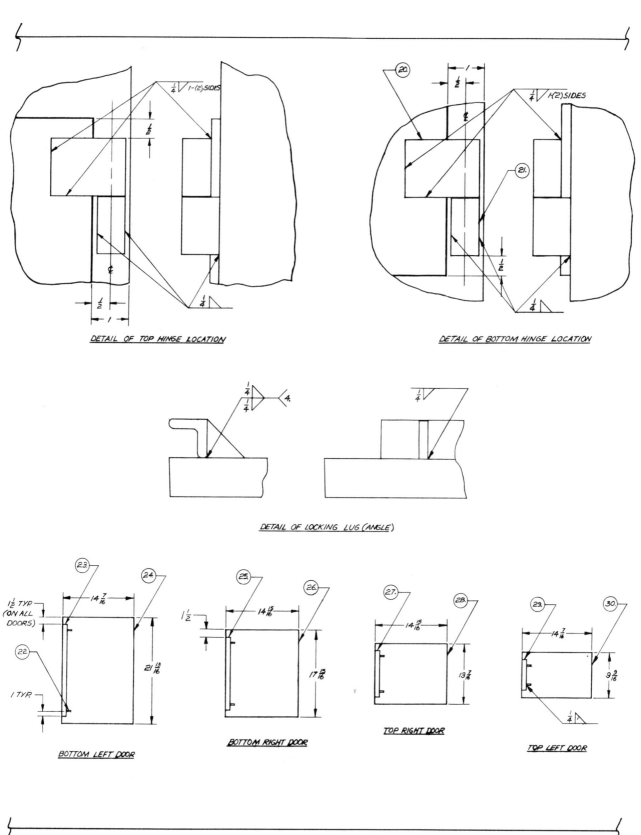

DETAIL OF TOP HINGE LOCATION

DETAIL OF BOTTOM HINGE LOCATION

DETAIL OF LOCKING LUG (ANGLE)

BOTTOM LEFT DOOR

BOTTOM RIGHT DOOR

TOP RIGHT DOOR

TOP LEFT DOOR

FABRICATION DWG. A-21
(Details, Doors)

ITEM	SIZE, MATERIAL, DESCRIPTION	QTY.
1	$\frac{1}{2}$ X 39 $\frac{3}{8}$ X 19 $\frac{1}{2}$ STL. PLT., SIDE	2
2	$\frac{1}{2}$ X 31 $\frac{13}{16}$ X 18 $\frac{11}{16}$ STL. PLT., VERT. DIVIDER	1
3	$\frac{1}{2}$ X 15 $\frac{7}{8}$ X 18 $\frac{11}{16}$ STL. PLT., HORIZ. DIVIDER	1
4	$\frac{1}{2}$ X 31 $\frac{7}{8}$ X 30 $\frac{7}{8}$ STL. PLT., BACK	1
5	$\frac{3}{8}$ X 3 X 3 X 12 ANGLE, MNTG. ANGLE	2
6	SHELF STD. X 17	4
7	$\frac{1}{2}$ X $\frac{1}{2}$ X 17 $\frac{13}{16}$ SQR. C.R.S., STOP BLOCK	2
8	$\frac{1}{2}$ X 1 $\frac{1}{2}$ X 17 $\frac{13}{16}$ F.B. C.R.S., HINGE BAR	1
9	$\frac{1}{2}$ X $\frac{1}{2}$ X 13 $\frac{7}{16}$ SQR. C.R.S., STOP BLOCK	2
10	$\frac{1}{2}$ X 1 $\frac{1}{2}$ X 13 $\frac{7}{16}$ F.B. C.R.S., HINGE BAR	1
11	$\frac{1}{2}$ X $\frac{1}{2}$ X 13 $\frac{13}{16}$ SQR. C.R.S., STOP BLOCK	4
12	$\frac{1}{8}$ X 19 $\frac{1}{2}$ X 30 $\frac{15}{16}$ STL. PLT., TOP & BOTT.	2
13	$\frac{1}{2}$ X 1 $\frac{1}{2}$ X 9 $\frac{3}{16}$ F.B. C.R.S., HINGE BAR	1
14	$\frac{1}{2}$ X $\frac{1}{2}$ X 9 $\frac{3}{16}$ SQR. C.R.S., STOP BLOCK	2
15	$\frac{1}{2}$ X $\frac{1}{2}$ X 21 $\frac{13}{16}$ SQR. C.R.S., STOP BLOCK	2
16	$\frac{1}{2}$ X $\frac{1}{2}$ X 13 $\frac{7}{16}$ SQR. C.R.S., STOP BLOCK	4
17	$\frac{1}{2}$ X 1 $\frac{1}{2}$ X 21 $\frac{13}{16}$ F.B. C.R.S., HINGE BAR	1
18	$\frac{1}{4}$ X 2 X 2 X 30 $\frac{7}{8}$ ANGLE	1
19	$\frac{1}{2}$ X 14 $\frac{1}{2}$ X 18 $\frac{11}{16}$ STL. PLT., HORIZ. DIVIDER	1
20	MINI CINCI. HINGE HASP	8
21	MINI CINCI. HINGE TAB	8
22	$\frac{1}{4}$ GUSSET	8
23	$\frac{1}{4}$ X 1 X 1 X 18 $\frac{7}{8}$ ANGLE, INTERLOCK	1
24	1 X 21 $\frac{13}{16}$ X 14 $\frac{7}{16}$ STL. PLT., DOOR	1
25	$\frac{1}{4}$ X 1 X 1 X 15 ANGLE, INTERLOCK	1
26	1 X 17 $\frac{15}{16}$ X 14 $\frac{15}{16}$ STL. PLT., DOOR	1
27	$\frac{1}{4}$ X 1 X 1 X 10 $\frac{5}{8}$ ANGLE, INTERLOCK	1
28	1 X 13 $\frac{7}{16}$ X 14 $\frac{15}{16}$ STL. PLT., DOOR	1
29	$\frac{1}{4}$ X 1 X 1 X 6 $\frac{5}{8}$ ANGLE, INTERLOCK	1
30	1 X 9 $\frac{5}{16}$ X 14 $\frac{7}{16}$ STL. PLT., DOOR	1

WELDING SYMBOL NOTES:

1.) 6" SPACE, FREE OF WELD, MUST BE LEFT BEHIND STOP FOR BOLT BAR, 3" ABOVE & BELOW HORIZONTAL CENTER LINE. (TOP COMPTS. ONLY)

2.) TYPICAL AT SAME LOCATION IN ALL FOUR COMPTS.

3.) TYPICAL OF ALL BUTTED SURFACES ON FRONT OF SAFE.

4.) ON SIDE OPPOSITE GUSSETS, WELD IN 6" FROM BOTH ENDS, ON GUSSET SIDE, WELD DISTANCE BETWEEN 6" WELDS.

5.) ON BOTT. LEFT COMPT., LEAVE (3) 1" SPACES FREE OF WELD, ON 8" ₵'S ABOVE & BELOW HORIZ. ₵ OF COMPT. FOR BOTT. RIGHT COMPT. LEAVE (3) 1" SPACES ON 3" ₵'S ABOVE & BELOW HORIZ. ₵ OF COMPT.

6.) WELD STARTS BEHIND DOOR STOPS.

7.) WELDS START BEHIND ANGLE.

1	ITMS. 2,3,19-18 $\frac{11}{16}$ WAS 18 $\frac{3}{4}$, ITMS. 11 & 16, DIM. CHNGE.	2-28-92	K.A.
REV.	DESCRIPTION	DATE	BY

CHANGE

DRAWN BY: K.A.	DATE 2-26-92	**ALLIED/GARY SAFE & VAULT CO., INC.**
APPROVED BY:	DATE:	SPOKANE, WASHINGTON - CINCINNATI, OHIO
QUOTE: REF. S-599-85		
SPECIAL NOTE:		REF. DWG. NO. - C-000542-S
UNLESS OTHERWISE SPECIFIED DIMENSIONS ARE IN INCHES. TOLERANCES ARE: FRACTIONS DECIMALS ANGLES ± .XX ± ± .XXX ±		FABRICATION DWG. A-21
DO NOT SCALE DWG.	SCALE NONE	D	DRAWING NO. D-007001-S REV.-1

FABRICATION DWG. A-21
(Welding Symbol Notes, Bill of Materials, and Title Block)

APPENDIX

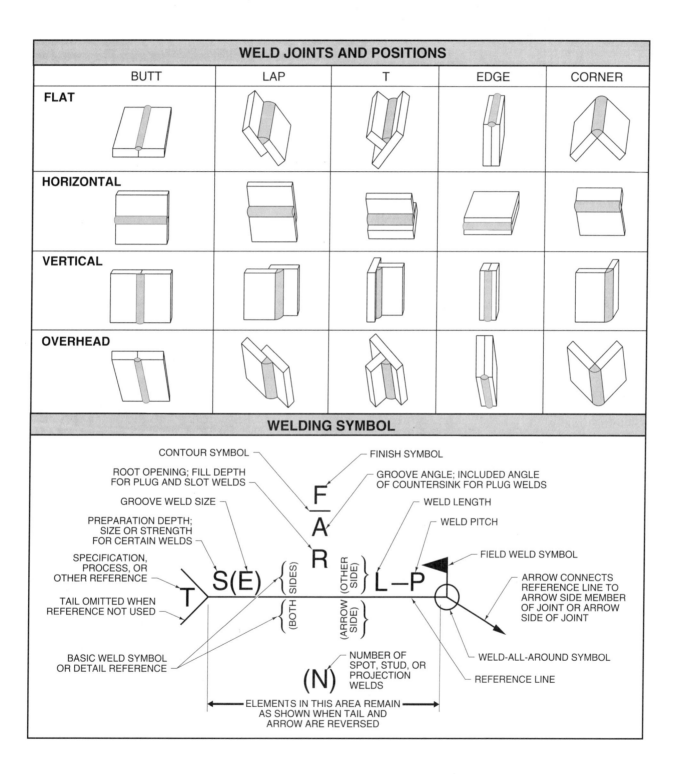

WELD JOINTS AND POSITIONS				
	BUTT	LAP	T	EDGE	CORNER
FLAT					
HORIZONTAL					
VERTICAL					
OVERHEAD					

WELDING SYMBOL

CONTOUR SYMBOL

FINISH SYMBOL

ROOT OPENING; FILL DEPTH FOR PLUG AND SLOT WELDS

GROOVE ANGLE; INCLUDED ANGLE OF COUNTERSINK FOR PLUG WELDS

GROOVE WELD SIZE

WELD LENGTH

PREPARATION DEPTH; SIZE OR STRENGTH FOR CERTAIN WELDS

WELD PITCH

SPECIFICATION, PROCESS, OR OTHER REFERENCE

FIELD WELD SYMBOL

TAIL OMITTED WHEN REFERENCE NOT USED

ARROW CONNECTS REFERENCE LINE TO ARROW SIDE MEMBER OF JOINT OR ARROW SIDE OF JOINT

T S(E) (SIDES) (BOTH SIDES) (OTHER SIDE) (ARROW SIDE) R A F L–P

BASIC WELD SYMBOL OR DETAIL REFERENCE

NUMBER OF SPOT, STUD, OR PROJECTION WELDS

WELD-ALL-AROUND SYMBOL

(N)

REFERENCE LINE

ELEMENTS IN THIS AREA REMAIN AS SHOWN WHEN TAIL AND ARROW ARE REVERSED

WELD JOINTS AND TYPES						
APPLICABLE WELDS	WELD SYMBOL	BUTT	LAP	T	EDGE	CORNER
SQUARE-GROOVE			—			
BEVEL-GROOVE						
V-GROOVE			—	—		
U-GROOVE			—	—		
J-GROOVE						
FLARE-BEVEL-GROOVE						
FLARE-V-GROOVE			—	—		
FILLET		—			—	
PLUG		—			—	
SLOT		—			—	
EDGE-FLANGE			—	—		—
CORNER-FLANGE		—				
SPOT		—			—	
PROJECTION		—			—	
SEAM		—				
BRAZE	BRAZE				—	

STRUCTURAL STEEL SHAPES

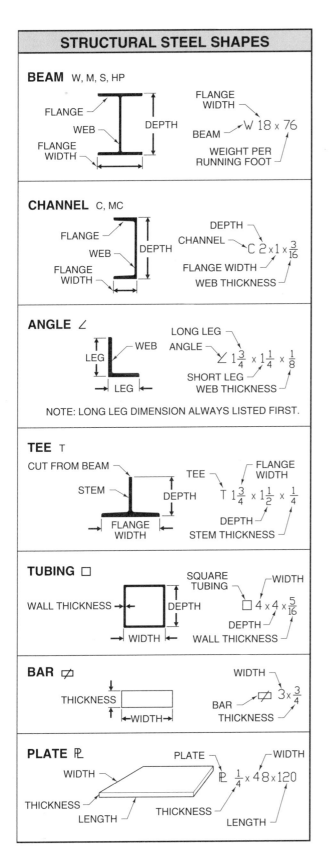

BEAM W, M, S, HP

FLANGE
WEB
FLANGE WIDTH
DEPTH
FLANGE WIDTH
BEAM W 18 x 76
WEIGHT PER RUNNING FOOT

CHANNEL C, MC

FLANGE
WEB
FLANGE WIDTH
DEPTH
DEPTH
CHANNEL C 2 x 1 x $\frac{3}{16}$
FLANGE WIDTH
WEB THICKNESS

ANGLE ∠

LONG LEG
WEB
LEG
ANGLE ∠ $1\frac{3}{4}$ x $1\frac{1}{4}$ x $\frac{1}{8}$
SHORT LEG
WEB THICKNESS
LEG

NOTE: LONG LEG DIMENSION ALWAYS LISTED FIRST.

TEE T

CUT FROM BEAM
STEM
DEPTH
FLANGE WIDTH
TEE
FLANGE WIDTH
DEPTH
T $1\frac{3}{4}$ x $1\frac{1}{2}$ x $\frac{1}{4}$
DEPTH
STEM THICKNESS

TUBING □

SQUARE TUBING
WIDTH
WALL THICKNESS
DEPTH
WIDTH
□ 4 x 4 x $\frac{5}{16}$
DEPTH
WALL THICKNESS

BAR ⊿

WIDTH
THICKNESS
WIDTH
BAR
⊿ 3 x $\frac{3}{4}$
THICKNESS

PLATE ℙ

PLATE
WIDTH
WIDTH
THICKNESS
LENGTH
THICKNESS
ℙ $\frac{1}{4}$ x 48 x 120
LENGTH

GREEK SYMBOLS

Alpha	A α	Nu	N ν
Beta	B β	Xi	Ξ ξ
Gamma	Γ γ	Omicron	O o
Delta	Δ δ	Pi	Π π
Epsilon	E ε	Rho	P ρ
Zeta	Z ζ	Sigma	Σ σ
Eta	H η	Tau	T τ
Theta	Θ θ	Upsilon	Y υ
Iota	I ι	Phi	Φ φ
Kappa	K κ	Chi	X χ
Lambda	Λ λ	Psi	Ψ ψ
Mu	M μ	Omega	Ω ω

METRIC/ENGLISH CONVERSIONS

LENGTH

Units	m	mm	ft	in.
1 m	1	1000	3.281	39.37
1 mm	0.001	1	$3.281(10^{-3})$	0.03937
1 ft	0.3048	304.8	1	12
1 in.	0.0254	25.4	0.00694	1

AREA

Units	m^2	cm^2	ft^2	in^2
1 m^2	1	104	10.764	1549.9
1 cm^2	10^{-4}	1	0.00108	0.155
1 ft^2	0.0929	929	1	144
1 in^2	$6.452(10^{-4})$	6.452	0.00694	1

VOLUME

Units	m^3	cm^3	ft^3	in^3
1 m^3	1	10^6	35.31	61.023
1 cm^3	107	10^4	$3.531(10^{-5})$	0.061023
1 ft^3	0.028317	1	1	1,728

UNIT PREFIXES

PREFIX	UNIT	SYMBOL	NUMBER
Other larger multiples			
Mega	Million	M	$1,000,000 = 10^6$
Kilo	Thousand	k	$1,000 = 10^3$
Hecto	Hundred	h	$100 = 10^2$
Deka	Ten	d	$10 = 10^1$
			Unit $1 = 10^0$
Deci	Tenth	d	$0.1 = 10^{-1}$
Centi	Hundreth	c	$0.01 = 10^{-2}$
Milli	Thousandth	m	$0.001 = 10^{-3}$
Micro	Millionth	μ	$0.000001 = 10^{-6}$
Other smaller multiples			

MILLIMETER AND DECIMAL INCH EQUIVALENTS*

mm	in.	mm	in.	mm	in.	mm	in.	mm	in.	mm	in.
1/50	.00079	25/50	.01969	1	.03937	26	1.02362	51	2.00787	76	2.99212
2/50	.00157	26/50	.02047	2	.07874	27	1.06299	52	2.04724	77	3.03149
3/50	.00236	27/50	.02126	3	.11811	28	1.10236	53	2.08661	78	3.07086
4/50	.00315	28/50	.02205	4	.15748	29	1.14173	54	2.12598	79	3.11023
		29/50	.02283								
5/50	.00394	30/50	.02362	5	.19685	30	1.18110	55	2.16535	80	3.14960
6/50	.00472	31/50	.02441	6	.23622	31	1.22047	56	2.20472	81	3.18897
7/50	.00551	32/50	.02520	7	.27559	32	1.25984	57	2.24409	82	3.22834
8/50	.00630	33/50	.02598	8	.31496	33	1.29921	58	2.28346	83	3.26771
9/50	.00709	34/50	.02677	9	.35433	34	1.33858	59	2.32283	84	3.30708
10/50	.00787	35/50	.02756	10	.39370	35	1.37795	60	2.36220	85	3.34645
11/50	.00866	36/50	.02835	11	.43307	36	1.41732	61	2.40157	86	3.38582
12/50	.00945	37/50	.02913	12	.47244	37	1.45669	62	2.44094	87	3.42519
13/50	.01024	38/50	.02992	13	.51181	38	1.49606	63	2.48031	88	3.46456
14/50	.01102	39/50	.03071	14	.55118	39	1.53543	64	2.51968	89	3.50393
15/50	.01181	40/50	.03150	15	.59055	40	1.57480	65	2.55905	90	3.54330
16/50	.01260	41/50	.03228	16	.62992	41	1.61417	66	2.59842	91	3.58267
17/50	.01339	42/50	.03307	17	.66929	42	1.65354	67	2.63779	92	3.62204
18/50	.01417	43/50	.03386	18	.70866	43	1.69291	68	2.67716	93	3.66141
19/50	.01496	44/50	.03465	19	.74803	44	1.73228	69	2.71653	94	3.70078
20/50	.01575	45/50	.03543	20	.78740	45	1.77165	70	2.75590	95	3.74015
21/50	.01654	46/50	.03622	21	.82677	46	1.81102	71	2.79527	96	3.77952
22/50	.01732	47/50	.03701	22	.86614	47	1.85039	72	2.83464	97	3.81889
23/50	.01811	48/50	.03780	23	.90551	48	1.88976	73	2.87401	98	3.85826
24/50	.01890	49/50	.03858	24	.94488	49	1.92913	74	2.91338	99	3.89763
				25	.98425	50	1.96850	75	2.95275	100	3.93700

*Based on 1/100 mm = .003973"   10 mm = 1 centmeter = 0.3937"   25.4 mm = 1"

DECIMAL EQUIVALENTS OF AN INCH

Fraction	Decimal	Fraction	Decimal	Fraction	Decimal	Fraction	Decimal
1/64	0.015625	17/64	0.265625	33/64	0.515625	47/64	0.765625
1/32	0.03125	9/32	0.28125	17/32	0.53125	25/32	0.78125
3/64	0.046875	19/64	0.296875	35/64	0.546875	51/64	0.796875
1/16	0.0625	5/16	0.3125	9/16	0.5625	13/16	0.8125
5/64	0.078125	21/64	0.328125	37/64	0.578125	53/64	0.828125
3/32	0.09375	11/32	0.34375	19/32	0.9375	27/32	0.84375
7/64	0.109375	23/64	0.359375	39/64	0.609375	55/64	0.859375
1/8	0.125	3/8	0.375	5/8	0.625	7/8	0.875
9/64	0.140625	25/64	0.390625	41/64	0.640625	57/64	0.890625
5/32	0.15625	13/32	0.40625	21/32	0.65625	29/32	0.90625
11/64	0.171875	27/64	0.421875	43/64	0.671875	59/64	0.921875
3/16	0.1875	7/16	0.4375	11/16	0.6875	15/16	0.9375
13/64	0.203125	29/64	0.453125	45/64	0.703125	61/64	0.953125
7/32	0.21875	15/32	0.46875	23/32	0.71875	31/32	0.96875
15/64	0.234375	31/64	0.484375	47/64	0.734375	63/64	0.984375
1/4	0.250	1/2	0.500	3/4	0.750	1	1.000

PIPE							
NOMINAL ID (IN.)	OD (BW GAUGE)	INSIDE DIAMETER (BW GAUGE)			NOMINAL WALL THICKNESS		
		STD	XS	XXS	SCHEDULE40	SCHEDULE60	SCHEDULE80
1/8	0.405	0.269	0.215		0.068	0.095	
1/4	0.540	0.364	0.302		0.088	0.119	
3/8	0.675	0.493	0.423		0.091	0.126	
1/2	0.840	0.622	0.546	0.252	0.109	0.147	0.294
3/4	1.050	0.824	0.742	0.434	0.113	0.154	0.308
1	1.315	1.049	0.957	0.599	0.133	0.179	0.358
1 1/4	1.660	1.380	1.278	0.896	0.140	0.191	0.382
1 1/2	1.900	1.610	1.500	1.100	0.145	0.200	0.400
2	2.375	2.067	1.939	1.503	0.154	0.218	0.436
2 1/2	2.875	2.469	2.323	1.771	0.203	0.276	0.552
3	3.500	3.068	2.900	2.300	0.216	0.300	0.600
3 1/2	4.000	3.548	3.364	2.728	0.226	0.318	
4	4.500	4.026	3.826	3.152	0.237	0.337	0.674
5	5.563	5.047	4.813	4.063	0.258	0.375	0.750
6	6.625	6.065	5.761	4.897	0.280	0.432	0.864
8	8.625	7.981	7.625	6.875	0.322	0.500	0.875
10	10.750	10.020	9.750	8.750	0.365	0.500	
12	12.750	12.000	11.750	10.750	0.406	0.500	

STANDARD SERIES THREADS — GRADED PITCHES

NOMINAL DIAMETER	UNC		UNF		UNEF	
	TPI	TAP DRILL	TPI	TAP DRILL	TPI	TAP DRILL
0 (.0600)			80	3/64		
1 (.0730)	64	No. 53	72	No. 53		
2 (.0860)	56	No. 50	64	No. 50		
3 (.0990)	48	No. 47	56	No. 45		
4 (.1120)	40	No. 43	48	No. 42		
5 (.1250)	40	No. 38	44	No. 37		
6 (.1380)	32	No. 36	40	No. 33		
8 (.1640)	32	No. 29	36	No. 29		
10 (.1900)	24	No. 25	32	No. 21		
12 (.2160)	24	No. 16	28	No. 14	32	No.13
1/4 (.2500)	20	No. 7	28	No. 3	32	7/32
5/16 (.3125)	18	F	24	I	32	9/32
3/8 (.3750)	16	5/16	24	Q	32	11/32
7/16 (.4375)	14	U	20	25/64	28	13/32
1/2 (.5000)	13	27/64	20	29/64	28	15/32
9/16 (.5625)	12	31/64	18	33/64	24	33/64
5/8 (.6250)	11	17/32	18	37/64	24	37/64
11/16 (.6875)					24	41/64
3/4 (.7500)	10	21/32	16	11/16	20	45/64
13/16 (.8125)					20	49/64
7/8 (.8750)	9	49/64	14	13/16	20	53/64
15/16 (.9375)					20	57/64
1 (1.000)	8	7/8	12	59/64	20	61/64

PIPE FITTINGS AND VALVES

	WELDED*	FLANGED	SCREWED		WELDED*	FLANGED	SCREWED		WELDED*	FLANGED	SCREWED
BUSHING				REDUCING FLANGE				AUTOMATIC BY-PASS VALVE			
CAP				BULL PLUG							
REDUCING CROSS				PIPE PLUG				AUTOMATIC REDUCING VALVE			
STRAIGHT-SIZE CROSS				CONCENTRIC REDUCER				STRAIGHT CHECK VALVE			
CROSSOVER				ECCENTRIC REDUCER				COCK			
45° ELBOW				SLEEVE				DIAPHRAGM VALVE			
90° ELBOW				STRAIGHT-SIZE TEE				FLOAT VALVE			
ELBOW—TURNED DOWN				TEE—OUTLET UP				GATE VALVE			
ELBOW—TURNED UP				TEE—OUTLET DOWN				MOTOR-OPERATED GATE VALVE			
BASE ELBOW				DOUBLE-SWEEP TEE				GLOBE VALVE			
DOUBLE-BRANCH ELBOW				REDUCING TEE				MOTOR-OPERATED GLOBE VALVE			
LONG-RADIUS ELBOW				SINGLE-SWEEP TEE				ANGLE HOSE VALVE			
REDUCING ELBOW				SIDE OUTLET TEE—OUTLET DOWN				GATE VALVE			
SIDE OUTLET ELBOW—OUTLET DOWN				SIDE OUTLET TEE—OUTLET UP				GLOBE VALVE			
SIDE OUTLET ELBOW—OUTLET UP				UNION				LOCKSHIELD VALVE			
STREET ELBOW				ANGLE CHECK VALVE				QUICK-OPENING VALVE			
CONNECTING PIPE JOINT				ANGLE GATE VALVE—ELEVATION				SAFETY VALVE			
EXPANSION JOINT				ANGLE GATE VALVE—PLAN							
LATERAL				ANGLE GLOBE VALVE—ELEVATION				GOVERNOR-OPERATED AUTOMATIC VALVE			
ORIFICE FLANGE				ANGLE GLOBE VALVE—PLAN							

* A • may be used instead of the "X" to represent a welded joint.

The American Society of Mechanical Engineers

Form A.7.1

**SUGGESTED
WELDING PROCEDURE SPECIFICATION (WPS)**

Identification _____

Date _____ Revision _____

Company name _____

Supporting PQR no.(s) _____ Type - Manual ( ) Semi-Automatic ( )

Welding process(es) _____ Machine ( ) Automatic ( )

Backing: Yes ( ) No ( )

Backing material (type) _____

Material number _____ Group _____ To material number _____ Group _____

Material spec. type and grade _____ To material spec. type and grade _____

Base metal thickness range: Groove _____ Fillet _____

Deposited weld metal thickness range _____

Filler metal F no. _____ A no. _____

Spec. no. (AWS) _____ Flux tradename _____

Electrode-flux (Class) _____ Type _____

Consumable insert: Yes ( ) No ( ) Classifications _____

Shape _____

Position(s) of joint _____ Size _____

Welding progression: Up ( ) Down ( ) Ferrite number (when reqd.) _____

PREHEAT: **GAS:**

Preheat temp., min _____ Shielding gas(es) _____

Interpass temp., max _____ Percent composition _____
(continuous or special heating, where
applicable, should be recorded) Flow rate _____
Root shielding gas _____

POSTWELD HEAT TREATMENT: Trailing gas composition _____

Temperature range _____ Trailing gas flow rate _____

Time range _____

Tungsten electrode, type and size _____

Mode of metal transfer for GMAW: Short-circuiting ( ) Globular ( ) Spray ( )

Electrode wire feed speed range: _____

Stringer bead ( ) Weave bead ( ) Peening: Yes ( ) No ( )

Oscillation _____

Standoff distance

Multiple ( ) or single electrode ( )

Other _____

Filler metal				Current			
Weld layer(s)	Process	Class	Dia.	Type & polarity	Amp range	Volt range	Travel speed range

e.g., Remarks, comments, hot wire addition, technique, torch angle, etc.

Approved for Production by _____
Employer

Note: Those items that are not applicable should be marked N.A.

Form A.7.2 **SUGGESTED** **Page 1 of 2**
PROCEDURE QUALIFICATION RECORD (PQR)

WPS no. used for test _____ Welding process(es) _____

Company _____ Equipment type and model (sw) _____

JOINT DESIGN USED (2.6.1)

WELD INCREMENT SEQUENCE

Single ( )   Double weld ( )

Backing material _____

Root opening _____ Root face dimension _____

Groove angle _____ Radius (J-U) _____

Back gouging: Yes ( ) No ( ) Method _____

BASE METALS (2.6.2)

Material spec. _____ To _____

Type or grade _____ To _____

Material no. _____ To material no. _____

Group no. _____ To group no. _____

Thickness _____

Diameter (pipe) _____

Surfacing: Material _____ Thickness _____

Chemical composition _____

Other _____

FILLER METALS (2.6.3)

Weld metal analysis A no. _____

Filler metal F no. _____

AWS specification _____

AWS classification _____

Flux class _____ Flux brand _____

Consumable insert: Spec. _____ Class. _____

Supplemental filler metal spec. _____ Class. _____

Non-classified filler metals _____

Consumable guide (ESW) Yes ( ) No ( )

Supplemental deoxidant (EBW) _____

POSITION (2.6.4)

Position of groove _____ Fillet _____

Vertical progression: Up ( )   Down ( )

PREHEAT (2.6.5)

Preheat temp., actual min _____

Interpass temp., actual max _____

POSTWELD HEAT TREAMTENT (2.6.6):

Temp. _____

Time _____

Other _____

GAS (2.6.7)

Gas type(s) _____

Gas mixture percentage

Flow rate _____

Backing gas _____ Flow rate _____

Root shielding gas

EBW vacuum ( )   Absolute pressure ( )

ELECTRICAL CHARACTERISTICS (2.6.8)

Electrode extension _____

Standoff distance

Transfer mode (GMAW) _____

Electrode diameter tungsten _____

Type tungsten electrode _____

Current: AC ( ) DCEP ( ) DCEN ( ) Pulsed ( )

Heat input _____

EBW: beam focus current _____ Pulse freq. _____

Filament type _____ Shape ____ Size _____

Other _____

TECHNIQUE (2.6.9)

Oscillation frequency _____Weave width _____

Dwell time _____

String or weave bead _____ Weave width _____

Multi-pass or single pass (per side) _____

Number of electrodes _____

Peening _____

Electrode spacing _____

Arc timing (SW) _____ Lift ( )

PAW: Conventional ( )   Key hole ( )

Interpass cleaning:

Pass no.	Filler metal size	Amps	Volts	Travel speed (ipm)	Filler metal wire (ipm)	Slope induction	Special notes (process, etc.)

Note: Those items that are not applicable should be marked N.A.

Form A.7.2 **Page 2 of 2**

TENSILE TEST SPECIMENS: SUGGESTED PROCEDURE QUALIFICATION RECORD PQR No. _____

Type: _____ Tensile specimen size: _____ Area: _____

Groove ( ) Reinforcing bar ( ) Stud welds ( )

Tensile test results: (Minimum required UTS _____ psi)

Specimen no.	Width, in.	Thickness, in.	Area, in.^{2}	Max load lbs	UTS, psi	Type failure and location

GUIDED BEND TEST SPECIMENS - SPECIMEN SIZE: _____

Type	Result	Type	Result

MACRO-EXAMINATION RESULTS: Reinforcing bar ( ) Stud ( )

1. _____ 4. _____
2. _____ 5. _____
3. _____

SHEAR TEST RESULTS - FILLETS:
1. _____ 3. _____
2. _____ 4. _____

IMPACT TEST SPECIMENS

Type: _____ Size: _____

Test temperature: _____

Specimen location: WM = weld metal; BM = base metal; HAZ = heat-affected zone

Test results:

Welding position	Specimen location	Energy absorbed (ft.-lbs.)	Ductile fracture area (percent)	Lateral expansion (mils)

IF APPLICABLE **RESULTS**

Hardness tests: ( ) Values _____ Acceptable ( ) Unacceptable ( )

Visual (special weldments 2.4.2) ( ) Acceptable ( ) Unacceptable ( )

Torque ( ) psi Acceptable ( ) Unacceptable ( )

Proof test ( ) Method _____ Acceptable ( ) Unacceptable ( )

Chemical analysis ( ) Acceptable ( ) Unacceptable ( )

Non-destructive exam ( ) Process _____ Acceptable ( ) Unacceptable ( )

Other _____ Acceptable ( ) Unacceptable ( )

Mechanical Testing by (Company) _____ Lab No. _____

We certify that the statements in this Record are correct and that the test welds were prepared, welded, and tested in accordance with the requirements of the American Welding Society Standard for Welding Procedure and Performance Qualification (AWS B2.1-83).

Qualifier: _____ Reviewed by: _____

Date: _____ Approved by: _____

Employer

Form A.7.3

**SUGGESTED
PERFORMANCE QUALIFICATION TEST RECORD**

Name _____ Identification _____ Welder ( ) Operator ( )

Social security number: _____ Qualified to WPS no. _____

Process(es) _____ Manual ( ) Semi-Automatic ( ) Automatic ( ) Machine ( )

Test base metal specification _____ To _____

Material number _____ To _____

Fuel gas (OFW) _____

AWS filler metal classification _____ F no. _____

Backing: Yes ( ) No ( ) Double ( ) or Single side ( )
Current: AC ( ) DC ( ) Short-circuiting arc (GMAW) Yes ( ) No ( )
Consumable insert: Yes ( ) No ( )
Root shielding: Yes ( ) No ( )

TEST WELDMENT   **POSITION TESTED**   **WELDMENT THICKNESS (T)**

GROOVE:
Pipe   1G ( ) 2G ( ) 5G ( ) 6G ( ) 6GR ( ) Diameter(s) _____ (T) _____
Plate   1G ( ) 2G ( ) 3G ( ) 4G ( )   (T) _____
Rebar   1G ( ) 2G ( ) 3G ( ) 4G ( )   Bar size _____ Butt ( )
  Spliced butt ( )

FILLET:
Pipe ( )   1F ( ) 2F ( ) 3F ( ) 4F ( ) 5F ( ) Diameter _____ (T) _____
Plate ( )   1F ( ) 2F ( ) 3F ( ) 4F ( )   (T) _____

Other (describe) _____

Test results: Remarks

Visual test results	N/A ( )	Pass ( )	Fail ( )	
Bend test results	N/A ( )	Pass ( )	Fail ( )	
Macro test results	N/A ( )	Pass ( )	Fail ( )	
Tension test	N/A ( )	Pass ( )	Fail ( )	
Radiographic test results	N/A ( )	Pass ( )	Fail ( )	
Penetrant test	N/A ( )	Pass ( )	Fail ( )	

QUALIFIED FOR:
PROCESSES
GROOVE: **THICKNESS**
Pipe   1G ( ) 2G ( ) 5G ( ) 6G ( ) 6GR ( ) (T) Min _____ Max _____ Dia _____
Plate   1G ( ) 2G ( ) 3G ( ) 4G ( )   (T) Min _____ Max _____
Rebar   1G ( ) 2G ( ) 3G ( ) 4G ( )   Bar size Min _____ Max _____

FILLET:
Pipe   1F ( ) 2F ( ) 4F ( ) 5F ( )   (T) Min _____ Max _____
Plate   1F ( ) 2F ( ) 3F ( ) 4F ( )   (T) Min _____ Max _____
Rebar   1F ( ) 2F ( ) 3F ( ) 4F ( )   Bar size Min _____ Max _____

Weld cladding ( )   Position(s) _____ T Min _____ Max _____ Clad Min _____

Consumable insert ( ) Backing type ( )
Vertical Up ( ) Down ( )
Single side ( ) Double side ( ) No backing ( )
Short-circuiting arc ( ) Spray arc ( ) Pulsed arc ( )
Reinforcing bar - butt ( ) or Spliced butt ( )

The above named person is qualified for the welding process(es) used in this test within the limits of essential variables including materials and filler metal variables of the AWS Standard for Welding Procedure and Performance Qualification (AWS B2.1).

Date tested _____ Signed by _____
  Qualifier

GLOSSARY

Terms in this glossary are defined as they relate to welding.

A

absortivity: Fraction of light absorbed.

acetylene: Colorless gas that is highly combustible when mixed with oxygen. Stable at pressures above 15 psi. Used in oxyacetylene welding. See *oxyacetylene welding.*

actual throat: Shortest distance from the face of a fillet weld to the weld root after welding. See *weld face, fillet weld,* and *weld root.*

acute angle: Angle with less than 90°. See *angle.*

addition: Process of uniting two or more numbers to make one number.

adhesion: Joining together of dissimilar metals by capillary action. See *capillary action.*

aligned section: Section view in which the cutting plane line is bent to pass through detailed features, and the section view is revolved. See *section view* and *cutting plane line.*

allowance: Difference between the design size and the basic size of a thread.

alloy: Metal that consists of more than one chemical element, with at least one of the elements being a pure metal.

ampere (amp or A): Unit of measure for electricity that expresses the quantity or number of electrons flowing through a conductor per unit of time. See *conductor.*

angle: 1. In plane figures, the intersection of two lines. **2.** In building steel, L-shaped structural steel of two equal or unequal widths. See *structural steel.*

angular dimension: Dimension that measures angles and is commonly expressed as degrees and decimal parts of a degree, or as degrees, minutes, and seconds. See *dimensions* and *angle.*

Arabic numerals: Numerals expressed by the ten digits 0, 1, 2, 3, 4, 5, 6, 7, 8, and 9.

arc: Portion of the circumference of a circle. See *circumference* and *circle.*

arc length: Distance from the electrode to the molten pool of the base metal. See *electrode* and *base metal.*

area: Number of unit squares equal to the surface of an object.

arrow: Welding symbol that identifies the location where the welding operation is to be performed. See *welding symbol.*

arrowhead: Symbol that indicates the extent of a dimension. See *dimensions.*

axis: Straight line around which a geometric figure is generated.

axonometric: Pictorial drawing showing three sides of an object with horizontal and vertical dimensions. Drawn to scale and containing no true view of any side. See *true view.*

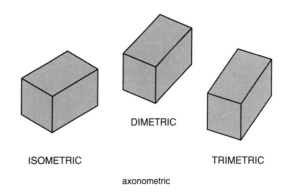

ISOMETRIC   DIMETRIC   TRIMETRIC

axonometric

B

back (transverse) pitch: Distance from the center of one row of rivets to the center of the adjacent row of rivets. See *rivet.*

backstep welding: Welding in which individual passes are made in the opposite direction of the weld.

back weld: Weld deposited in the weld root opposite the face of the weld on the other side of the joint member. Deposited after the weld on the opposite side of the part. See *backing weld.*

back weld

backing symbol: Supplementary symbol indicated by a rectangle on the opposite side of the groove weld symbol on the reference line. See *supplementary symbol.*

backing weld: Weld deposited in the weld root opposite the face of the weld on the other side of the joint member. Deposited before the weld on the opposite side of the part. See *back weld.*

bar: Round-, square-, or rectangular-shaped structural steel. See *structural steel.*

base metal: Material to be welded.

beam: I-shaped structural steel. See *structural steel.*

bending stress: Stress caused by equal forces acting perpendicular to the horizontal axis of an object. See *stress* and *axis.*

bevel: Sloped edge of an object running from surface to surface.

blind hole: Drilled hole that does not pass through.

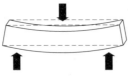
bending stress

blind rivet: Rivet with a hollow shank that joins two parts with access from one side. See *rivet* and *shank*.

brazing: Group of welding processes in which metal is joined by heating the filler metal at temperatures greater than 840°F, but less than the melting point of the base metal. See *filler metal* and *base metal*.

brazing symbol: Graphic symbol that shows braze locations and specifications on prints.

break line: Line that shows internal features or avoids showing continuous features.

brittleness: Lack of ductility in a metal. See *ductility*.

broken-out section: Partial section view which appears to have been broken out of the object. See *section view*.

butt joint: Weld joint formed when two joint members, located approximately in the same plane, are positioned edge to edge. See *weld joint*.

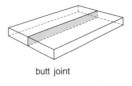
butt joint

C

cabinet: Oblique with receding lines drawn to one-half the scale of lines in the true view. See *oblique* and *true view*.

capillary action: Force by which a liquid in contact with a solid is distributed between faying surfaces. See *faying surface*.

carburizing (reducing) flame: Oxyfuel flame with an excess of fuel.

cast: Metal heated to its liquid state and poured into a mold, where it cools and resolidifies.

cavalier: Oblique with receding lines drawn to the same scale as the lines in the true view. See *oblique* and *true view*.

center line: Line that locates the centerpoints of arcs and circles. See *arc* and *circle*.

central processing unit: Control center of a computer.

chamfer: Sloped edge of an object running from surface to side. See *edge*.

channel: C-shaped structural steel used in conjunction with other structural shapes as support members or combined to serve as an I beam. See *structural steel*.

chemical properties: Properties of metals that are directly related to molecular composition and pertaining to the chemical reactivity of metals and the surrounding environment.

chemical test: Metal identification test using chemicals which react when placed on certain types of metals.

chip test: Metal identification test that identifies metal by the shape of its chips.

chord: Line from circumference to circumference not through the centerpoint of a circle. See *circumference* and *circle*.

circle: 1. In plane figures, a figure generated around a centerpoint. See *plane figure*. **2.** In solid figures, a conic section that has the plane perpendicular to the axis. See *conic section* and *axis*.

circumference: 3.1416 times the diameter of a circle. See *diameter* and *circle*.

coefficient of thermal expansion: Unit change in the length of a material caused by changing the temperature 1°F.

color test: Metal identification test that identifies metals by their color.

combined weld symbols: Weld symbols used when the weld joint, weld type, and welding operation require more information than can be specified with one weld symbol. See *weld symbol*.

complementary angles: Angles equaling 90°. See *angle*.

compressive stress: The stress caused by two equal forces acting on the same axial line to crush an object. See *stress* and *axis*.

compressive stress

computer-aided design (CAD): Software program used for generation and reproduction of line drawings and prints with computers. See *software*.

concave: Curved inward.

concentric circles: Circles that have different diameters and the same centerpoint. See *circle* and *diameter*.

conductor: Any material through which electricity flows easily.

conic section: Curve produced by a plane intersecting a right circular cone. See *right circular cone*.

constant pitch: Standard screw thread series with a set number of threads per inch regardless of diameter. See *standard series*.

consumable insert: Spacer that provides proper opening of the weld joint and becomes part of the filler metal during welding. See *weld joint* and *filler metal*.

consumable insert symbol: Supplementary symbol indicated by a square on the opposite side of a groove weld on the reference line. See *supplementary symbol*.

contour symbol: Supplementary symbol indicated by a horizontal line or arc parallel to the weld symbol, which specifies the shape of the completed weld. See *supplementary symbol*.

conventional drafting practices: Language of standard lines, symbols, and abbreviations used in conjunction with drafting principles to ensure drawings are consistent and easy to read.

convention break: Standard method of showing shortened views of elongated objects.

convex: Curved outward.

corner: Angular space at the intersection of surfaces.

corner-flange weld: Flange weld with one joint member bent. See *flange weld.*

corner joint: Weld joint formed when two joint members are positioned at an approximate 90° angle with the weld joint at the outside of the joint members. See *weld joint.*

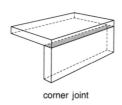

corner joint

corrosion: Combining of metals with elements in the surrounding environment that leads to the deterioration or wasting away of the metal.

counterbored hole: Enlarged and recessed hole with square shoulders.

counterdrilled hole: Hole with a cone-shaped opening below the outer surface.

countersink: Tool that produces a countersunk hole. See *countersunk hole.*

countersunk hole: Hole with a cone-shaped opening or recess at the outer surface. See *countersink.*

cover pass: Final weld pass deposited. See *weld pass.*

creep: Slow progressive strain that causes metal to fail. See *strain.*

cryogenic properties: Abilities of metals to resist failure when subjected to very low temperatures.

crystal: Solid composed of atoms arranged in a pattern that is repetitive in three dimensions.

cubic foot: 1'-0″ × 1'-0″ × 1'-0″ or 1728 cu in.

cubic inch: 1″ × 1″ × 1″ or its equivalent.

current: Movement of electrons through a conductor. See *conductor.*

cursor: Solid or flashing pointer indicating position of work on a video display terminal. See *video display terminal.*

cutting plane line: Line that shows where an object is imagined to be cut in order to view internal features.

D

decimal: Fraction with a denominator of 10, 100, 1000, etc. See *fraction* and *denominator.*

decimal point: The period in a decimal number. See *decimal.*

denominator: Number of parts the whole number has been divided into. Lower (or right-hand) number of a fraction. See *whole number.*

depth of fusion: Distance from the fusion face to the weld interface. See *fusion face* and *weld interface.*

destructive testing: Any type of testing that damages the test part (specimen).

detailed representation: Method of thread representation in which the thread profiles are connected by helices. See *thread representation* and *helix.*

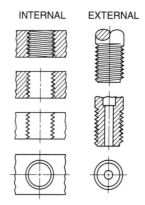

INTERNAL   EXTERNAL

detailed representation

diagonal pitch: Distance between the centers of rivets nearest each other in adjacent rows. See *rivet.*

diameter: Distance from circumference (outside) to the circumference through the centerpoint of a circle. See *circumference* and *circle.*

dimension line: Line that is used with dimensions to show size or location. See *dimensions.*

dimensions: 1. In welding, part of a welding symbol that specifies weld size, number, and location. See *welding symbol.* **2.** Numerical values that give size, form, or location of objects.

dimetric: Axonometric drawing with two axes drawn on equal angles and one axis containing either fewer or more degrees. See *axonometric* and *axis.*

direct current electrode negative (DCEN): Flow of current from electrode (–) to work (+). See *electrode.*

direct current electrode positive (DCEP): Flow of current from work (–) to electrode (+).

distortion: Change in the original shape of the metal as metal expands when heated and contracts when cooled.

dividend: Number to be divided. See *division.*

division: Process of finding how many times one number contains the other number.

divisor: Number by which division is done. See *division.*

double-bevel-groove weld: Groove weld having joint members beveled on both sides with the weld made from both sides. See *groove weld.*

double-fillet weld: Fillet weld that has filler metal deposited on both sides. See *fillet weld.*

double-flare-bevel-groove weld: Groove weld having two radiused joint members with the weld made from both sides. See *groove weld.*

double-flare-V-groove weld: Groove weld having radiused joint members with the weld made from both sides. See *groove weld*.

double-J-groove weld: Groove weld having joint members grooved in a J shape on both sides with the weld made from both sides. See *groove weld*.

double-square-groove weld: Groove weld having square-edged joint members with the weld made from both sides. See *groove weld*.

double-U-groove weld: Groove weld having joint members grooved in a U shape on both sides with the weld made from both sides. See *groove weld*.

double-V-groove weld: Groove weld having joint members angled on both sides with the weld made from both sides. See *groove weld*.

drill: Round hole in a material produced by a twist drill.

ductility: Ability of a metal to stretch, bend, or twist without breaking or cracking.

duty cycle: Length of time (expressed as a percentage) that a welding machine can operate at its rated output within a ten minute period. See *output*.

E

eccentric circles: Circles that have different diameters and different centerpoints. See *circle* and *diameter*.

edge: Intersection of two surfaces.

edge-flange weld: Flange weld with both joint members bent. See *flange weld*.

edge joint: Weld joint formed when the edges of two joint members are joined. See *weld joint*.

edge joint

effective throat: Shortest distance from the face of a fillet weld to the weld root, minus any convexity after welding. See *weld face, fillet weld,* and *weld root*.

elastic deformation: Ability of a metal to return to its original size and shape after loading and unloading.

elastic limit (yield): Last point at which a material can be deformed and still return to its original shape.

electrical properties: Abilities of a metal to conduct or resist electricity or the flow of electrons.

electrode: Coated metal wire which forms and cleans the weld bead. See *weld bead*.

electrode holder: Hand-held device that holds the electrode securely at the required angle for maximum access to the weld area. See *electrode*.

ellipse: 1. In plane figures, a curve with two focal points. **2.** In solid figures, a conic section that has the plane oblique to the axis but making a greater angle with the axis than with the elements of the cone. See *conic section* and *axis*.

equation: Means of showing that two numbers or two groups of numbers are equal to the same amount.

equilateral triangle: Triangle that has three equal sides and three equal angles. See *triangle* and *angle*.

equilibrium: State of balance between opposing forces.

even number: Number that can be divided by 2 an exact number of times.

extension line: Line that extends from surface features and terminates dimension lines.

F

face reinforcement: Filler metal which extends above the surface of the joint member on the side of the joint on which welding was done. See *filler metal*.

faying surface: Part of the joint member which is in full contact prior to welding. See *capillary action*.

ferrous metal: Any metal with iron as a major alloying element.

field rivet: Rivet placed in the field. See *rivet*.

field weld symbol: Supplementary symbol indicated by a triangular flag rising from the intersection of the arrow and reference line, which specifies the welding operation is to be completed in the field at the location of final installation. See *supplementary symbol*.

file test: Metal identification test in which a file is used to indicate the hardness of steel compared with that of the file.

filler metal: Metal deposited during welding process.

filler pass: Weld pass that fills remaining portion of the weld after the root pass and hot pass. See *weld pass*.

fillet: Rounded interior corner. See *corner*.

fillet weld: Weld type made in the cross-sectional shape of a triangle. See *weld type*.

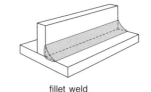

fillet weld

fillet weld leg: Distance from the joint root to the weld toe. See *joint root* and *weld toe*.

fillet weld leg size: Dimension from the root of a weld to the toes of a weld after welding. See *fillet weld leg*.

fit-up: Positioning of pipe with other pipe or fittings before welding.

fitting: Standard connection used to join two or more pieces of pipe.

fixture: Device used to maintain the correct positional relationship between joint members required by print specifications.

flange height: Distance from the point of tangency on the flange of a flange weld to the edge of the flange before welding.

flange radius: Radius of the joint member(s) requiring edge preparation in a flange weld.

flange weld: Weld type made of light-gauge metal with one or both joint members bent at approximately 90°. See *weld type*.

flange weld thickness: Cross-sectional distance of a flange weld from the weld face to the weld root. See *flange weld, weld face,* and *weld root.*

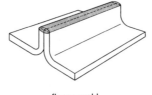

flange weld

flux: Coating on the electrode. See *electrode.*

flux cored arc welding (FCAW): Welding process that uses an arc shielded by gas from within the electrode. See *electrode.*

forged: Metal formed by a mechanical or hydraulic press with or without heat.

formula: Mathematical equation which contains a fact, rule, or principle. See *equation.*

fraction: Less than 1. One part of a whole number. See *whole number.*

fractional dimension: Dimension that measures lines and is commonly expressed as inches and fractional parts of an inch. See *dimensions.*

fracture test: Metal identification test that breaks the metal sample to check for ductility and grain size.

full section: Section view created by passing the cutting plane line completely through the object. See *section view* and *cutting plane line.*

fusion: Melting together of filler metal and base metal. See *filler metal* and *base metal.*

fusion face: Surface of the base metal that is melted during welding. See *fusion.*

G

gas metal arc welding (GMAW): Welding process with a shielded gas arc between a continuous wire electrode and the weld metal. See *electrode.*

gas tungsten arc welding (GTAW): Welding process in which shielding gas protects the arc between a tungsten electrode and the weld area.

general notes: Notes that apply a given specification to all items on the prints.

generator: Welding machine that produces DC only.

globular transfer: Metal transfer in which molten metal from a consumable electrode is spread across the arc in large drops.

graded pitch: Standard screw thread series with a different number of threads per inch based on the diameter. See *standard series.*

grain: Individual crystal in a metal that has multiple crystals. See *crystal.*

grip: Effective holding length of a rivet. See *rivet.*

groove face: Surface of the joint member included in the groove of the weld.

groove weld: Weld type made in the groove of the pieces to be welded. See *weld type.*

grounding device (ground): Connection between welding cable and weld parts in the welding circuit.

H

half section: Section view created by passing the cutting plane line halfway through the object. See *section view* and *cutting plane line.*

hardfacing: Applying filler metals, which provide a coating to protect the base metal from wear caused by impact, abrasion, erosion, or from other wear. See *filler metal.*

hardness: Ability of a metal to resist indentation.

hardware: Physical components of a computer system, including the input devices, central processing unit (CPU), and output devices.

heat-affected zone: Area of base metal in which the mechanical properties and structure are affected by the welding process. See *base metal* and *mechanical properties.*

helix: Curve formed by a line angular to the axis of a cylinder and in a plane wrapped around the cylinder. See *axis.*

heptagon: Polygon that has seven sides. See *polygon.*

hexagon: Polygon that has six sides. See *polygon.*

hidden line: Line that represents shapes which cannot be seen.

horizontal line: Line parallel to the horizon.

hot pass: Weld pass that penetrates deeply into the root pass and the root face of the joint. See *weld pass, root pass,* and *root face.*

hyperbola: Conic section that has the plane making a smaller angle with the axis than with the elements of the cone. See *conic section, angle,* and *axis.*

hypotenuse: Side of a right triangle opposite the right angle. See *right triangle* and *right angle.*

I

impact load: Load that is applied suddenly or intermittently. See *load.*

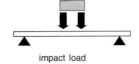

impact load

inclined (slanted) line: Line that is neither horizontal nor vertical.

inclined surface: Plane surface perpendicular to one plane of projection and inclined to the remaining two orthographic views.

incomplete fusion: Discontinuance of a weld where complete fusion does not occur between the weld metal and the fusion faces of the joint members. See *fusion.*

inert gas: Gas that will not readily combine with other elements.

input: Electrical requirements for operating a welding machine.

input device: Hardware used to enter information into a computer system. See *hardware.*

intermittent fillet welds: Short sections of fillet welds applied at specified intervals on the weld part. See *fillet weld.*

interpass temperature: Weld area temperature between passes of a multiple-pass weld. See *weld pass.*

intersecting surface: When one surface meets another surface.

irregular plane figure: Plane figure that does not have equal angles and sides. See *plane figure* and *angle.*

irregular polygon: Polygon that has unequal sides and unequal angles. See *polygon* and *angle.*

isometric: Axonometric drawing with the axes drawn 120° apart. See *axonometric* and *axis.*

isosceles triangle: Triangle that has two equal sides and two equal angles. See *triangle* and *angle.*

J

joint root: Part of a joint to be welded where the members are the closest to each other.

joystick: Electromechanical device used to control the cursor on the display screen and enter information into a computer. See *cursor.*

K

keyboard: Electronic device that sends signals to the central processing unit (CPU) of a computer.

L

lap joint: Weld joint formed when two joint members are lapped over one another. See *weld joint.*

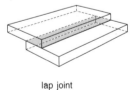

lap joint

large rivets: Rivets with a shank of ½″ or greater in diameter. See *rivet* and *shank.*

leader: Line that connects a dimension, note, or specification with a particular feature of the drawn object. See *dimensions* and *specifications.*

light pen: Photosensitive device used to enter data into a computer.

linear dimension: Dimension that measures lines and is commonly expressed as decimal inches or millimeters. See *dimensions.*

linetype library: CAD file that contains dashed, hidden, center, phantom, dot, dotdash, border, and divide lines. See *computer-aided design (CAD).*

load: External force applied to an elastic body that causes stress in a material. See *stress.*

lowest common denominator (LCD): Highest number that will divide equally into the denominator and numerator. See *denominator* and *numerator.*

M

magnetic test: Metal identification test that checks for the presence of iron in a metal.

malleability: Ability of a metal to be deformed by compressive forces without developing defects.

margin: Distance from the edge of a plate to the center line of the nearest row of rivets. See *plate* and *rivet.*

matte finish: Dull finish on film that will accept and hold pencil and ink lines well.

mechanical properties: Properties that describe the behavior of metals under applied loads.

melting point: Amount of heat required to melt a given amount of metal.

melt-through symbol: Supplementary symbol indicated by a darkened radius on the reference line opposite the weld symbol specified. Filler metal deposited on one side must completely penetrate through to the other side of weld. See *supplementary symbol.*

metal: Material consisting of one or more chemical elements having crystalline structure, high thermal and electrical conductivity, the ability to be deformed when heated, and high reflectivity.

microstructure: Microscopic arrangement of the components within a metal. See *metal.*

millions period: Third period (1,000,000 through 999,999,999). See *period.*

minuend: Number from which the subtraction is made. See *subtraction.*

modulus of elasticity: Ratio of stress to strain within the elastic limit. See *stress, strain,* and *elastic limit.*

monitor: Video display terminal of a computer. See *video display terminal.*

mouse: Electronic device used to input information into a computer.

multiplicand: Any number which is multiplied. See *multiplication.*

multiplication: Process of adding one number as many times as there are units in another number.

multiplier: Number by which multiplication is done. See *multiplication.*

multiview: Orthographic projection in which each view shows two dimensions of an object. See *orthographic projection.*

N

neutral flame: Oxyfuel flame with a balanced mixture of oxygen and fuel.

nondestructive examination (NDE) symbol: Symbol that specifies examination methods and requirements to verify weld quality.

nondestructive testing: Any type of testing that leaves the test part undamaged.

nonferrous metal: Pure metal, other than iron or metals with iron as a major alloying element.

non-threaded fasteners: Devices that join or fasten parts together without threads.

normal surface: Plane surface parallel to a plane of projection.

numerator: Number of parts in a fraction. Upper (or left-hand) number of a fraction. See *fraction*.

O

object line: Line that defines the visible shape of an object.

oblique: 1. In pictorial drawing, a drawing that shows one surface of an object as a true view. See *true view*. **2.** In plane surfaces, any plane that is not perpendicular to the base. See *perpendicular*.

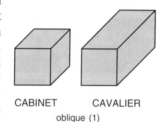

CABINET   CAVALIER

oblique (1)

oblique surface: Plane surface not parallel to any plane of projection.

obtuse angle: Angle with more than 90°. See *angle*.

octagon: Polygon that has eight sides. See *polygon*.

odd number: Any number that cannot be divided by 2 an exact number of times.

offset section: Section view created by an offset cutting plane line that shows internal features not in a straight line. See *section view* and *cutting plane line*.

optical properties: Color of a metal and how it reflects light.

orthographic projection: Drawing at right angles.

output: Maximum amperage and voltage of a welding machine. See *ampere* and *volt*.

output device: Hardware of a computer that either displays or generates drawings. See *hardware*.

overlapping: Extending weld metal beyond the weld toes or weld root. See *weld toes* and *weld root*.

oxidation: Combination of metal and oxygen into metal oxides.

oxidizing flame: An oxyfuel flame with an excess of oxygen.

oxyacetylene welding (OAW): Oxyfuel welding with acetylene. See *oxyfuel welding* and *acetylene*.

oxyfuel welding (OFW): Welding process that uses oxygen combined with a fuel to sustain a flame that generates the heat necessary for welding.

P

parabola: Conic section that has the plane oblique to the axis and making the same angle with the axis as with the elements of the cone. See *conic section, axis,* and *angle*.

parallel lines: Lines that remain the same distance apart.

parallelogram: Four-sided plane figure that has opposite sides parallel and equal. See *plane figure*.

pentagon: Polygon that has five sides. See *polygon*.

period: Group of three digits separated from other periods by a comma.

perpendicular: Straightness of a line making a right angle with another line. See *right angle*.

physical properties: Thermal, electrical, optical, magnetic, and general properties of metal.

pipe: Round-shaped structural steel. See *structural steel*.

pipe-jig: Device which holds sections of pipe or fittings before tack welding. See *tack weld*.

pitch: Distance between corresponding points on adjacent thread forms.

plane figure: A flat figure with only two dimensions.

plastic deformation: Failure of a metal to return to its original size and shape after being loaded and unloaded. See *load*.

plate: $3/16''$ or more thick structural steel used to cover large expanses of a structure. See *structural steel*.

plotter: Output device of a computer that generates finished drawings with pens. See *output device*.

plug weld: Weld type made in the cross-sectional shape of a hole in one of the joint members. See *weld type*.

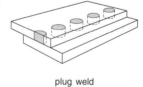

plug weld

plug weld size: Diameter of the hole through the joint member at the faying surface of the weld joint.

plumb: Exact verticality (determined by a plumb bob and line) with Earth's gravity.

polygon: Many-sided plane figure. See *plane figure*.

porosity: Cavity or cavities in the weld metal or weld interface caused by trapped gas.

positioner: Mechanical device that supports and moves joint members for maximum loading, welding, and unloading efficiency.

postheating: Application of heat to the weld part after welding to facilitate a controlled cooling rate.

preheating: Application of heat to the base metal before welding to reduce the temperature difference between the weld metal and the surrounding base metal.

primary weld: Weld that is an integral part of a structure and that directly transfers the load. See *load*.

prime number: Number that can be divided an exact number of times only by itself and the number 1.

prints: Reproductions of working drawings.

product: Result of multiplication. See *multiplication*.

projection weld: Weld type produced by confining fusion of molten base metal using heat and pressure with a preformed dimple or projection in one joint member prior to welding. See *weld type* and *base metal*.

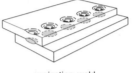

projection weld

proportional limit: Maximum stress a material can withstand without permanent deformation. See *stress*.

pure metal: Metal that consists of one chemical element. See *metal*.

Pythagorean Theorem: Theorem that states the square of the hypotenuse of a right triangle is equal to the sum of the squares of the other two sides. See *hypotenuse*, *right triangle*, and *sum*.

Q

quadrant: One-fourth of a circle. See *circle*.

quadrilateral: Four-sided plane figure. See *plane figure*.

quotient: Result of division. See *division*.

R

radiographic examination: Testing of welds for weld defects and strength requirements using X rays. See *X ray*.

radius: One-half the length of the diameter of a circle. See *diameter* and *circle*.

reaming: Enlarging and improving the surface quality of a hole.

rectangle: Plane figure that has opposite sides equal and four 90° angles. See *plane figure* and *angle*.

rectifier: Welding machine that produces AC or DC.

reducing flame: See *carburizing flame*.

reduction: Loss or removal of oxygen during the welding process.

reference line: Part of the welding symbol that identifies the side of the joint to be welded. See *welding symbol*.

reflectivity: Fraction of light reflected.

regular plane figure: Plane figure that has equal angles and equal sides. See *plane figure* and *angle*.

regular polygon: Polygon that has equal sides and equal angles. See *polygon* and *angle*.

remainder: 1. In subtraction, the difference between the minuend and the subtrahend. See *minuend* and *subtrahend*. **2.** In division, the part of the quotient leftover whenever the quotient is not a whole number. See *quotient* and *whole number*.

resistance welding (RW): Welding processes in which welding occurs from the heat obtained by resistance to the flow of current through the workpieces.

revolved section: Cross-sectional shape of elongated objects.

rhomboid: Plane figure that has opposite sides equal with opposite angles equal and no 90° angles. See *plane figure* and *angle*.

rhombus: Plane figure that has four equal sides with opposite angles equal and no 90° angles. See *plane figure* and *angle*.

right angle: Angle that contains 90°. See *angle*.

right circular cone: Cone with the axis at a 90° angle to the circular base. See *axis* and *angle*.

right triangle: Triangle that contains one 90° angle. See *triangle* and *angle*.

rivet: Cylindrical metal pin with a preformed head.

rivet pitch: Distance from the center of one rivet to the center of the next rivet in the same row. See *rivet*.

Roman numerals: Numerals expressed by the letters I, X, L, C, D, and M.

root edge: Weld face that comes to a point and has no width. See *weld face*.

root face: Surface of the groove next to the root.

root opening: Distance between joint members at the root of the weld before welding.

root pass: Initial weld pass that provides complete penetration through the thickness of the joint member. See *weld pass*.

root reinforcement: Filler metal which extends above the surface of the joint on the opposite side of the joint on which welding was done. See *filler metal*.

root surface: Surface of the weld on the opposite side of the joint on which welding was done.

round: Rounded exterior corner. See *corner*.

runout: Curve produced by a plane surface tangent to a cylindrical surface.

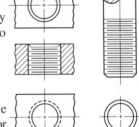

INTERNAL EXTERNAL

S

scalene triangle: Triangle that has no equal sides or equal angles. See *triangle* and *angle*.

schematic representation:

schematic representation

Method of thread representation in which solid lines

perpendicular to the axis represent roots and crests. See *thread representation* and *axis*.

screw thread series: Groups of diameter-pitch combinations. See *pitch*.

seam weld: Weld type produced by confining fusion of molten base metal using heat and pressure for a series of continuous or overlapping successive spot welds on joint members. See *weld type* and *spot weld*.

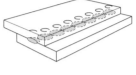
seam weld

secant: Straight line touching the circumference of a circle at two points. See *circumference* and *circle*.

secondary weld: Weld used to hold joint members and subassemblies together.

section line: Line that identifies the internal features cut by a cutting plane line. See *cutting plane line*.

section view: Interior view of an object through which a cutting plane has been passed. See *cutting plane line*.

sector: Pie-shaped piece of a circle. See *circle*.

segment: Portion of a circle set off by a chord. See *circle* and *chord*.

segregation: Separation of elements comprising the base metal. See *base metal*.

semicircle: One-half of a circle. See *circle*.

shank: Cylindrical body of a rivet. See *rivet*.

sheared plate: Plate that is rolled between horizontal and vertical rollers and trimmed on all edges.

shearing stress: Stress caused by two equal and parallel forces acting upon an object from opposite directions. See *stress*.

shearing stress

sheet: 3/16″ or less structural steel used to cover large expanses of a structure. See *structural steel*.

shielded metal arc welding (SMAW): Arc welding process in which the arc is shielded by the decomposition of the electrode covering. See *electrode*.

shop rivet: Rivet placed in the shop. See *rivet*.

short-circuit transfer: Metal transfer in which molten metal from a consumable electrode is deposited during repeated short circuits.

simplified representation: Method of thread representation in which hidden lines are drawn parallel to the axis at the approximate depth of the thread. See *thread representation* and *axis*.

single-bevel-groove weld: Groove weld having one joint member beveled with the weld made from that side. See *groove weld*.

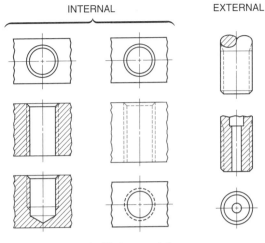
INTERNAL   EXTERNAL

simplified representation

single-fillet weld: Fillet weld having filler metal deposited on one side. See *fillet weld* and *filler metal*.

single-flare-bevel-groove weld: Groove weld having one straight and one radiused joint member with the weld made from one side. See *groove weld*.

single-flare-V-groove weld: Groove weld having radiused joint members with the weld made from one side. See *groove weld*.

single-J-groove weld: Groove weld having joint members grooved in a J shape on one side with the weld made from that side. See *groove weld*.

single-square-groove weld: Groove weld having square-edged joint members with the weld made from one side. See *groove weld*.

single-U-groove weld: Groove weld having joint members grooved in a U shape on one side with the weld made from that side. See *groove weld*.

single-V-groove weld: Groove weld having both joint members angled on the same side with the weld made from that side. See *groove weld*.

sketching: Drawing without instruments.

slag inclusions: Small particles of slag (cooled flux) trapped in the weld metal which prevent complete penetration.

slot weld: Weld type made in the cross-sectional shape of a slot (elongated hole) in one of the joint members. See *weld type*.

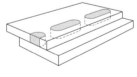
slot weld

small rivet: Rivet with a shank of 7/16″ or less in diameter. See *rivet* and *shank*.

software: Operating system of a computer, on magnetic tape or disks, that provides operational instructions for capturing and formatting keystrokes and generating lines.

soldering: Group of welding processes in which metal is joined by heating the filler metal at temperatures less than 840°F and less than the melting point of the base metal. See *filler metal* and *base metal.*

space lattice: Uniform pattern produced by lines connected through the atoms.

spacer symbol: Supplementary symbol indicated by a rectangle centered on reference line. See *supplementary symbol.*

spark test: Metal identification test that identifies metals by the shape, length, and color of a spark emitted from contact with a grinding wheel.

special series: Screw thread series with combinations of diameter and pitch not in the standard screw thread series. See *screw thread series* and *pitch.*

specifications: Documents that supplement working drawings with written instructions giving additional information.

specific notes: Notes that apply a given specification to specific items.

spotface: Flat surface machined at a right angle to a drilled hole. See *right angle.*

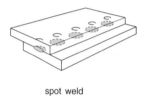

spot weld: Weld type produced by confining the fusion of molten base metal using heat and

spot weld

pressure without preparation to the joint members. See *weld type* and *base metal.*

spray transfer: Metal transfer in which molten metal from a consumable electrode is sprayed across the arc in small drops. See *electrode.*

square: Plane figure that has four equal sides and four 90° angles. See *plane figure* and *angle.*

staggered intermittent fillet welds: Intermittent fillet welds that have a staggered pitch and are applied to both sides of a weld joint. *See intermittent fillet welds.*

standard series: Screw thread series of coarse (UNC/UNRC), fine (UNF/UNRF), and extra-fine (UNEF/UNREF) graded pitches and eight series with constant pitches. See *screw thread series* and *pitch.*

static load: Load that remains constant. See *load.*

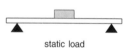

static load

straight angle: Angle that contains 180°. See *angle.*

straight line: Shortest distance between two points.

strain: Deformation per unit length of a solid under stress. See *stress.*

stress: Effect of an external force applied upon a solid material.

structural steel: Steel used in the erection of structures.

stud weld: Weld type made by joining threaded studs with other parts using heat and pressure. See *weld type.*

stud weld

stylus: Electromechanical device used to input information into a computer.

subtraction: Process of taking one number away from another number.

subtrahend: Any number subtracted. See *subtraction.*

sum: Result obtained from adding two or more numbers. See *addition.*

supplementary angles: Angles which equal 180°. See *angle.*

supplementary symbol: Symbol used on welding symbols to further define the operation to be completed.

surface feature: A part of a surface where change occurs.

surfacing: Applying filler metals which have similar characteristics to the base metal. See *filler metal* and *base metal.*

surfacing weld: Weld type in which weld beads are deposited on a surface to increase the dimensions of the part or to add special properties to the weld part. See *weld type.*

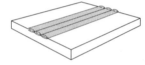

surfacing weld

T

tack weld: Weld that joins the joint members at random points to keep the joint members from moving out of their required positions.

tail: Part of a welding symbol included when a specific welding process, specification, or procedure must be indicated. See *welding symbol.*

tangent: Straight line touching the circumference of a circle at only one point. See *straight line, circumference,* and *circle.*

tee: T-shaped structural steel made of I beams cut to specifications by mill or suppliers. See *structural steel.*

tensile stress: Stress that is caused by two equal forces acting on the same axial line to pull an object apart.

tensile stress

theoretical throat: Distance from the face of a fillet weld to the root before welding. See *fillet weld, weld face,* and *weld root.*

thermal conductivity: Rate which metal transmits heat.

thermal expansion: Expansion of a metal when subjected to heat.

thermal properties: One of the physical properties of metal. Includes melting point, thermal conductivity, and thermal expansion and contraction. See *physical properties.*

thousands period: Second period (1,000 through 999,999). See *period.*

threaded fasteners: Devices such as nuts and bolts that join or fasten parts together with threads.

thread representation: Method of drawing used to show a threaded part.

through hole: Drilled hole passing completely through the material.

thumbwheel: Electromechanical device used to control the position of a computer cursor in horizontal and vertical planes.

T-joint: Weld joint formed when two joint members are positioned approximately 90° to one another in the form of a T. See *weld joint.*

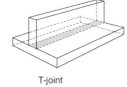

T-joint

tolerance: Amount of variation allowed above or below a dimension. See *dimensions.*

torch test: Metal identification test that can be used to identify a metal by its color change with the application of heat, its melting point, and its behavior in the molten state.

torque: Product of the applied force (P) times the distance (L) from the center of application.

torsional stress: Stress caused by two forces acting in opposite twisting motions. See *stress.*

toughness: Combination of strength and ductility of metals. See *ductility.*

torsional stress

trackball: Electromechanical device used to control the position of a computer cursor in horizontal, vertical, and diagonal planes. See *cursor.*

transformer: Welding machine that produces AC only.

translucent paper: Paper that allows light to pass through.

transmissivity: Fraction of light transmitted.

transverse pitch: See *back pitch.*

trapezium: Plane figure that has no sides parallel. See *plane figure.*

trapezoid: Plane figure that has two sides parallel. See *plane figure.*

travel angle: Angle less than 90° of the electrode in relation to a perpendicular line from the weld and the direction of the weld. See *electrode.*

travel speed: Speed at which the electrode is moved across the weld area. See *electrode.*

triangle: Three-sided plane figure. See *plane figure.*

trimetric: Axonometric drawing with all axes drawn at different angles. See *axonometric* and *axis.*

true view: View in which the line of sight is perpendicular to the surface.

tubing: Round-, square-, or rectangular-shaped structural steel. See *structural steel.*

U

undercutting: Creating a groove that is not completely filled by weld metal in the base metal during the welding process.

union: Fitting consisting of three parts having threads and flanges which draw together when tightened.

units period: First period (000 through 999). See *period.*

universal plate: Plate that is rolled between horizontal and vertical rollers and trimmed only on the ends.

V

variable load: Load that varies with time and rate, but without the sudden change that occurs with an impact load. See *impact load.*

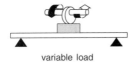

variable load

vertical: Line in a straight, upward position.

vertical line: Line that is perpendicular to the horizon. See *perpendicular.*

video display terminal: Monitor or screen of computer.

volt (V): Unit of measure for electricity that expresses the electrical pressure differential between two points in a conductor. See *conductor.*

volume: Three-dimensional size of an object measured in cubic units.

W

weld-all-around symbol: Supplementary symbol indicated by a circle at intersection of arrow and reference line, which specifies that the weld extends completely around the joint. See *supplementary symbol.*

weld bead: Weld that results from a weld pass. See *weld pass.*

weld contour: Cross-sectional shape of the completed weld face. See *weld face.*

weld cracks: Linear discontinuities that occur in the base metal, weld interface, or the weld metal. See *base metal* and *weld interface.*

weld defects: Undesirable characteristics of a weld which may cause the weld to be rejected.

weld face: Exposed surface of weld, bounded by the weld toes of the side on which welding was done. See *weld toe.*

weld finish: Method used to achieve the surface finish.

welding symbol: Graphic symbol that shows weld locations and specifications on prints.

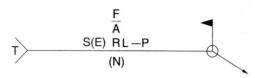

welding symbol

weld interface: Area where filler metal and base metal mix together. See *filler metal* and *base metal.*

weld joint: Physical configuration of the joint members to be joined.

weld pass: Single progression of welding along a joint.

weld root: Area where filler metal intersects base metal opposite weld face. See *filler metal* and *base metal.*

weld symbol: Graphic symbol which defines the cross-sectional shape of a weld.

weld toe: Intersection of the base metal and the weld face. See *base metal* and *weld face.*

weld type: Cross-sectional shape of the filler metal after welding. See *filler metal.*

whole number: Number that has no fractional or decimal parts. See *fraction* and *decimal.*

work angle: Angle less than 90° of the electrode in relation to the workpiece. See *electrode.*

working drawings: Set of plans that contains the information necessary to complete a job.

X

X ray: Electromagnetic radiation with a very short wavelength.

Y

yield: See *elastic limit.*

Z

INDEX